The Chemistry of Biology

Armin Börner · Juliana Zeidler

The Chemistry of Biology

Basis and Origin of Evolution

 Springer

Armin Börner
Leibniz-Institut für Katalyse e. V.
Rostock, Germany

Juliana Zeidler
Institut für Chemie, Universität Rostock
Rostock, Germany

ISBN 978-3-662-66520-6   ISBN 978-3-662-66521-3   (eBook)
https://doi.org/10.1007/978-3-662-66521-3

This Springer imprint is published by the registered company Springer-Verlag GmbH, DE, part of Springer Nature.
The registered company address is: Heidelberger Platz 3, 14197 Berlin, Germany

The beginning of all sciences is the surprise that things are as they are.

Aristotle

Give me only matter, and I will build you a world from it.

Immanuel Kant

Only apparently does a thing have a color, only apparently is it sweet or bitter, in reality there are only atoms in empty space.

Democritus

Preface

In modern knowledge societies, there is probably no greater contradiction than in the public perception of biology and chemistry. While biological findings are considered with great sympathy, chemistry is often negatively connoted. This fact, which is also empirically confirmed, is popularized by the poorly informed reporting in almost all mass media, which focuses primarily on a few products of synthetic chemistry such as plastic, agrochemicals, food additives, etc. and their effects on humans and their environment. The chemical didactics in schools and universities also concentrate mainly on non-biological teaching content. In this way, the animated nature is explicitly separated from chemistry. This state must cause amazement, since not only biology and physics, but also chemistry have always been counted among the so-called natural sciences, that is, those sciences that deal with the investigation of nature.

The gap between chemistry and biology could close biochemistry. However, it has adapted to its own, very diverse forms of representation and perspectives of the complexity of the subject of investigation and has partly removed itself from chemistry with its strict concentration on the chemical formula language as the most important tool. However, chemical knowledge can contribute to reducing this biochemical complexity. A central conclusion that results from this is the actually trivial finding that the material basis of all living organisms is chemical compounds. Organisms differ or resemble each other due to the underlying chemistry!

With the knowledge of chemical structures, not only the physical properties of chemical compounds become clear, but also the possibilities and the determinacy of the interactions of molecules with other molecules in the form of chemical reactions emerge. Only chemistry provides answers to the question of why, out of the more than one hundred elements of the periodic table of elements (PSE), it is precisely carbon and not silicon that is the dominant element in biochemistry and thus in biology, although both are found in large quantities on Earth. Only the laws of the PSE allow conclusions to be drawn about why phosphoric acid and not sulfuric acid or even perchloric acid acts as a bridge building block in DNA, and why precisely the D-glucose and not the D-mannose plays such a central role in the construction of biological scaffold building blocks such as cellulose, starch or chitin. At the same time, only chemistry shows why the citrate cycle proceeds as we know it, and how determined it is in itself and in interaction with other biochemical cycles.

Since the material basis of biology can only be described by chemistry, the question arises as to whether the central guiding principle of modern biology, the theory of evolution, which goes back to Charles Darwin and Alfred Russel Wallace, also has a counterpart in the underlying chemistry. The (positive) answer to this question is a central concern of this book. Based on the laws of the PSE and the analysis of the formation and interaction of natural products in the context of biochemical reactions, it is actually possible to find a corresponding overarching narrative. As a result, Darwinism is put from the (biological) head onto the (chemical) feet. By focusing on chemical formulas, it becomes clear once again that biochemistry is also a synthetic chemistry, which "only" differs from the "man-made" synthetic chemistry in the framework conditions. Biochemical transformations are embedded in a higher-level complex context. Out of the multitude of elements of the PSE and the almost infinite number of chemical compounds that result from this, individual ones are selected. The way of selection is formed by the environmental conditions on Earth, such as moderate temperatures, preferably atmospheric pressure, the solvent water and, as the primary reaction partner, oxygen. Directed selection means evolution, thus establishing a connection to biology.

Since the approach to chemistry presented in this book is limited to biogenic ("natural") processes, a significant and necessary increase in reputation for the natural science of chemistry is to be expected in public perception. Living nature, up to chemical processes in humans, offers itself as an experimental and completely safe demonstration with almost unlimited scope and variety. This not only results in a desirable increase in general education, but also in a changed focus on the understanding of the nature of so-called "natural" processes. In order to enable easy access to the thoughts presented here, basic knowledge of the school material of the higher classes is briefly recapitulated before consequences are derived for biochemistry and biology. This book can therefore also serve as a guide and orientation aid, especially for teaching at higher schools and in basic studies at academic institutions. At the same time, the content is suitable for a deeper understanding in biochemical metabolic pathways, as they are taught, for example, in medicine and biochemistry studies.

We would like to thank Oliver Zeidler for the design of the graphics. A.B. would like to thank his big daughter Anna and Robert Franke (Marl) for many interesting discussions outside of chemistry during the conception of this book. Our joint thanks go to the colleagues of the LIKAT Dilver Peña Fuentes and Baoxin Zhang for the careful reading of the manuscript and for many valuable suggestions for content and design. The colleagues of the University of Rostock Wolfgang Schulz and Wolfram Seidel are thanked for stimulating professional hints on special chemical topics. We would like to thank the Springer-Verlag, in particular Desirée Claus and Carola Lerch, for their constant support and the courage to explore chemical questions on less beaten paths. We would like to thank Claus-Dieter Bachem and Maximilian Rittelmann for the translation of the German original book into English by means of artificial intelligence (AI). Without the organizational support of the Leibniz Institute for Catalysis e. V. (LIKAT) in Rostock, this book would not have been possible, for which, by way of example, its managing director Matthias Beller is thanked.

Armin Börner
Juliana Zeidler

Introduction

In modern societies, the increasing diversification and specialization of chemistry and biology is leading to a separation of both natural sciences, which not only has considerable disadvantages for the reputation of chemistry in the public, but also represents an enormous blockade for the gain of knowledge. Due to an accumulation of knowledge unique in educational history in the twenty-first century, a situation has arisen that makes it almost impossible for the modern human being to assess the importance of learned "fragments of education" and thus to classify them. This affects above all the assessment of so-called "natural" processes.

Biochemistry could build a bridge between chemistry and biology. But it has increasingly moved away from the mother science of chemistry. A world of terms, abbreviations and representations has arisen that hardly give an impression that the basis of biochemistry is still inorganic and organic chemistry. This is even more true for biology, which in social reception has almost completely detached itself from chemistry. Already when writing this book, the different levels of precision in the description of chemical phenomena became apparent.

Below an attempt should be made to bring chemistry and biology back together and to give biochemistry that connecting role that it can actually play. For this purpose, the properties of the periodic system of the elements (PSE) should be placed at the beginning as well as characteristic functional groups and typical reactions of inorganic and organic chemistry in the center of biochemical processes.

In order not to exceed the limited scope of an introduction and in view of the enormous number of natural substances and biochemical transformations, a selection had to be made. But it will show that with the tools of chemistry, the evolution and structure of natural substances as well as the processes of biochemical processes can be intellectually penetrated. With the often used term of evolution, which comes from biology, it is also signaled that chemical evolution is the prerequisite for biological evolution.

The proof for this thesis is transported almost exclusively through the formula language of chemistry. Formulas and reaction equations offer the singular advantage in contrast to other forms of human communication that the facts fit into a physically determined framework. Chemical formulas are also "only" symbols of human communication from an epistemological point of view, but they correspond very closely to the real conditions at an atomic level. The crystal structure analysis by means of X-rays has been possible for a long time to determine the position of atoms in space and their position to each other exactly. If atoms come particularly and permanently close, chemistry speaks of a bond. By enlarging the crystallographic result by a factor of 10^9 or 10^{10} and visualizing it on paper or on the computer screen, a chemical formula results. Chemical formulas thus describe unshakeable and thus objective relationships. Focusing on chemical formulas disciplines thinking. Structural formulas have the invaluable advantage of revealing the property and reaction potential of chemical compounds and their relations to each other. At the same time, the complexity of biochemical and biological systems is reduced. In order not to cause confusion, the nomenclature used in biochemistry is often used in parallel here.

This approach forces in the following introduction to a selection. When dealing with element properties that arise from the PSE, only those will be discussed that are necessary for understanding the structure of biogenic compounds and the course of biotic metabolic pathways. Such relationships are memorized in the form of cartoon-like excerpts from the PSE. For further explanations, the broad knowledge base of general and inorganic chemistry is referred to. In contrast to textbooks of organic chemistry, the traditional classification system, which places the individual functional groups with their numerous homologous representatives in the center, is only peripherally mentioned. A wider treatment would exceed the scope of the present representation. For more detailed explanations, the possibility has only been available for a few years to retrieve further explanations via electronic information services in parallel to reading the book. Corresponding terms and relationships, which are, for example, treated in Internet encyclopedias or specialized books in a wider chemical or physical context, are marked in the text. A short list of introductory and further literature can be found at the end of the book.

Natural products and associated biochemical reactions, often referred to as "chemistry of life", are characterized by special properties that span a uniform and clearly delineated framework, which serves as a red thread in the present book:

1. Reactivity requirement

Only a small part of the PSE is relevant in biochemistry and thus in living nature. The special role of these elements, which include carbon, hydrogen, oxygen, nitrogen, phosphorus, sulfur, some non-noble metals and few semi-precious metals, can be derived from intrinsic properties of the PSE. Such properties concern atomic radii, electron affinities, ionization energies, electronegativities of elements as well as oxidation numbers and basic geometries of the corresponding compounds. They give the respective elements a singular evolutionary potential that can be connected to higher levels of biochemical compounds and biological phenomena. This deductive analysis also leads to logical explanations why a large part of the elements, such as the noble metals, play no role in the chemistry of life, which simplifies the considerations considerably.

2. Concentration requirement

In addition to these intrinsic properties, extrinsic properties such as the frequency of occurrence and the type of distribution of these elements on Earth play a central role.

3. Adaptation requirement

The basis for life in an aerobic atmosphere, i.e. in the presence of oxygen, is the insertion of oxygen into X–H bonds (X = C, N, P, S). Water not only serves as a solvent for all reactions, but also reacts as a reaction partner. Only those compounds that are sufficiently stable and react in a graded manner under these conditions are relevant in biochemistry. Reactions in biochemical systems take place in a very limited environment with regard to solvent, pH, temperature and pressure. Strongly varying conditions, as they are constitutive for synthetic chemistry in research laboratories and for the chemical industry, do not play a role and therefore do not have to be discussed.

4. Carbon-centeredness

4a. General properties of carbon

Due to its singular physical and chemical properties, carbon occupies a position of exception with respect to all other elements and thus has by far the widest biochemical potential. In modern industrial societies, this potential is further extended by organic synthetic chemistry beyond the relatively limited number of natural products.

4b. Formation of carbon-carbon bonds and the formation of functional groups

The formation of long and branching chains over carbon-carbon bonds is unique. By insertion of oxygen into C–H bonds, first functional groups are formed. If oxygen-containing functional groups are replaced by other groups with heteroatoms (nitrogen, sulfur, phosphorus, halogens, etc.), this leads to the formation of further functional groups. In total, this results in an almost infinite number of stable compounds with different physical and chemical properties, into which biochemical and biological selection "intervenes" evolutionarily.

For the basics of chemistry treated here, the following cognitive framework results with regard to biochemistry:

I. The genesis of organic from inorganic chemistry

Organic chemistry as the basis of biochemistry can be seen as an extreme extension of inorganic chemistry, with countless new compounds being added, some of which have the properties of their inorganic base molecules, but even more new properties. Basic qualitative trends can be derived from the laws of PSE.

II. Life as a slowed-down total oxidation of energy-rich carbon

Carbon takes on all oxidation states from -4 (methane) to $+4$ (carbon dioxide) in natural products. Based on these oxidation numbers, every natural product, no matter how complex, can be assigned a place. At the same time, based on this red thread, basic principles of biochemical mechanisms and relations can be illustrated. By focusing on the oxidation numbers, an exclusively chemical system of order is applied to biochemistry, which makes it possible to dispense with the biochemical distinction between catabolic metabolism (breakdown of metabolic products from complex to simple molecules) and anabolic metabolism (synthesis of body-own substances) to a large extent.

It can be seen from this procedure that only very few reaction paths in the biochemical context (in the presence of oxygen!) lead to compounds with lower oxidation states. This includes, for example, methanogenesis, which is coupled to certain bacteria, and photosynthesis, which takes place in cyanobacteria and green plants. In contrast, processes predominate in which compounds of carbon in low oxidation state are converted into those with higher oxidation state. The result is the basic fact that the majority of all life is the slowed-down total oxidation of energy-rich carbon. Slowing down is a time phenomenon. Slowing down results from the existence of mutually penetrating biochemical mechanisms and structurally increasingly complex natural products. This does not exclude the short-term achievement of lower oxidation states. The focus on oxidation numbers makes it clear that even highly complex molecules, such as polysaccharides, proteins or polynucleic acids, are part of this "slowed-down" degradation process. The resulting natural products and their interactions form the material basis for the richness of biological species and their individuality.

III. Why- and comparison questions as cognitive aids

From an epistemological point of view, the reception of facts is not very sustainable, it is important to relate these facts to each other. Questions about why are particularly enlightening in the case of natural phenomena. By answering decision-making questions, why certain elements and chemical structures are present in biochemistry and biology and others are not, a deeper understanding of the nature and determinism of "natural" processes results. This of course includes the functioning of the human body itself.

Contents

The Periodic Table of Elements and Basic Consequences for the Structure of Natural Substances and the Course of Biochemical Processes

1

1.1 Elements of the Periodic Table

The periodic table of elements (PSE) currently comprises 118 elements. The elements of the order numbers 1 to 94 occur in nature. Elements beyond the order number 94 have to be synthetically generated by nuclear physics and are not stable.

Hydrogen as the lightest element was created during the Big Bang about 13.8 billion years ago together with helium and traces of lithium. The elements following hydrogen in the PSE up to carbon were produced in light stars, including our sun, by fusion reactions. Stars that were much heavier produced the elements up to iron. The remaining elements were generated by the explosion of particularly massive stars (supernovae). These processes are still taking place in space. Of the 94 naturally occurring elements, 83 have existed since the earth was formed about 4.3 billion years ago.

In Fig. 1.1 those elements are highlighted in green that have so far been found in living beings.

Elements highlighted in blue have so far only been found in organisms in traces or there is no evidence yet for a direct involvement in biochemical processes. Apparently they were present in the environment of the organism and were randomly absorbed by constantly taking place exchange processes. Perhaps they will be integrated into biochemistry in the course of further evolutionary steps, which explains the provisional character of this representation.

This relatively small number of chemical elements is responsible for the diversity of organic life that developed over time on Earth for about 1.5 billion years. This was also not fundamentally changed by numerous geological (volcanic eruptions, meteorite impacts) or biologically caused (emergence of land plants, increasing concentrations of "toxic" oxygen) environmental disasters. These disasters had *a priori* always chemical causes. These include, first and foremost, changing concentrations of free oxygen, the

© The Author(s), under exclusive license to Springer-Verlag GmbH, DE, part of Springer Nature 2023
A. Börner and J. Zeidler, *The Chemistry of Biology*,
https://doi.org/10.1007/978-3-662-66521-3_1

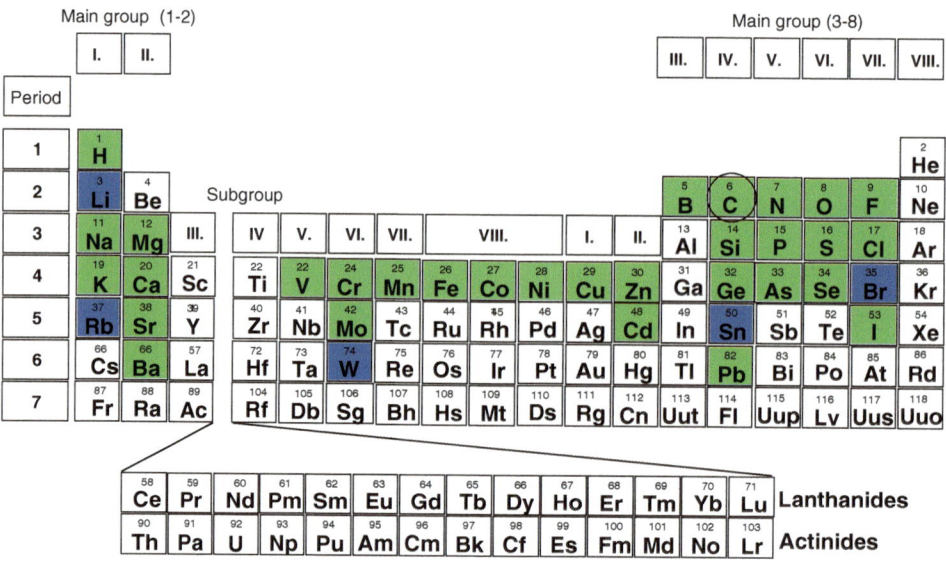

Fig. 1.1 The periodic table of elements (PSE) with emphasis on biologically relevant elements. *Source* Börner A (2019) Chemistry—Connections for Life, WBG, Darmstadt

temporary production of large amounts of methane, as well as changes in the carbon dioxide content in the atmosphere. As a result, they occasionally led to the almost complete extinction of microorganisms, plants and animals. However, an approximately constant chemical reservoir of biochemically relevant elements always served as a source for the re-emergence of conservative or, above all, for the emergence of modified biochemical reaction mechanisms and resulting natural products. In comparison to the predecessor worlds, partly more complex biological structures were formed. The information carriers DNA or RNA intervene in this limited reservoir of chemical compounds, combine them in biochemical processes and generate an immense number of biological species, genera, families and individual organisms.

The limited pool of chemical elements is not only restricted to all phylogenetic processes, but also forms the chemical basis for the emergence of every biological individual in ontogenesis. So we are always witnesses of this fascinating and never-ending development, while chemical compounds always form the material basis.

Despite an almost limitless number of conceivable and stable inorganic and especially organic compounds, however, only a relatively small number is realized in biological organisms. Even more, many uniform structures and mechanisms are found in very different organisms, which proves that life is subject to limited laws of chemical composition of compounds and the course of biochemical mechanisms. This is a big difference to the "man-made" synthesis chemistry, which has a much greater range of variation. Therefore, some of their products are referred to as "plastics", an attribute that is often used in a derogatory way, since the creativity of man is particularly expressed here.

Life means *a priori* adaptation to the conditions on the planet Earth. This includes the physical framework, moderate temperatures and, with the exception of some deep-sea residents, an air pressure of about 1 bar. The chemical framework conditions are water as a ubiquitous solvent and oxygen as the main reaction partner.

First of all, some typical properties of the periodic system should be considered, always with a view to the suitability, subsequently referred to as evolutionary potential, of the respective elements in biochemical mechanisms. Similar to biological processes, the selection of chemical elements probably also underwent and still undergoes directed evolution, which was and is characterized by the occurrence of increasingly complex molecules and their interactions with each other in the biochemical context. In parallel with the emergence of molecules with increasingly large molecular masses (e.g. cellulose, starch, chitin), molecules in the past also emerged that contained information and passed it on to this day (e.g. RNA or DNA, proteins). Life will also take place in the future within these framework conditions.

Evolution means that higher levels of complexity select alternatives on lower levels by feedback effects, which are thereby stabilized. This requires a sufficient number of alternatives on lower levels, from which selection is made in the context of a *Top-down* principle. At the same time, the structures of the lower levels determine the structures on the next or even higher level, thus a *Bottom-up* principle. It seems legitimate to apply Darwin's theory of evolution *survival of the fittest* also to the chemistry of life, which can be described in the context of the pre-living, i.e. with natural substances and biochemical reaction mechanisms, as the conditioned continuation of the formed.

Elements   Chemical   Biochemistry   Biology   Social
                compounds   structures

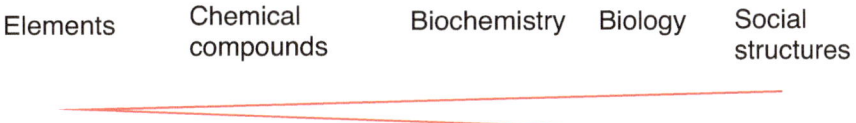

Evolution from the simple to the complex

The chemical evolution is determined by the physical properties of the elements and their occurrence frequency or distribution on Earth. Depending on the conditions on Earth, chemical compounds with special properties are built from the elements and create new, namely the biochemical selection level, with the natural substances. Chemical compounds are then selected by their suitability, in biology the term *Fitness* is used, in higher biochemical and biological structures and mechanisms. In contrast to biological evolution, where often non-adapted species are sorted out, many of the rare natural substances survive this selection process and ultimately give all biological species their characteristic.

It is extremely important to realize that biological properties thus exclusively depend on the chemical properties of selected inorganic, but especially organic compounds. This is a triviality that has not yet played a role in the public discussion and thus also denies chemistry the place it deserves as a life science. Biochemical processes are chemical processes that have undergone and still undergo biochemical and biological selection pressure during the evolution of life on Earth. This determines these chemical processes and the chemical compounds they are based on through similar selection effects as biological processes, up to the formation of social structures of organisms. The latter reach a new culmination point in human societies.

Nevertheless, it must be warned against deriving simple "if-then-relationships", as they are pursued, for example, in synthetic chemistry in the laboratory in order to achieve high yields of the desired products. This also applies to teleological "in-order-to-explanation patterns" that concern biologically naive arguments that are not based on the theory of evolution (in this context, the often circulated story of the giraffes, which, according to this hypothesis, got long necks to reach the leaves in tall trees, should be mentioned.). Some chemical properties lead directly to a biological property, others not. As an example of a direct correlation, astaxanthin is mentioned (see Fig. 4.21). Astaxanthin is red due to the many $C=C$ double bonds, which represent a chemical structural property. These double bonds are responsible for the reaction with oxygen radicals and thus have a protective effect on other compounds from a biochemical point of view. Astaxanthin is found in the legs and wings of flamingos and is a color indicator for sexual fitness and thus also a biological phenomenon. Examples of compounds for which the color does not play a role in biology are red hemoglobin and green chlorophyll (see Fig. 4.16). Their color results from the chemical structure as metal complexes for oxygen transport. However, since other, usually also colored compounds can take over oxygen transport, the color is a property without meaning for biology. They are only connectable on the cultural level of humans (e.g. for the labeling of political parties). However, it seems daring to derive an evolutionary advantage for the respective organisms from this within the framework of an "extended phenotype".

1.2   Biological Relevant Elements, the Singular Roles of Which can be Derived from Intrinsic Properties within the Periodic System of Elements (PSE)

Certain elements occur in biochemical processes, others do not. Apparently there are intrinsic physical properties that are responsible for this remarkable selection. In the periodic table of the elements (PSE), tendencies become clear that allow logical conclusions as to why this is so. At the same time, each biochemically relevant element has individual properties that ultimately decide the function in biological structures.

Therefore, the knowledge of important trends of the PSE and individual properties is indispensable. The following should remind these relationships in the form of catchy rules of thumb and cartoons and examine them with a focus on biochemical relevance. Completeness is not intended.

1.2.1   Atom Radii

In general, the atomic radii decrease with increasing atomic number within a period in the PSE (Fig. 1.2).

   This trend can be explained by the increasing number of positively charged particles (protons) in the atomic nucleus, which cause an increasing attraction to the electrons (negatively charged!) in the outer shell of the atom. According to classical ideas about the structure of atoms (**shell model**) the electrons are on orbits or shells around the atomic nucleus. As a result, the atomic radius of lithium is larger than that of fluorine. The atomic radii increase with increasing atomic number within a main group. For example, the radius of lithium is smaller than that of sodium or potassium, i.e. elements that are directly below in the PSE. This fact can be explained by an increase in the number of electron shells around the nucleus within the main group. This shields the outer electron shells from the electron-attracting effect of the atomic nucleus by underlying ones. Carbon in the 4th main group and in the first 8th period occupies a central position.

Carbon versus Nitrogen
Even small differences in atomic radii can have a dramatic effect on chemical properties, even of adjacent elements in the PSE. This is impressively shown by the comparison of compounds with triple bonds on the basis of carbon and nitrogen, two of the most important elements in biochemistry. In the molecular nitrogen N_2, the distance between the two nitrogen atoms, which are held together by a bond energy of 945 kJ/mol, is only 110 pm.

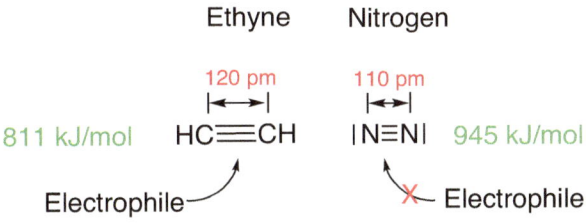

Although the triple bond is formed by a total of six electrons and thus represents a high concentration of negative charge in this area of the molecule, N_2 is an extremely stable molecule. The attack of positively charged particles (**electrophiles**) is prevented by the close proximity of both nitrogen atoms, which is equivalent to a shielding effect.

Fig. 1.2 Tendencies of the atomic radii in the PSE and selected elements

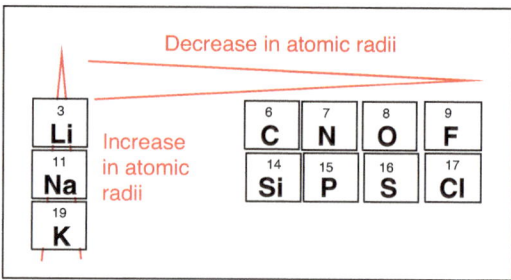

Even protons, that is, ions with the smallest radius, are not able to do this. In contrast, the distance between the two C atoms in ethyne, which belongs to the class of **alkynes**, is 120 pm. It is therefore larger and the bond energy between the two carbon atoms of 811 kJ/mol is weaker. Electrophiles, even much larger than protons, can therefore attack the C≡C triple bond.

Alkynes are rarely found in nature due to their high reactivity with the dynamic processes. There are only about a thousand natural substances with a C≡C triple bond. They are produced by marine organisms, special cyanobacteria, mollusks and species of lizards. Examples are neocarzinostatin and (−)-histrionicotoxin (Fig. 1.3). The former is produced by bacteria of the type *Streptomyces macromomyceticus*. Histrionicotoxin is a poison that occurs in the skin of tree frogs and protects them from reptiles and mammals. Both compounds are exotics in nature. In contrast, the stable and inert N_2 is the dominant molecule in the atmosphere with a 78% share of volume.

Carbon versus Silicon

Another illustrative example of the influence of atomic radii on binding relationships is the comparison between carbon and silicon. Both elements are directly above each other in the 4th main group. The atomic radius of silicon is slightly larger than that of carbon according to the trend described above in the PSE, which is a significant reason why carbon and not silicon has prevailed for the evolution of life on Earth. One consequence of different atomic radii is the effect of the so-called **"Erlenmeyer rule"**, an empirical rule that states that two hydroxy groups (or generally oxy groups) are not stable on a small central atom. The space requirement of two HO groups at the carbon atom of acetic acid is too great, so it is unstable. By splitting off the energy-poor water molecule, the situation relaxes.

Erlenmeyer
rule

O=C(OH)(OH) ⇌ H_2O + O=C=O↑

Carbonic acid   Carbon dioxide
                        (gaseous)

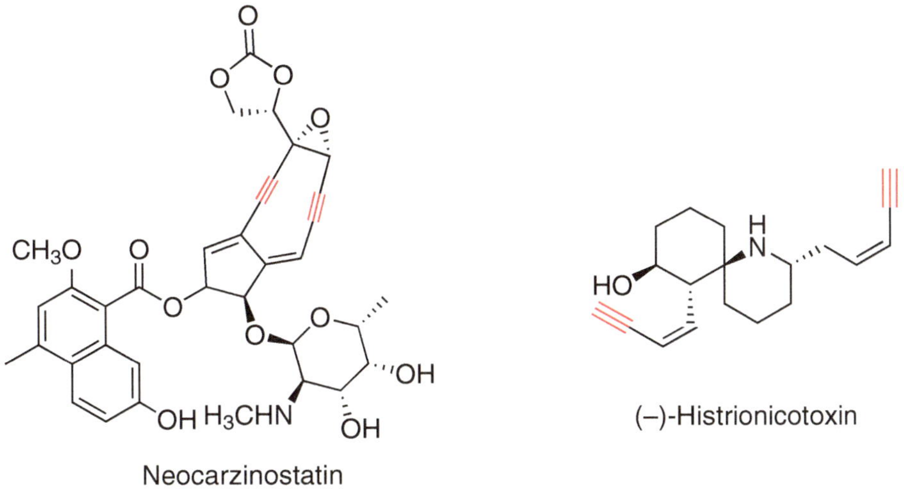

Fig. 1.3 Examples of alkyne structures in natural substances

The balance of the reaction lies far on the side of the also very stable carbon dioxide (CO_2). If you take the entire reaction into account, it becomes clear that the effect described by the Erlenmeyer rule, the reactant, the carbonic acid, destabilizes. At the same time, the formation of CO_2, which is a gas under biotic conditions and escapes, shifts the balance of the reaction in favor of this product. This situation is by the **principle of Le Chatelier-Braun** (the "principle of least constraint") described, which in its simple version says: If you exert a force on a chemical system in equilibrium, it reacts so that the effect of the force is minimized.

Silicon behaves quite differently than carbon due to its slightly larger atomic radius. Silicon is no longer subject to the Erlenmeyer rule. In orthosilicic acid, even several hydroxy groups can fit on the silicon atom, which is larger than the carbon atom, resulting in a compound with the formula $Si(OH)_4$ called orthosilicic acid (Fig. 1.4).

However, orthosilicic acid only occurs in small concentrations in nature. In contrast to the comparable carbonic acid, no products with one or two Si=O bonds are formed by water cleavage, but by **intermolecular condensation** with a second molecule, disilicic acid is formed. With further silicic acid molecules and cleavage of water, which already happens in the weakly acidic milieu, the hydrogen content becomes smaller and smaller, and at the end the three-dimensional cross-linked product also has the general formula SiO_2 in terms of composition. However, silicon dioxide is described exactly with the formula $(SiO_2)_n$. The compound has a very large molar mass and is a solid, most impressively materialized in sand, an extremely stable and unreactive material. The compound also differs in terms of the aggregate state under biotic conditions from the gaseous carbon dioxide CO_2. Since carbon dioxide is always at the end of all oxidative degradation reactions of carbon compounds, all life processes are shifted out of equilibrium in the direction of this gas according to the principle of Le Chatelier-Braun, which causes the dynamics of life processes with food intake, energy generation, formation of biological structures and continuous degradation processes. Life as it exists on Earth would not be possible on the basis of silicon, specifically SiO_2. Of course, this does not exclude the possibility that completely different life forms based on static systems, such as semiconductors, as pursued by artificial intelligence (AI), will become significant at some point in the future.

In carbon dioxide, two double bonds emanate from the carbon atom. Their formation is only possible by the spatial proximity of C and O, which in turn is made possible by the small atomic radii of both binding partners. In contrast, the slightly larger silicon atom does not have a pronounced tendency to form double bonds with oxygen due to its larger atomic radius. Even the average distance of a Si–O bond is much longer than that of a C–O single bond. Compounds with Si=O bonds can only be produced and stabilized under laboratory conditions. They do not occur in nature.

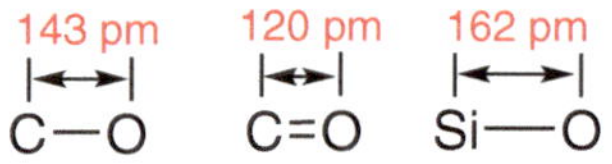

$$
\begin{array}{ccc}
143\ pm & 120\ pm & 162\ pm \\
|\!\leftarrow\!\rightarrow\!| & |\!\leftrightarrow\!| & |\!\leftarrow\!\longrightarrow\!| \\
C\!-\!O & C\!=\!O & Si\!-\!\!-\!O
\end{array}
$$

This fact, in addition to the suspension of the Erlenmeyer rule, is the reason that, as shown above, two orthosilicic acids react by water elimination to form disilicic acid. This situation is subject to the so-called **double bond rule**, which states that elements of the 3rd period of the PSE are hardly able to form stable chemical compounds with

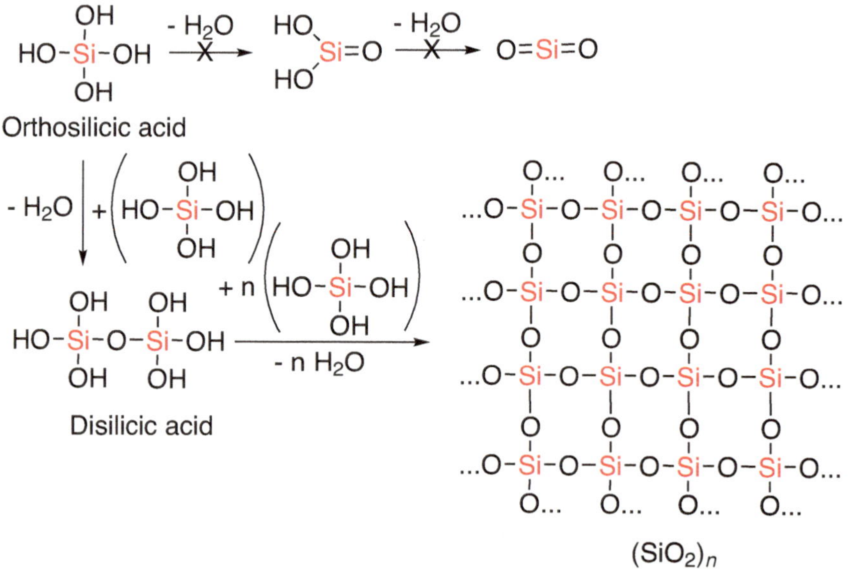

Fig. 1.4 The condensation behavior of orthosilicic acid

multiple bonds. In the end, only small differences in atomic radii are responsible for the fundamental difference between carbon and the adjacent silicon.

The instability of silicic acid is also the reason why silicon and its compounds do not find their way into the biochemistry of carbon. Nevertheless, there are a number of organisms that use Si–O-containing structures outside this framework to improve their own stability, for example in the form of an exoskeleton. Diatoms, sponges and radiolarians are known. Plants such as horsetails and bamboo also contain silicon dioxide in their stems and leaves.

Unsaturated Compounds

The "mutual penetration" of elements to form multiple bonds is particularly characteristic of carbon compounds. Only on the basis of the mean atomic radius of carbon as an element of the 4th main group are organic compounds with both C=C double and C≡C triple bonds possible. In general, a distinction is made between σ-bonds and π-bonds. The primary cohesion between two carbon residues is ensured by a σ-bond, while π-bonds, which are formally drawn above or below the σ-bond, contribute additional binding interactions.

$$\sigma\text{Binding} \qquad \pi\text{-Bindings}$$

$$C\!-\!C \qquad C\!=\!C \qquad C\!\equiv\!C$$

The distance between the atoms decreases with increasing degree of bonding. On average, a C–C single bond is much longer than a C=C double bond. The latter is again longer than a C≡C triple bond.

$$154\ \text{pm} \qquad 134\ \text{pm} \qquad 120\ \text{pm}$$

$$C\!-\!C \qquad C\!=\!C \qquad C\!\equiv\!C$$

Such π-bonds are generally weaker than σ-bonds. This has the consequence that in organic compounds π-bonds can be cleaved without the more stable C–C single bonds being affected. This is the most important prerequisite for the existence of long hydrocarbon chains at all. This ultimately results in an almost infinite number of individual compounds from which biochemical processes can "select" evolutionarily. This is another difference to molecular nitrogen N_2, where the N≡N triple bond is much more stable than an N=N double bond as in the metastable diimine, which decomposes already at room temperature. Even N–N single bonds, as in hydrazine, are less robust.

$$N\!\equiv\!N \quad \gg \quad HN\!=\!NH \quad < \quad H_2N\!-\!NH_2$$

Nitrogen   Diimine   Hydrazine

Stability

The bond relationships influence the reactivity. Organic compounds with C–C single bonds, in the simplest case alkanes, are usually less reactive than compounds containing multiple bonds. The former are therefore also referred to as **saturated compounds**. The rarity of the very reactive C≡C triple bond in natural products has already been mentioned. C=C double bonds, the characteristic of **alkenes,** are however slightly more stable than triple bonds and are located in the reactivity between triple and single bonds. Also alkenes are referred to as **unsaturated compounds**. They can be found in numerous biologically important structures. Due to the negative charge concentration (four electrons) in the area of the double bond and the sufficiently large distance between the two carbon atoms, they are easily attacked by polar reagents of various kinds and converted into saturated compounds by addition reactions. By elimination reactions, alkenes can again be converted into alkenes, thus unsaturated compounds.

$$C=C \quad \underset{\substack{\text{Elimination}\\\text{reactions}}}{\overset{\substack{\text{Addition}\\\text{reactions}}}{\rightleftharpoons}} \quad C-C$$

$$\text{unsaturated} \qquad\qquad \text{saturated}$$

From these properties it follows that compounds with alkene substructures are important transit and switching stations in almost all biochemical processes.

Phosphoric Acid and Phosphoric Anhydride
The suspension of the Erlenmeyer rule for hydroxy compounds with larger central atoms is not only a characteristic of silicic acid, but also occurs in phosphoric acid. In phosphoric acid, three hydroxy groups are gathered around the central phosphorus atom.

$$\begin{array}{c} \text{O} \\ \| \\ \text{HO}-\text{P}-\text{OH} \\ | \\ \text{OH} \end{array}$$

Phosphoric acid

Phosphoric acid is a stable compound and does not have a pronounced tendency to intramolecular water splitting; a monomeric structure with the molecular formula HPO_3 with two P=O double bonds does not exist (Fig. 1.5). However, when energy is supplied, water is split intermolecularly from two molecules of phosphoric acid; diphosphoric acid anhydride, also called pyrophosphoric acid, is formed. If another phosphoric acid molecule is condensed onto the pyrophosphoric acid, triphosphoric acid anhydride is formed. As with the silicic acids, further phosphoric acid molecules can be added to polyphosphoric acids, with the degree of polymerization being several thousand.

Phosphorus oxyacids do not occur in the animate nature as free acids, but in the form of their corresponding salts, which are the phosphates or the corresponding salts of the polyphosphoric acids, the polyphosphates. They are found in all eukaryotic and

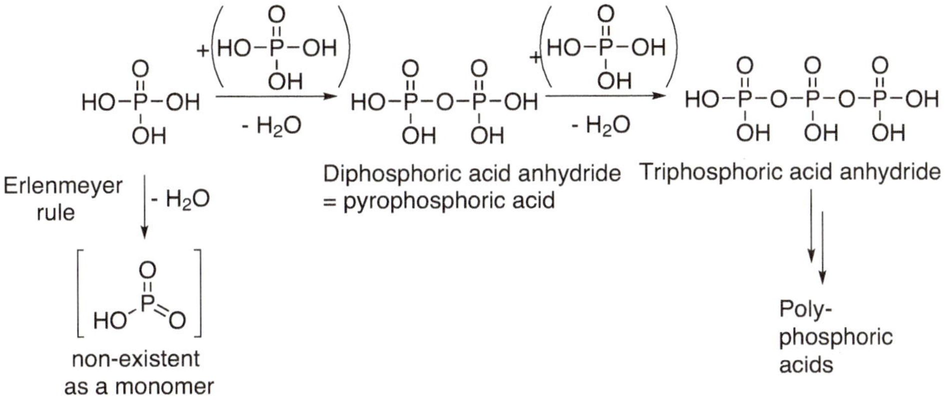

Fig. 1.5 The condensation properties of phosphoric acid

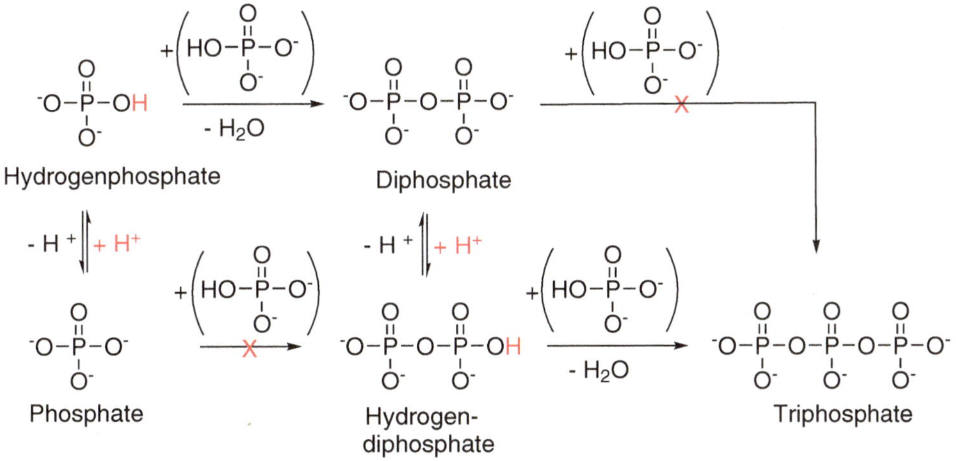

Fig. 1.6 The pH dependence of the condensation properties of phosphates

prokaryotic cells. In biochemistry, the inorganic salts of phosphoric acid are often abbreviated as P_i (inorganic phosphate).

It should be noted that the phosphate anion does not condense with another phosphate, but also not with a hydrogen phosphate to the diphosphate (Fig. 1.6). The intermolecular water cleavage only takes place after protonation of one of the three anionic oxygen atoms to OH. This is comparable to the formation of polysilicic acids in acidic environments. Consequently, the diphosphate must first be converted into the hydrogen diphosphate by stoichiometric amounts of protons before the reaction proceeds to the triphosphate. The pH-dependence of these condensation reactions is significant in connection with the formation of ATP (see below).

Since phosphorus as an element of the 5th main group has a smaller atomic radius than silicon (4th main group!), the repulsive forces between the oxygen substituents in the di- or triphosphate are greater than in the disilicate or in the polysilicates. Phosphoric acid anhydrides and their salts are thus characterized by a considerable amount of energy. When reacted with water (hydrolysis), this energy is released (Fig. 1.7). The resulting hydrogen phosphates are converted into phosphate by releasing the proton. By protonation—and thus pH-dependent—phosphate is in equilibrium with hydrogen phosphate and hydrogen diphosphate.

This unique property of phosphoric anhydrides has central consequences for the energy balance of almost all biological organisms. Unlike synthetic chemistry in the laboratory or in an industrial plant, a biochemical transformation cannot be forced by changing the pressure, temperature or solvent. These parameters are always approximately constant or completely constant. The required energy must therefore always be provided by other, usually chemical processes. Pairs of **coupled reactions**, in which one reaction provides the energy that is consumed in the other reaction in parallel. The energy-providing reaction is termed **exergonic** and the energy-consuming termed **endergonic**.

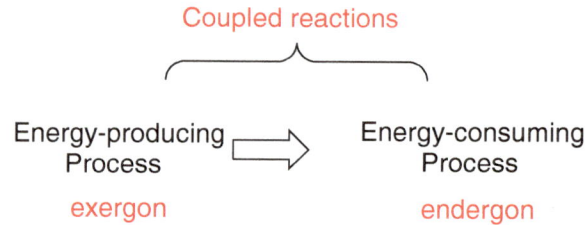

Fig. 1.7 The hydrolysis of triphosphate

In most cases, the energy in biological systems comes from salts of phosphoric acid anhydrides with their high energy density. In particular, monoesters of di- and triphosphates serve as universal energy storage and -suppliers in all living cells. The prototype is realized in adenosine triphosphate, abbreviated ATP (Fig. 1.8). The human body builds up an equivalent of about half its body weight in ATP every day. However, due to its high reactivity towards water, the current concentration in cells is very low.

In ATP there are two different types of P–O bonds (Fig. 1.9): The phosphoric acid anhydride groups already described above with the characteristic P–O–P unit, and a phosphoric acid ester group. The latter is characterized by a low-energy P–O–C bond, which is not cleaved by water in the absence of catalysts. The bond with the organic

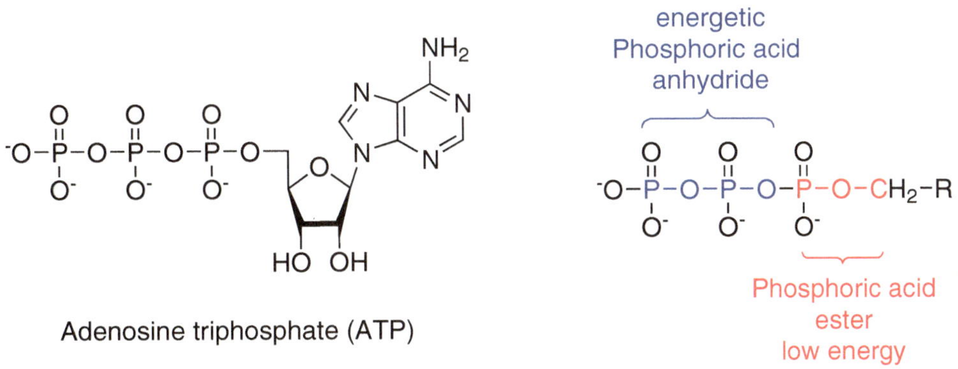

Fig. 1.8   Adenosine triphosphate (ATP) as the most important energy supplier in biochemical processes

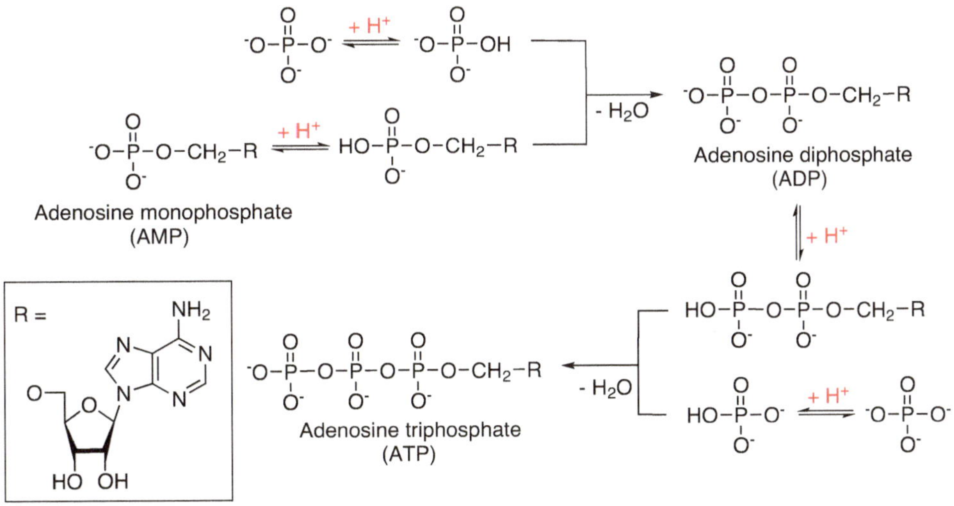

Fig. 1.9   Successive synthesis of ATP from AMP

moiety stops the attachment of further phosphoric acid molecules to this end of the molecule, thus preventing the formation of insoluble polyphosphates during biomineralization (Sect. 2.1). In contrast, esterification does not take place under biotic conditions with analogous silicic acids, which prevents their access to bioorganic processes. Therefore, their polycondensation only ends with highly molecular and insoluble silicon dioxide. This property and the comparatively low energy density of polysilicates are the reasons why silicic acid is not a biochemical alternative to phosphoric acid.

The formation of ATP from AMP or ADP requires a stoichiometric amount of protons. Therefore, it is logical that this step is closely coupled with processes in which a large number of protons are generated. These protons usually come from parallel dehydrogenation reactions of organic compounds (carbohydrates, fats). In the course of this process, in addition to electrons, an equal number of protons are generated, which in the context of a **chemiosmosis,** that is the migration through a semipermeable membrane, drives the production of ATP. Therefore, the two processes are not only chemically, but also spatially linked in mitochondria.

Energy-providing processes based on ATP run successively via hydrolysis over ADP and stop at the AMP level (Fig. 1.10). Alternatively, the pyrophosphate unit can be cleaved off from ATP and then decomposed separately, also a highly exergonic process.

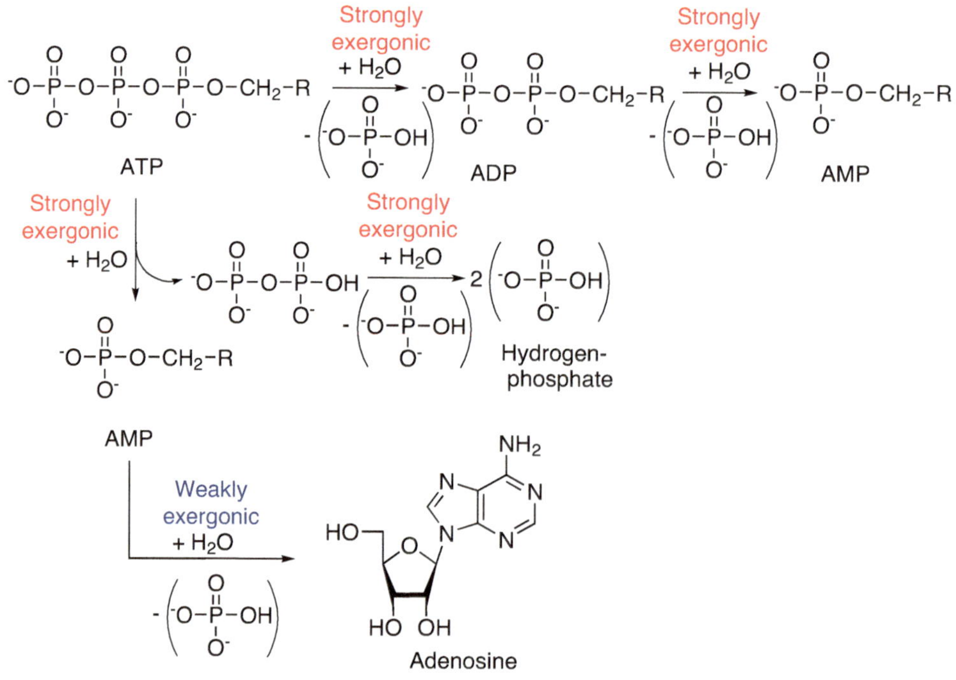

Fig. 1.10 Energy contents of different P–O bonds in ATP during hydrolysis

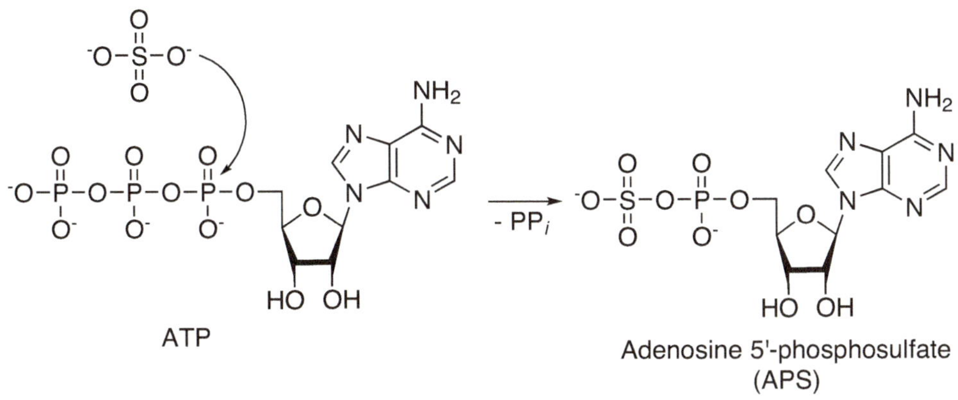

Fig. 1.11 Activation of sulfate by anhydride formation with ATP

In contrast, the cleavage of the ester bond to adenosine hardly provides any energy and therefore plays no role in this context. The differences in reactivity arise from the repulsive forces between the negatively charged oxygen atoms. Between those of the pyrophosphate, they are permanently present, while they do not exist for the ester group.

ATP not only provides the energy for numerous parallel biochemical reactions, but also serves as an activating carrier. For example, the mixed anhydride of AMP and sulfate is present at the beginning of the entire sulfur metabolism in microorganisms and plants (Fig. 1.11). The "activated sulfate" (APS) is formed by condensation between ATP and sulfate with pyrophosphate (PP_i) as the leaving group.

Numerous anhydrides with organic carboxylic acids or carbonic acid are formed according to the same pattern (Chap. 4), while comparable anhydrides of phosphoric acid with orthosilicic acid are not known due to the high tendency of the latter to self-condensation, which in turn emphasizes the difference between the two elements silicon and phosphorus for the chemistry of life.

1.2.2   Ion Radii and Ionization Energies

In the biochemical context, in addition to the atomic radii, the corresponding **ion radii** are also significant. Ion radii follow the same trend for the same charge number (Fig. 1.2). The **ionization energy** denotes the amount of energy that must be supplied to an atom in order to remove an electron from it and generate the corresponding cation Na^+ or K^+. A cation is formed.

Elements forming cations are located on the left side of the PSE (Fig. 1.12). From the simplified scheme of the PSE it can be seen that only little energy is required to remove an electron from an alkali metal like sodium or potassium and generate the corresponding cations Na^+ or K^+. However, the removal of an electron from an element at the top

Fig. 1.12 Trends in ionization energies in the PSE and selected elements

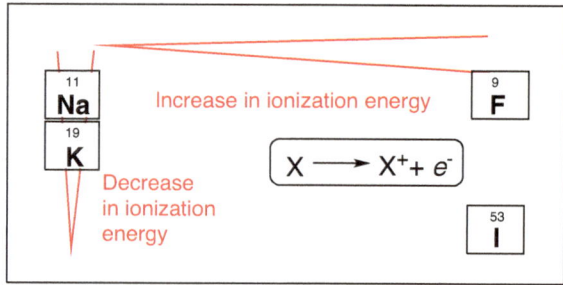

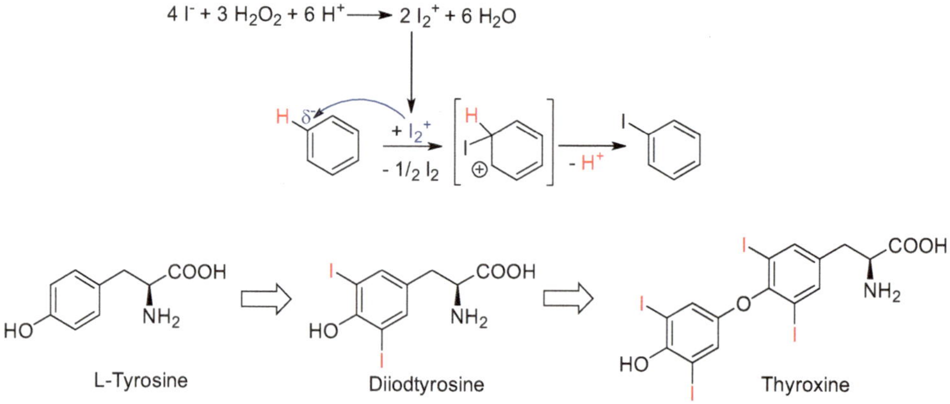

$$4\ I^- + 3\ H_2O_2 + 6\ H^+ \longrightarrow 2\ I_2^+ + 6\ H_2O$$

Fig. 1.13 The iodonium ion. Generation and reaction with L-tyrosine

right of the PSE, such as molecular fluorine, requires much more energy. F_2^+ does not exist in biogenic systems. On the other hand, the formation of an iodonium ion I^{2+} from I_2 is less energy-intensive and also of biological relevance. The iodonium ion, which is formed, for example, from the unreactive iodide by reaction with hydrogen peroxide, plays a role in iodinations of **aromatics** (Fig. 1.13). Aromatics, a simple case is benzene, represent accumulation of electrons (Sect. 4.1.2) and are therefore attacked by **electrophiles,** that is, positive particles. Via a positively charged intermediate state, the system stabilizes again, and at the end iodine has replaced a hydrogen atom in benzene. This type of reaction is generally referred to as **electrophilic substitution**.

Electrophilic substitutions with iodine form the basis of the biosynthesis of diiodo-tyrosine from the amino acid L-tyrosine, a precursor of the thyroid hormone thyroxin. Since thyroxin can only be built up using external iodine sources, the supply of inorganic iodide (I^-) or organic iodine compounds is essential for mammals and thus also for humans. Deficiency periods can be countered by "stockpiling" in the thyroid. This is an illustrative example of how biological evolution follows or supplements chemical evolution.

1.2.3   Electron Affinity

As the counterpart to ionization energy, the electron affinity can be understood. It quantifies the energy that must be expended to add an electron to a neutral atom, resulting in an anion. Elements that form anions are located on the right side of the PSE (Fig. 1.14). Particularly large electron affinities (with the exception of the noble gases, which play no role in the biological context) characterize the halogens, with fluorine at the top.

Halides, in particular Cl^-, Br^-, or I^-, are taken up by water and then integrated into biochemical processes. Due to different solubilities in water, their availability to organisms is also different. Fluoride F^- plays almost no role in this context because it only occurs in very small concentrations in natural sources (soils or water).

1.2.4   Solubility in Water

Care must be taken when making statements about the solubility properties of the corresponding salts in water based on the ionic radii of cations and anions. Small cations such as Na^+ or K^+ are always more effectively surrounded by a water shell (hydrated) than large ones. As a result, their salts are often soluble in water. However, a generalization is not possible, since the **hydration,** that is, the surrounding of ions by polar water molecules, depends not only on the cations, but also on the anions throughout the entire solution process. Lithium, sodium or potassium salts with anions such as F^-, Cl^-, Br^- and I^- are soluble in water. But sulfates and carbonates with small alkali metal ions such as Li_2SO_4 or Na_2CO_3 are also characterized by a high water solubility. On the other hand, many salts with large ions, such as $CaCO_3$ (aragonite) or $BaSO_4$ (gypsum), are insoluble in water.

Fig. 1.14 Tendencies of the electron affinities in the PSE and selected elements

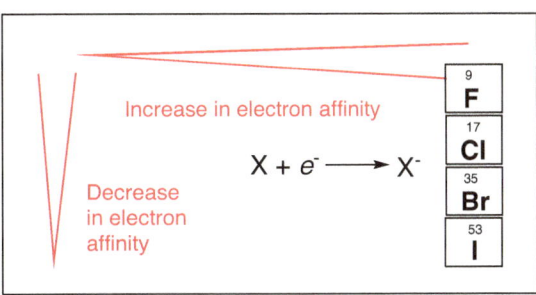

1.2.5   Reactivity Towards Oxygen: Oxides, Hydroxides and Oxoacids

Ionization energies and electron affinities refer to the isolated (gaseous) atom and describe tendencies to form cations or anions from the elements. But which of these ionic species is formed is not only a function of the atomic sizes, but also depends on the interaction with reaction partners. The most important partners under the conditions on Earth are oxygen and water. The corresponding element oxides are formed from the elements by oxidation with oxygen, which then react further with water to form the corresponding hydroxy compounds. In the PSE, the acid-forming compounds on the right are distinguished from the basic-reacting compounds on the left, which are separated from each other by a diagonal line, which is formed by the **amphoteric** (Greek: of both kinds) element oxides (Fig. 1.15).

For example, a series of (very unstable) oxides such as Cl_2O, Cl_2O_3, Cl_2O_5 or Cl_2O_7 is formed from the reaction of the non-metal chlorine with oxygen successively (Fig. 1.16). By reaction with water, acids such as hypochlorous acid, chloric acid, hydrochloric acid or perchloric acid are formed from them.

In contrast, the metal sodium reacts with oxygen to form sodium oxide Na_2O. By reaction with water, sodium hydroxide is formed, which is a strong base in water. Similar applies to lithium, potassium or calcium, which react to form LiOH, KOH or $Ca(OH)_2$.

Fig. 1.15 Different products of the reaction of elements with oxygen depending on the position in the PSE

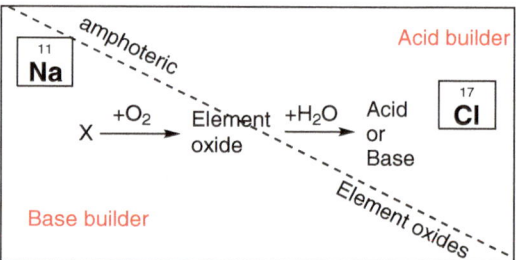

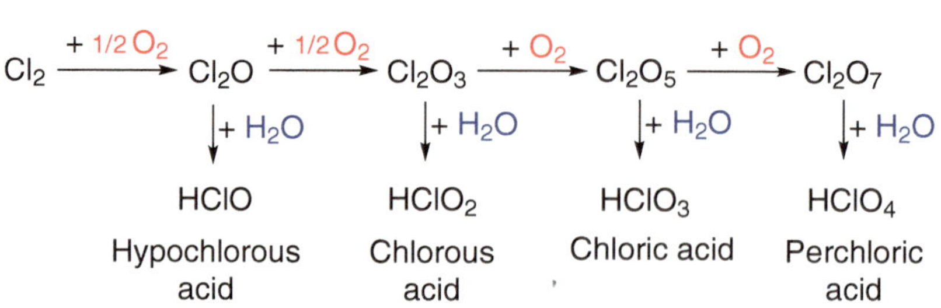

Fig. 1.16 Chlorine's stepwise and exhaustive oxidation

$$2\ Na \xrightarrow{+\ 1/2\ O_2} Na_2O$$
$$\downarrow + H_2O$$
$$2\ NaOH$$

Sodium hydroxide

Whether the formed HO bond gives off protons or hydroxide ions in water depends on the electron-shifting or -attracting properties of the central atom (Fig. 1.17). Electronegative elements (which are located in the PSE at the top right, Sect. 1.2.6) intensify the polarity of the H–O bond and lead to the release of H^+. They thus represent, in analogy to the **acid-base definitions** of **Arrhenius** or **Brønsted** acids. In the case of electropositive elements (metals), however, the bond between the central atom and the HO group is broken, HO− is released, and, according to the definition of Arrhenius, it is a base.

The influence of different **oxidation states** on acid strength can be demonstrated using the chloric acids described above. With increasing oxidation state of chlorine, acid strength increases from hypochlorous acid to perchloric acid. Perchloric acid is an extremely strong acid ($pK_S = -10$) and is therefore also referred to as a superacid. A highly charged central atom like Cl in oxidation state +7 leads to a particularly strong polarization of the H–O bond.

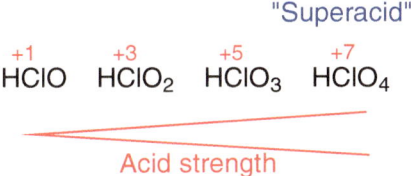

"Superacid"

$$\overset{+1}{HClO} \quad \overset{+3}{HClO_2} \quad \overset{+5}{HClO_3} \quad \overset{+7}{HClO_4}$$

Acid strength

The same trend can be seen in the large number of corresponding nitrogen, phosphorus and sulfuric acids. In this context, nitric acid HNO_3, sulfuric acid H_2SO_4 and phosphoric

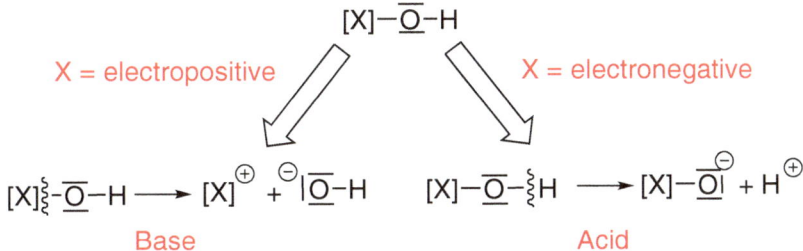

Fig. 1.17 Acid and base properties of inorganic HO compounds as a function of the electronegativity of the central atom

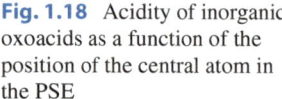

Fig. 1.18 Acidity of inorganic oxoacids as a function of the position of the central atom in the PSE

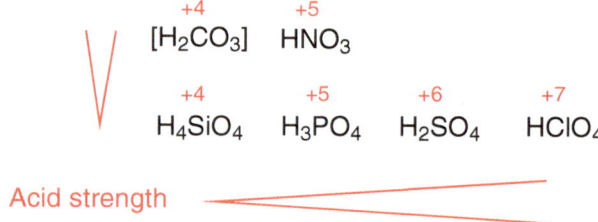

acid H_3PO_4 are not only the strongest acids in their series of oxoacids, but also occur most frequently in nature in the form of their salts.

The acidity of oxoacids increases from left to right in the same period in the PSE, which results from the increase in the oxidation number of the central atom (Fig. 1.18). Orthosilicic acid is the weakest and perchloric acid is the strongest acid. Within the main group, the acidity decreases from top to bottom, which can be explained by the weakening electron pull on the HO groups with increasing atomic radius of the elements. This means that nitric acid is more acidic than phosphoric acid. Carbonic acid would also be more acidic than ortho-hydroxybenzoic acid, at least in theory, and apart from its instability.

Salts of acids are formed by **neutralization** with bases. In water, these are HO^- ions. The formation of salts depends on the number of neutralizable HO groups. The salts of carbonic acid are hydrogen carbonate and carbonate (Fig. 1.19). Silicates of ortho-hydroxybenzoic acid and their condensation products are derived from silicon. There is only one anion of nitric acid, nitrate. The neutralization of sulfuric acid leads to

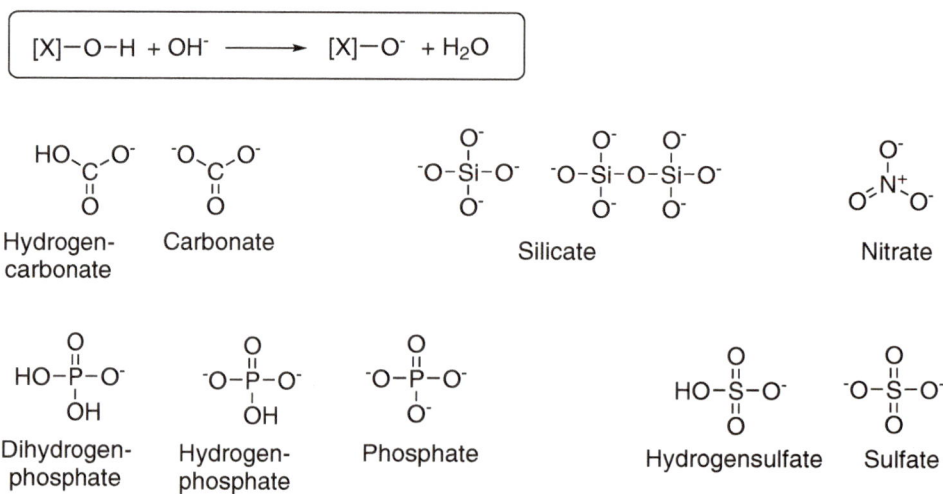

Fig. 1.19 Neutralization products of inorganic oxoacids

hydrogen sulfate and sulfate. Salts of phosphoric acid are dihydrogen phosphate, hydrogen phosphate and phosphate.

Whether free acids or the associated salts predominate in living nature depends on the stability of the acids and the acid strength. Carbonic acid is not stable under biotic conditions and decomposes into carbon dioxide and water. In contrast, the more stable orthosilicic acid occurs in small concentrations in waters and, as a result, in all plant and animal fluids. However, the main occurrence of silicon on Earth is represented by amorphous silicon dioxide and silicate minerals, i.e. salts. In comparison, nitric acid, phosphoric acid and sulphuric acid are only found in the form of their salts as increasingly stronger acids. The particularly strong perchloric acid is not found in the living world in the form of an acid or in the form of its salts due to its extremely oxidizing effect. In it, two tendencies of the PSE converge: high acid strength and extremely strong oxidizing effect. Both properties are counterproductive for the construction of living organisms.

1.2.6 Electronegativity

An important parameter for estimating the strength of individual bonds and the polarity of compounds is the **electronegativity**. It is a relative measure of the ability of an atom to attract the bonding electron pair to itself in a chemical bond of the type X–Y. Electronegativities are listed in most representations of the PSE (Fig. 1.20). The element with the highest electronegativity is fluorine. This results in the electronegativity increasing from left to right and decreasing from top to bottom. The partial charges in a bond of the type X–Y are indicated by δ^+ and δ^-, respectively.

The small concentration of available fluoride compared to other halogens on Earth has a significant impact on biochemistry. A greater supply of fluoride would cause a fundamental change in the properties of natural substances and numerous biochemical processes: The only slightly larger atomic radius of fluorine compared to hydrogen causes an exchange of both atoms to take place without any problems. However, since fluorine has a much higher electronegativity than hydrogen, the corresponding C–F bonds can form very strong **hydrogen bonds** (Sect. 1.2.11) to other electronegative elements in

Fig. 1.20 Tendencies of the electronegativities in the PSE and selected elements

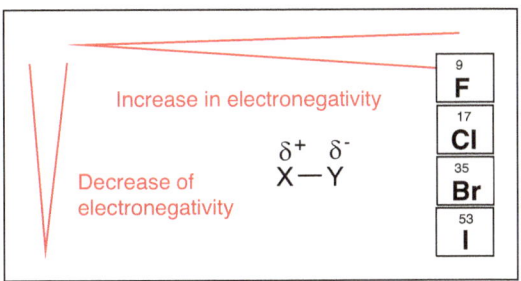

compounds, in contrast to the analogous C–H bonds. The resulting associates have different physical and chemical properties.

$$\delta^+ \; \delta^+ \qquad\qquad \delta^+ \; \delta^-$$
$$\text{C–H} \implies \text{C–F}$$

Hydrogen bonds

In addition to this effect transmitted across space, there is the electron-withdrawing effect of bound fluorine over C–C bonds. This not only affects the acidity—fluoroacetic acid FCH_2COOH is a stronger acid than acetic acid CH_3COOH—, but also the increased positive charge of the C atom of the acid group.

$$\text{H–C–C}\,\delta^+ \quad < \quad \text{F–C–C}\,\delta^{++}$$

Acid strength and positivation of the C=O-Group

As will be discussed later, acetic acid or its salts, the acetates, play a central role in many biochemical mechanisms, such as the citrate cycle (Sect. 4.1.1). Fluoroacetic acid displaces acetic acid from this context. This is the reason why, in most living organisms, fluoroacetic acid acts as a powerful poison. Only a few, like the poison leaf *(Dichapetalum cymosum)*, which grows on fluorine-rich soil, have adapted to this fluorine compound. Interestingly, even higher organisms, such as some kangaroo species in Australia, which feed on this plant, have become resistant over the course of evolution. It is obviously possible that some biochemical processes can adapt relatively quickly to new chemical compounds with different properties. This mainly affects compounds in very small concentrations, the so-called **micronutrients**, which originally developed from body-foreign compounds, sometimes even poisons. Micronutrients thus drive the evolution of biochemical mechanisms and, as a consequence, the evolution of new biological species.

No or only small differences in electronegativity are the hallmark of a **covalent bond**. Large differences are characteristic of polar bonds. In the extreme case, one partner almost completely draws the binding electron pair to itself and as a result becomes the carrier of the negative charge. Electronegativity differences of >1.7 characterize an **ionic bond**. The latter characterizes many inorganic salts such as NaCl or KBr. Salts are often water-soluble. Carbon atoms in unfunctionalized hydrocarbon chains, however, do not differ in terms of electronegativity. Adjacent carbon or silicon atoms are held together by a covalent bond.

$$X-X \Longleftarrow \overset{\delta^+ \ \delta^-}{X-Y} \Longrightarrow X^+ \ +|Y^-$$

Covalent bond   Ionic bond

e.g. C-C, Si-Si or C-H   e.g. NaCl, KI

The electronegativity difference between carbon and hydrogen is also very small. Therefore, C–C and C–H bonds in unfunctionalized hydrocarbons are non-polar. Pure hydrocarbons therefore do not dissolve in the polar solvent water, which is central to their function as boundary layers in biological systems.

Inactivated C–H bonds are not attacked by polar reagents. The cleavage of such C–C and C–H bonds proceeds radically, a property that is predestined for the attack of the oxygen molecule (a radical!). This reaction is the "opening reaction" for the functionalization of carbon chains and thus for the emergence of organic life and the dynamics of most life processes on Earth.

1.2.7   Element-Element Single Bonds

One of the most important requirements for evolutionary processes of any kind is a sufficiently large number of possibilities for variation, which enable selection and feedback at higher levels of complexity. In the biochemical context, a chemical element must have the potential to form very many different and sufficiently stable compounds that come into contact with each other through chemical reactions. These compounds can be formed by linking together same or different elements. In principle, in a polar solvent such as water, bonds between same elements are more robust than those between different elements due to the same electronegativity. In water, polarized bonds are easier to cleave, which is counterproductive to the construction of stable and long chains. On the other hand, polarized bonds give biochemical construction and decomposition processes the required dynamics.

Carbon versus Silicon
The bonds of a carbon atom to other carbon atoms, i.e. C–C bonds, are usually extremely stable. The simplest organic compound based on two carbon atoms is ethane and is a long-lived component of natural gas. By further chain extension by one CH_2 unit, other hydrocarbons can be formed, which can be arranged in an **homologous series**, starting with methane. Higher molecular weight hydrocarbons are obtained from petroleum and are used, for example, in the fuel sector. Their occurrence in petroleum also points to their enormous chemical stability.

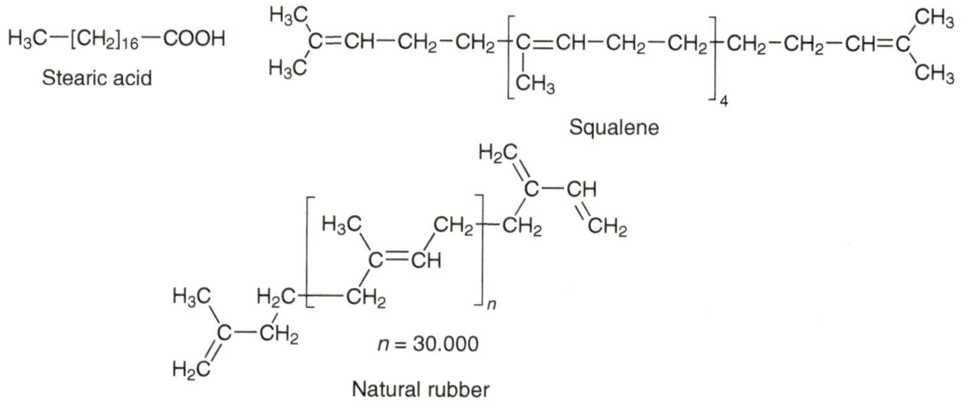

$H_3C-[CH_2]_{16}-COOH$

Stearic acid

Squalene

Fig. 1.21 Natural products with particularly long C–C chains

Organic carbon chains are almost unlimited in length. Macromolecules synthesized chemically, such as polyethylene, have average chain lengths of up to 400,000 CH_2 units.

CH_4 H_3C-CH_3 ... $H_3C-[CH_2]_n-CH_3$ $n = 400.000$

Methane   Ethane   Polyethylene

C–C chains in natural products can also reach considerable dimensions. It should be noted that these are never pure hydrocarbons, but the chains are always functionalized, either with double bonds or functional groups. Typical examples are stearic acid with a total of 18 carbon atoms, which occurs in almost all animal and vegetable fats and oils, and squalene, a natural product based on 30 carbon atoms, which is produced by all higher organisms (Fig. 1.21). Natural rubber from various rubber plants forms stable chains with up to 150,000 carbon atoms.

Another reason for the stability of hydrocarbon chains is the small electronegativity difference between C and H. The C–H bond is shorter than the C–C bond due to the small atomic radius of hydrogen, and the strength of a C–H bond even exceeds that of an average C–C bond.

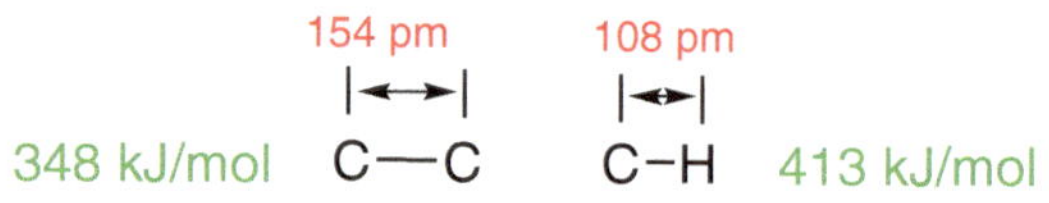

154 pm   108 pm

348 kJ/mol   C—C   C–H   413 kJ/mol

Long hydrocarbon chains thus represent C–C chains surrounded by a "shell" of hydrogen atoms. Due to the tetrahedral geometry (Sect. 1.2.10) of the four substituents around each carbon atom, the chain is presented in its most stable form as a zigzag pattern (Fig. 1.22). Other chains are aligned parallel to it. Attractive weak interactions, called van der Waals bonds, cause their cohesion. Van der Waals interactions arise due to

Fig. 1.22 The principle of van der Waals bonds

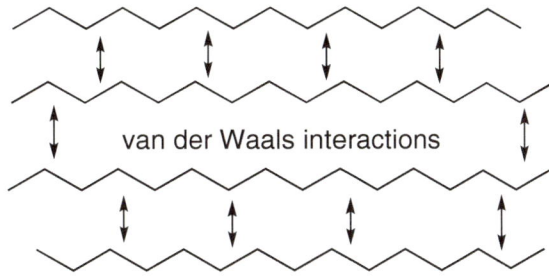

van der Waals interactions

temporary polarizations in individual atoms, which in turn induce charge differences in other atoms. They are weaker than most other types of bonds.

Hydrocarbon chains are non-polar on the outside. They do not dissolve in water; they are hydrophobic. Hydrocarbon chains organize themselves in the biochemical context after smaller chemical modifications at one end of the chain (e.g. fatty acid esters) at certain concentrations into ordered aggregates. These aggregates form the framework for more or less closed spaces in water. Such boundaries are called as **membranes** and form special microreaction spaces in biology, organelles and cells (Fig. 1.82). These compartments are the prerequisite for biochemical reactants to find each other in the context of reactions and not to be diluted immediately to infinity in the aqueous environment. After the boundary is built, it provides lasting protection. Due to the hydrophobic character of the aggregating alkyl chains, ions or polar compounds surrounded by a hydration shell must strip them off before crossing the membrane. Spontaneous diffusion is thus energetically unfavorable and the boundary for ions, as they occur in seawater, e.g. Cl^- or Na^+, almost insurmountable.

In comparison to carbon, silicon can also form chains (Fig. 1.23). Under laboratory conditions, it is possible to produce chains with up to eight silicon atoms or five- or six-membered rings such as cyclopentasilane or cyclohexasilane, i.e. Si–Si single bonds

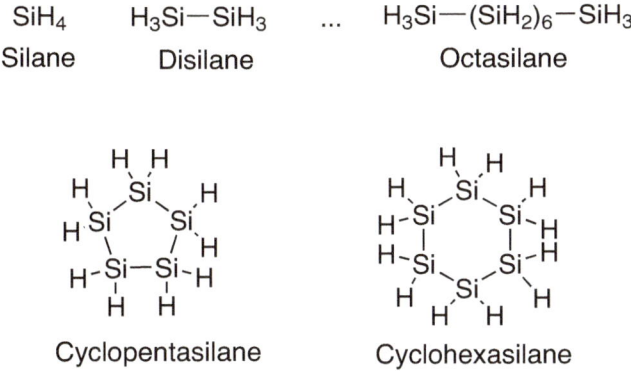

Fig. 1.23 Compounds based on Si–Si chains

are quite stable, although very limited. Only SiH$_4$ is unlimitedly stable at room temperature. The higher homologues are thermally less robust and decompose even in daylight, releasing hydrogen.

In an aerobic atmosphere, oxygen is immediately inserted into the Si–H bonds. Due to this high reactivity and the inability to form long Si–Si chains, silanes are consequently not involved in living nature, even though carbon and silicon are adjacent in the 4th main group and silicon is the second most common chemical element on Earth after oxygen.

Nitrogen

A single bond between two nitrogen atoms (nitrogen is directly adjacent to carbon in the PSE!) is significantly weaker than that between two carbon atoms. The only known stable compound with an N–N bond is hydrazine. Hydrazine is formally produced by combining two ammonia molecules by splitting off H$_2$ and can only be produced by synthetic chemical means. The bond energy in hydrazine is 159 kJ/mol. In the C-analogous ethane, the two methyl groups are held together with more than twice the energy (331 kJ/mol). At the same time, the N–N distance is significantly shorter than the comparable C–C distance, which further increases the repulsion effect.

153 pm   135 pm

331 kJ/mol   H$_3$C —— CH$_3$ H$_2$N—NH$_2$ 159 kJ/mol

Ethane   Hydrazine

The destabilization in hydrazine can be explained by the two adjacent "free" electron pairs that repel each other (Fig. 1.24). A compound with three nitrogen atoms, triazine, is already unstable for the same reason and decomposes immediately after synthesis. The structure of tetrazine has so far only been determined theoretically. The extremely limited number of nitrogen compounds is one of the main reasons why nitrogen was not considered as the basis for life during chemical evolution, even though N$_2$ with 78%

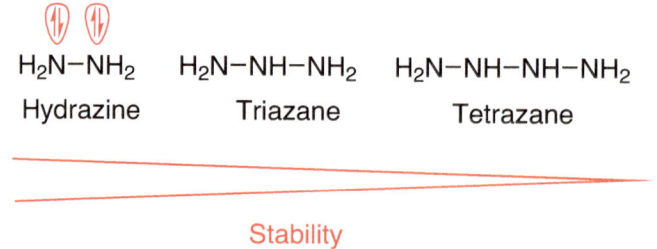

Fig. 1.24   The stability of compounds with N–N single bonds

Fig. 1.25 Atomic radii of elements of the 6th main group of the PSE

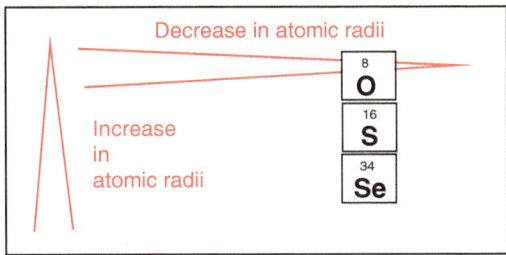

share in the Earth's atmosphere is a compound with an enormous occurrence and ubiquitous distribution.

Oxygen versus Sulfur versus Selenium

Oxygen, like sulfur and selenium, belongs to the 6th main group of the PSE. According to the periodic properties, the atomic radius of oxygen increases from oxygen to selenium, which affects the stability of chains (Fig. 1.25).

The bond between the two oxygen atoms in hydrogen peroxide is even weaker than the N–N bond in hydrazine. The bond length is also longer, which can be attributed to the repulsion of the four free electron pairs on the two oxygen atoms. Hydrogen peroxide H_2O_2 therefore decomposes very quickly into two hydroxyl radicals (Fig. 1.26).

The repulsive forces mentioned are also responsible for the fact that ozone (O_3) is an unstable molecule. The compound is part of the ozone layer of the Earth's atmosphere and decomposes under normal conditions within a few days to O_2. Compounds of oxygen with even more oxygen atoms, such as tetraoxygen (O_4, oxozone) or octaoxygen

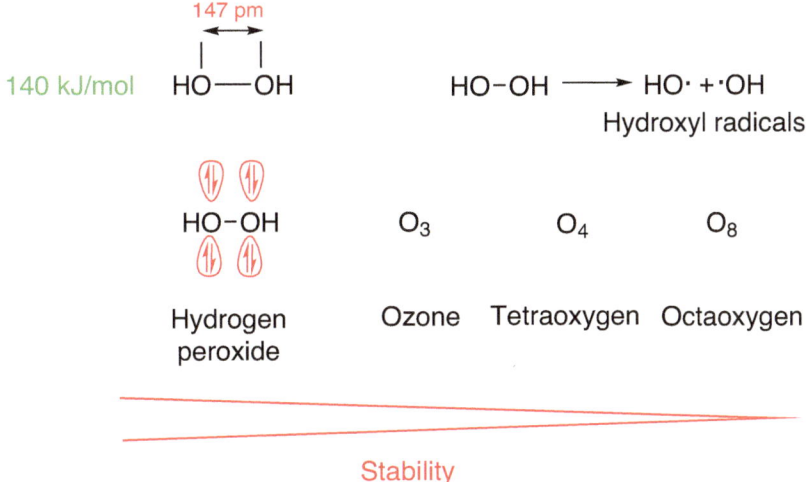

Fig. 1.26 On the stability of compounds with O–O bonds

(O_8), are of only theoretical importance. They are not stable, and their synthesis can only be achieved under laboratory conditions.

Another element that occurs in all biological systems without exception is sulfur. With an occurrence of 0.46%, sulfur ranks eighth in terms of abundance on Earth as a whole. In accordance with the periodic properties of the PSE, the radius of atomic sulfur is larger than that of oxygen (see Fig. 1.25). The repulsive forces due to the free electron pairs that were used to explain the lability of O–O bonds are therefore weaker. S–S Bonds in disulfides have a bond energy of 268 kJ/mol and are thus almost twice as strong as comparable O–O bonds. The typical occurrence of elemental sulfur is cyclooctasulfur. The fact that it occurs in nature and thus outside of a synthesis laboratory allows the conclusion that, in contrast to oxygen, S–S bonds are much more stable. This is yet another piece of evidence that the laws of the PSE explain the existence of such structures.

Cyclooctasulfur

Selenium, another element of the 6th main group, also occurs in nature. There are different modifications, including rings with eight or more selenate atoms. Although selenium is about 200 times less abundant on Earth than sulfur, it plays an important biochemical role. Selenium is an essential trace element for animals and many single-celled organisms. Just as sulfur can replace oxygen in many compounds, so can selenium. A biochemically relevant example can be found in L-selenocysteine, a selenium-analogous compound of the α-amino acids L-serine and L-cysteine. The latter two compounds belong to the **canonical proteinogenic amino acids,** i.e. those special 20 amino acids that make up proteins. L-Selenocysteine can partially take over the biochemical function of L-cysteine. It is therefore also referred to as the 21st proteinogenic amino acid. Selenocysteine can be found, for example, in the enzyme glutathione peroxidase and other selenoproteins. All three amino acids are biochemically related (Sect. 1.2.11).

L-Serine   L-Cysteine   L-Selenocysteine

Selenium occurs in yeast and plants as L-selenomethionine. It is incorporated nonspecifically into many proteins in place of L-methionine. This is an indication that some biochemical mechanisms react relatively insensitively to the exchange of elements when they are adjacent in the PSE and only offered in small amounts to the organism (micronutrients!). The switch from the sulfur-containing to the selenium-containing compound can thus be understood as part of a chemical evolution process that depends on environmental conditions (e.g. availability).

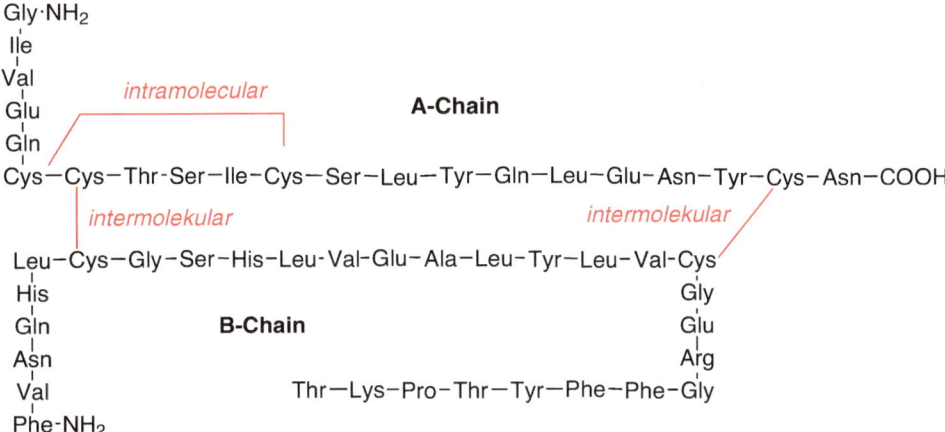

L-Methionine   L-Selenomethionine

As already mentioned, the atomic radius of oxygen is smaller than that of sulfur. While H_2O_2 and organic peroxides usually decompose immediately into radicals (Sect. 4.1.3), the analogous disulfides are more stable. The S–S bond cleavage only takes place with the help of a hydrogen donor, i.e. under reducing conditions, as shown below using the redox equilibrium between L-cysteine and L-cystine as an example. The formation of the stable L-cystine simultaneously prevents the cleavage of H_2S from L-cysteine. Since hydrogen sulfide is a gas, the disulfide bridge delays the degradation of this amino acid in the equilibrium.

L-Cystine   L-Cysteine

Disulfide bridges emanating from two cysteine residues are frequently found in natural products. For example, they stabilize the tertiary structure of many oligopeptides and proteins, with the reaction between L-cysteine/L-cystine playing a central role.

Proinsulin is a precursor of insulin, the hormone of the pancreas, and is found in all vertebrates. It receives its conformational stability through two intermolecular disulfide bridges between the A and B chains (Fig. 1.27). An intramolecular disulfide bridge in the A chain is also characteristic of this hormone.

Gly·NH₂

Fig. 1.27   The structure of proinsulin

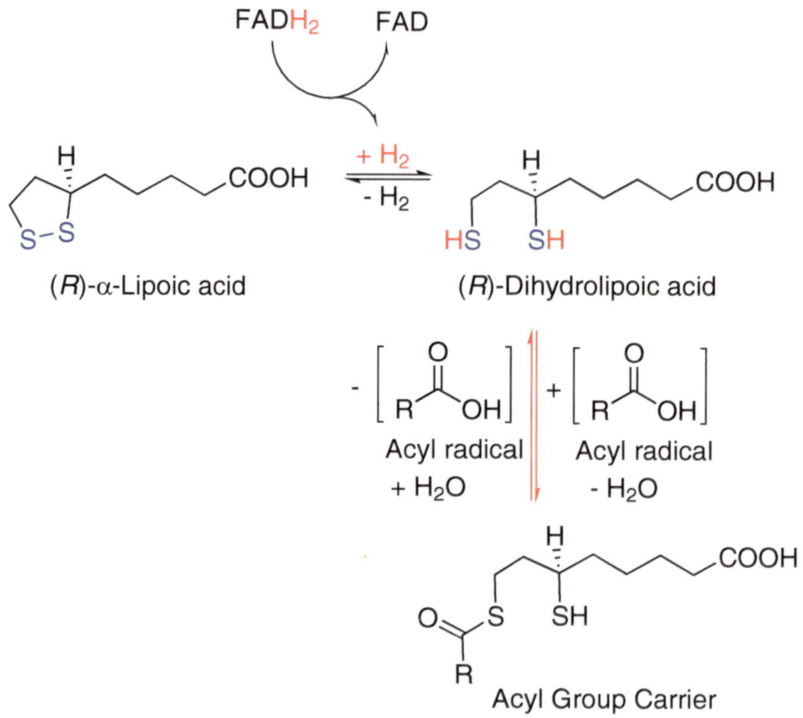

Fig. 1.28 The reversibility of the cleavage of S–S bridges

The *(R)*-α-lipoic acid, which is found in the mitochondria of almost all eukaryotes and plays a role in the oxidative decarboxylation (removal of CO_2) of pyruvate to acetyl-CoA, can only fulfill its function as an acyl group carrier in the cell in the reduced form as *(R)*-dihydrolipoic acid (Fig. 1.28). The required molecular hydrogen is supplied by the hydrogen donor $FADH_2$. Most reactions involving S–S bridges are reversible, which is not the case for unstable peroxides.

Hybrid structures based on S–Se bridges also play a role in biochemistry, e.g. in the synthesis of glutathione (GSH) (Fig. 1.29). Glutathione consists of three amino acids, including L-cysteine. By oxidation, glutathione changes from its monomeric form to the corresponding dimer (GSSG). The released hydrogen reacts to form water and thus acts as an oxygen radical scavenger [O]. GSSG is reduced again by high molecular weight hydrogen donors, such as (NADPH + H$^+$). Under biotic conditions, the balance is 90% on the left side, which explains the great protective effect against attacking oxygen radicals.

The balance between GSH and GSSG is only possible through the mediation of L-selenocysteine (Fig. 1.30). First, the oxyphilic HSe structure in L-selenocysteine is partially oxidized. Glutathione replaces the resulting HO group in the product by displacement of

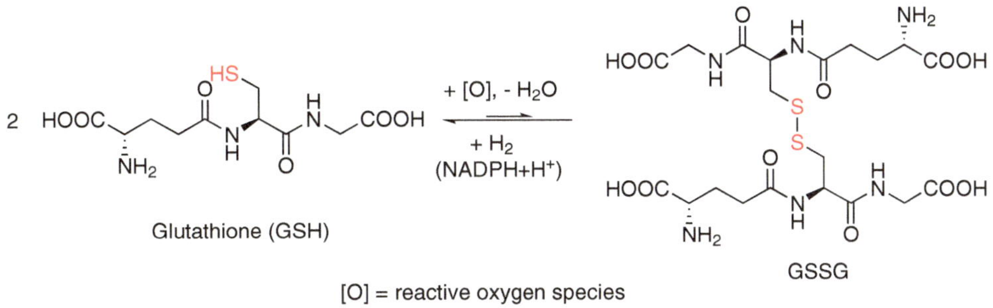

Fig. 1.29 The effect of gluthathione as an example of an oxygen scavenger

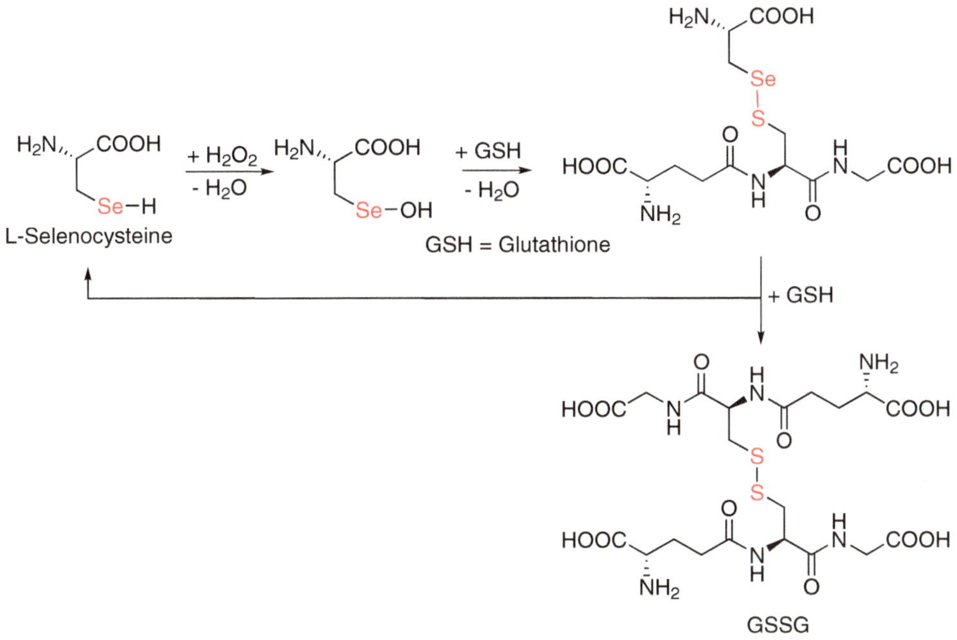

Fig. 1.30 S–Se and S–S bridges in a concerted biochemical reaction

water, forming a compound with an S–Se bridge. With a second equivalent of GSH, L-selenocysteine is displaced from it and the dimer GSSG is formed.

With the exception of some single-celled parasites, most eukaryotes are capable of GSH synthesis, which underscores the importance of disulfide radical scavengers in the "oxygen management" of living organisms. At the same time, the tripeptide GSH acts as a "reserve" of cysteine, which is continuously lost to the organism by oxidative processes or by elimination of H_2S.

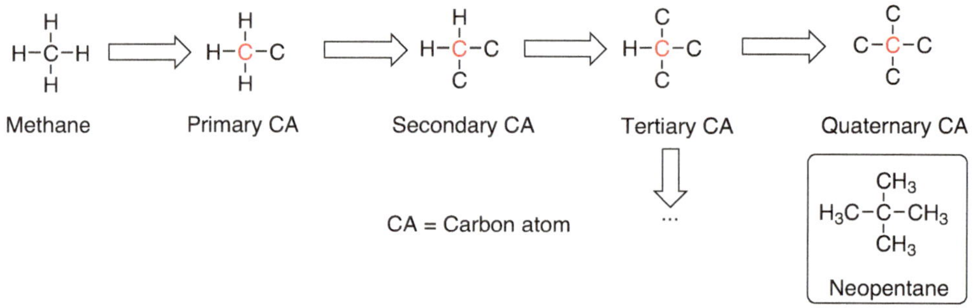

Fig. 1.31 By substituting H atoms with carbon radicals, branching possibilities arise

1.2.8 Branching of Chains

Not only the ability to form long chains gives an element an increased biochemical evolutionary potential, but it must also be able to form branched stable structures. In this respect, too, carbon shows its uniqueness compared to all other elements of the PSE. In principle, all C–H bonds can be replaced by carbon.

$$C-H \implies C-C$$

In the extreme case, in a four-valent carbon atom such as in methane, the replacement of all four hydrogen atoms by carbon atoms is possible (Fig. 1.31). In addition to chain formation, the branching potential multiplies the number of organic compounds. Carbon atoms are generally distinguished between primary, secondary, tertiary and quaternary depending on the number of bonded carbon atoms. Since no H atom is bonded to a quaternary carbon atom, oxygen cannot be inserted, as will be shown later. This brings the fundamental oxygen-initiated dynamics of biochemical mechanisms to a standstill. Quaternary carbon atoms are not only more expensive to construct, but also represent *dead-ends* in organic chemistry in terms of their oxidative breakdown. Even in crude oil, which has undergone a "maturing process" over several million years, compounds with quaternary carbon atoms, such as neopentane, are only present in small concentrations. Therefore, compounds with quaternary carbon atoms are also less common in biochemistry.

The property of branching of carbon chains enables the formation of rings with tertiary carbon atoms as bridgeheads. A biochemically relevant example is prostaglandin, the prototype of all prostaglandins (Fig. 1.32). Also possible are **annulated rings,** i.e. rings with a common bond, as they are found in bile acid. Bile acid is a end product of cholesterol metabolism.

It has already been mentioned that sulfur also forms longer stable S–S chains. However, only compounds with bridges between two sulfur atoms (disulfides) and in some exceptions S–Se bridges play a role in biochemistry. A branching of S–S chains with sulfur as terminal atom is not known in contrast to carbon. However, branching can be found in S–C compounds, i.e. with carbon as terminal atom. Biochemically relevant

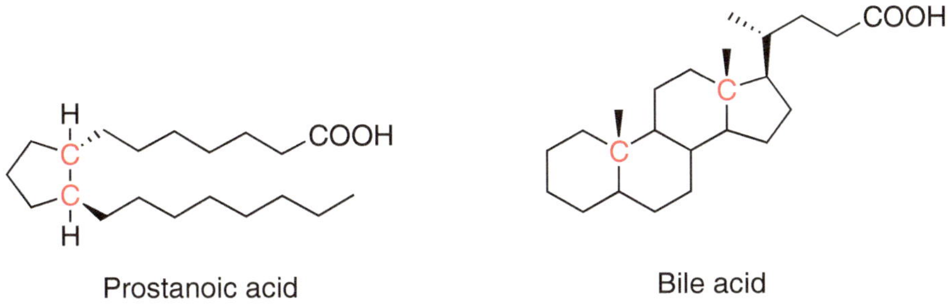

Prostanoic acid   Bile acid

Fig. 1.32 Ring formation and *annulated* rings increase the number of possible carbon compounds

is S-adenosylmethionine, which serves as a methyl group transfer agent, for example, on L-methionine. In both compounds, the sulfur atom is positively charged, which enables the reversible transfer of methyl groups (Fig. 1.33).

The comparison of the stability of element-element bonds such as C–C, N–N, O–O, S–S, Si–Si proves that of all the elements of the PSE, only carbon is suitable for forming almost an infinite number of stable element-element bonds. These structures in turn serve as the backbone for almost an infinite number of compounds that, due to their outstanding importance in the form of organic chemistry, have founded their own independent field of science. Biochemistry is for the most part organic chemistry!

1.2.9   Stoichiometric Valences, Element-Hydrogen Compounds, Molecule Geometries

Stoichiometric valences of the main group elements are also a periodic function of the atomic numbers. In hydrogen compounds, they increase along the period from one to four. After that, they decrease again to one (actually to zero in the case of the noble

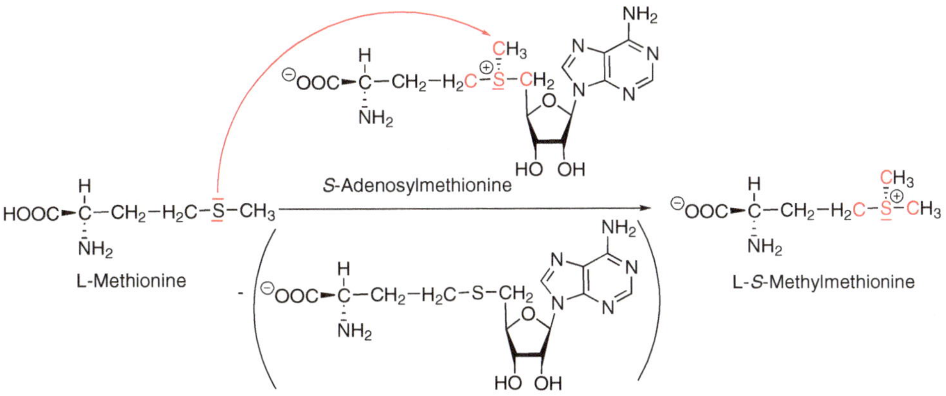

Fig. 1.33 Branching possibilities with S as bridge atom

gases). Therefore, the relevant lithium-hydrogen compound has the stoichiometric composition LiH, that of beryllium BeH_2, that of boron BH_3 etc.

2. Period $\boxed{LiH}$ $\boxed{BeH_2}$ $\boxed{BH_3}$ $\boxed{CH_4}$ $\boxed{NH_3}$ $\boxed{H_2O}$ $\boxed{HF}$

3. Period $\boxed{SiH_4}$ $\boxed{PH_3}$ $\boxed{H_2S}$ $\boxed{HCl}$

4. Period $\boxed{H_2Se}$ $\boxed{HBr}$

Carbon as an element of the 4th main group has four valence electrons that are available for forming up to four bonds. The simplest carbon-hydrogen compound of carbon is therefore methane CH_4. The nitrogen atom in ammonia NH_3 can, in contrast to carbon, only form three bonds to hydrogen due to the smaller atomic radius (Fig. 1.2). In water H_2O, the even smaller oxygen atom can only bind two hydrogen atoms. These relations can also be transferred to heavier elements. For example, the simplest element-hydrogen compounds of silicon, phosphorus and sulfur have the atomic compositions SiH_4, PH_3 or H_2S. In this way, it can also be explained why selenium-hydrogen must have the formula H_2Se. For the corresponding chlorine or bromine compound, the formula HCl or HBr results necessarily.

Carbon reaches the noble gas configuration of neon with eight valence electrons in CH_4, which explains the stability of the corresponding compounds (Fig. 1.34). A hypothetical structure of CH_5 is therefore not existent. Covalent bonds with their two binding electrons are places of increased electron density. Adjacent bonds therefore repel each other, which has a decisive influence on the geometry of the molecule. On the carbon atom in methane, four hydrogen atoms are bound with a total of eight valence electrons. Due to the mutual repulsion of the four electron pairs, the molecule is not planar, but takes on the geometric shape of an ideal tetrahedron, in which each H–C–H

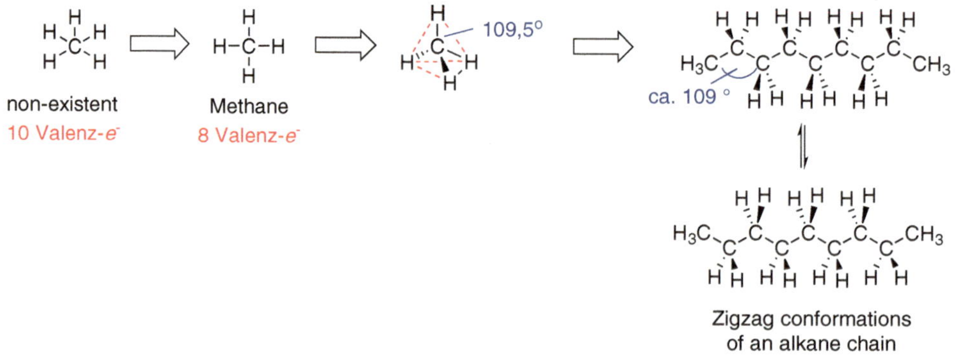

Fig. 1.34 The geometry of methane and higher alkane homologues

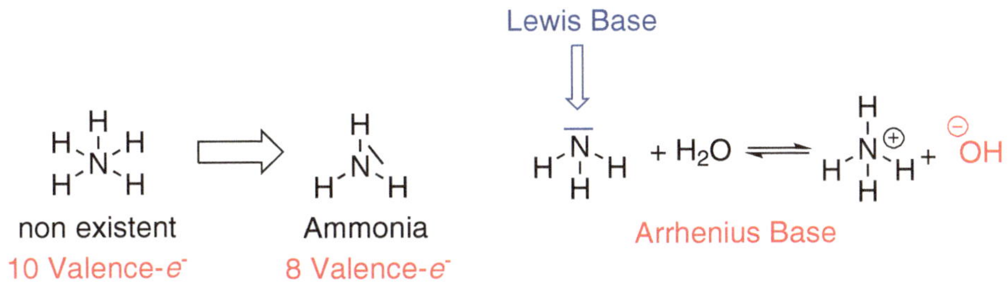

Fig. 1.35   The structural cause of the basicity of ammonia

angle measures 109.5°. This geometry around a carbon atom also changes only slightly in longer hydrocarbon chains. Even a C–C–C angle is approximately 109°. Such chains therefore take on zigzag conformations (Fig. 1.22), which merge into each other without barriers at room temperature. In this context, **conformations** describe the different spatial arrangements of atoms or groups of atoms around a carbon atom, which arise by rotation around single bonds.

If you combine this geometry-related analysis with a lawfulness already discussed above in the PSE, the size of the atomic radii in a period (Fig. 1.2), this results in a simple explanation for the basicity of ammonia: In contrast to carbon, the nitrogen atom is smaller, and thus the expected coordination number of five is not achieved in a hypothetical compound of the formula NH_5 (Fig. 1.35). This means that of the five **valence electrons** of the nitrogen atom, only three are available for the bonds with hydrogen, the remaining two remain as a "free" electron pair on nitrogen. According to the Arrhenius acid-base theory, bases are compounds that generate HO^- ions in water, which is actually the case for ammonia. In the formed ammonium ion NH_4^+, nitrogen indeed binds four protons, but still does not exceed the number of eight valence electrons. Ammonia also meets the definition of the **Lewis acid-base theory,** according to which bases have a free electron pair.

The explanation of the limited number of valence electrons is also suitable to understand the so-called **octet rule**, a rule that has already been classical, which says that elements of the first octave (Li, Be, C, N, O, F) only take up a maximum of eight valence electrons in all of their compounds. The central atoms have the tendency to adopt the next noble gas configuration, but not to exceed it. Also from this rule of the PSE it becomes clear why NH_5 with ten valence electrons does not exist, but NH_3 does. Ammonia reaches the noble gas configuration of neon with the eight valence electrons, comparable to carbon.

Free electron pairs—like covalent bonds—occupy space as locations of increased electron density, which affects the geometry of the overall compound. Thus, the free electron pair on the nitrogen atom is responsible for the fact that ammonia is not a planar molecule, but rather approximately assumes the geometric shape of a tetrahedron with

nitrogen in the center. A similar three-dimensional geometry is possessed by the ammonium ion, which represents the corresponding acid to the base ammonia.

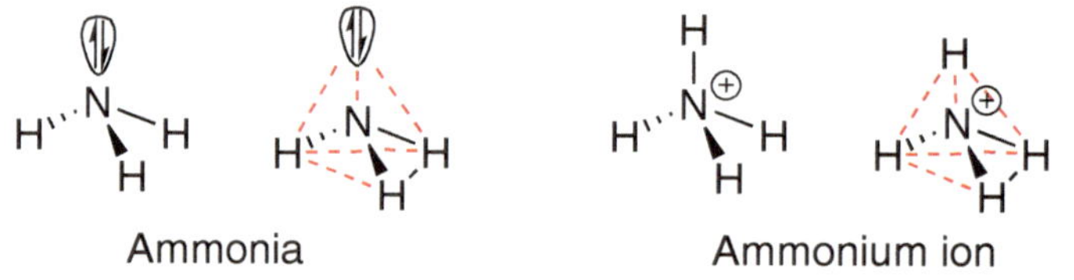

Ammonia   Ammonium ion

Oxygen has an even smaller atomic radius than nitrogen, so only two hydrogen atoms
"fit" around the central atom. Water therefore has the molecular formula H_2O. The molecule is not planar due to the two free, repelling electron pairs, which has grave consequences for the physical properties of water: Positive and negative charges do not come
together in the center of the molecule, but are separated. Thus, water is a dipole and
interacts with other water dipoles via attractive **electrostatic interactions**. In addition to
the formation of hydrogen bridges (Sect. 1.2.11), this is one of the reasons for the high
boiling point of water. If water were a gas under the conditions of Earth, biochemical
reactions in solution could not take place, which is one of the most important conditions
for any life on Earth.

Water   Water as a dipole   Binding interactions
                                                       due to dipoles

Free electron pairs, as noted above for ammonia, are, according to the acid-base theory
of Lewis, the hallmark of bases. Therefore, water also belongs to them. By means of the
free electron pairs on the oxygen atom, water molecules are able to attack carbon atoms
that carry a positive partial charge (δ^+), as demonstrated by the hydrolysis of an organic
ester to the carboxylic acid (Fig. 1.36).

Carboxylic ester   Carboxylic acid

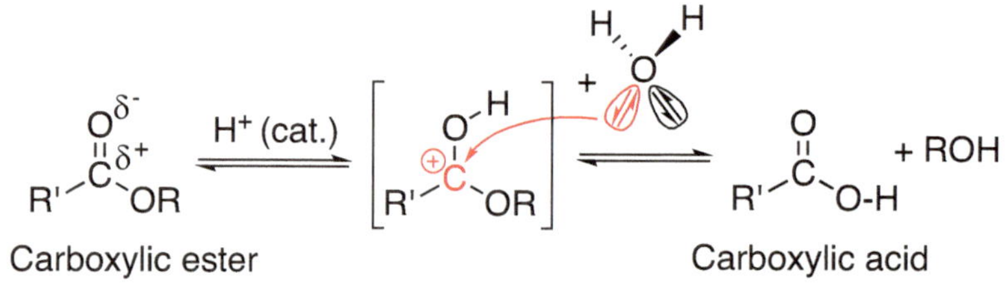

Fig. 1.36 Water as a Lewis base and reaction partner

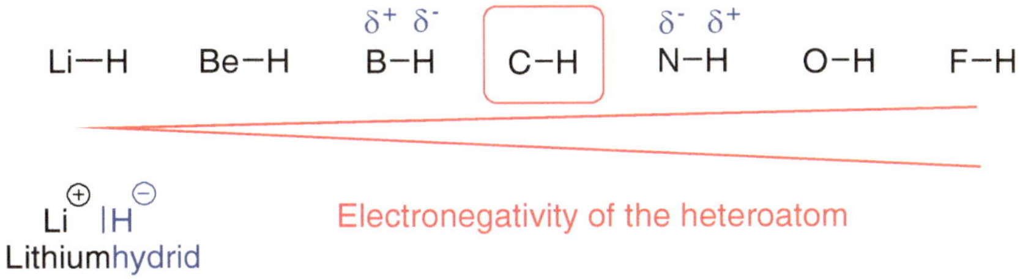

$$\text{Li}-\text{H} \quad \text{Be}-\text{H} \quad \overset{\delta^+ \;\; \delta^-}{\text{B}-\text{H}} \quad \boxed{\text{C}-\text{H}} \quad \overset{\delta^- \;\; \delta^+}{\text{N}-\text{H}} \quad \text{O}-\text{H} \quad \text{F}-\text{H}$$

$$\overset{\oplus}{\text{Li}} \; |\overset{\ominus}{\text{H}}$$
Lithiumhydrid

Electronegativity of the heteroatom

Fig. 1.37 Effects of electronegativity on the polarization of X–H bonds

In biotic systems, the reaction with water only takes place after the substrate has been activated by small (catalytic) amounts of acids or bases. In this way, numerous vital functional groups, including structure-forming (membranes), information molecules (nucleic acids and proteins), and storage molecules (cellulose, starch), are destroyed. For this reason, it is logical and imperative that almost all processes in living nature take place in an approximately neutral environment. Water has no way of bringing its properties as a Lewis base into play without external support, even though it is present in extremely large quantities as a solvent.

The direction of the polarity and the extent of the electronegativity differences in X–H bonds allow conclusions to be drawn about their polarization and allow estimates of the resulting reactivity. In aqueous systems, the small difference in electronegativity between C and H in a C–H bond serves as a reference (Fig. 1.37).

For elements in the PSE to the left and right of carbon, the bond to hydrogen is polarized in different directions. In X–H compounds with elements X to the left of carbon, the binding electron pair increasingly moves to the side of the hydrogen atom. In compounds with a B–H or a Be–H bond, the hydrogen atom is already the more electronegative partner. Compounds such as borane (B_2H_6) can therefore only be used in synthetic reactions in the absence of water. The bond is particularly polarized in lithium hydride, which is also reflected in the name "hydride" for a negatively charged hydrogen atom (H−). Lithium hydride is a salt based on lithium cations and negatively charged hydrogen ions. The interatomic cohesion is caused by **ionic bonds**. Due to the negatively charged hydrogen, LiH reacts violently with the protons of water, resulting in molecular hydrogen

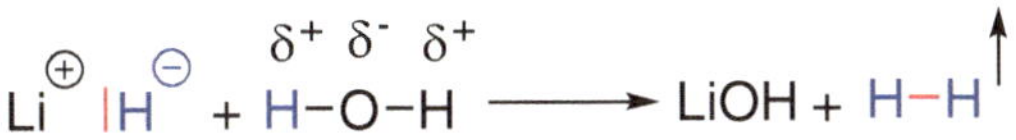

$$\overset{\oplus}{\text{Li}} \; |\overset{\ominus}{\text{H}} + \overset{\delta^+ \;\; \delta^- \;\; \delta^+}{\text{H}-\text{O}-\text{H}} \longrightarrow \text{LiOH} + \text{H}-\text{H}\uparrow$$

Since the electronegativity also decreases with increasing atomic mass within a main group, it becomes clear that, in comparison to a C–H bond in methane, the polarity is inverted in the Si–H bond of silane. In the Si–H bond, hydrogen is already the

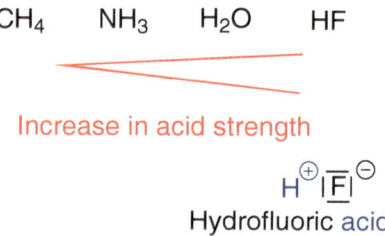

Fig. 1.38 Influence of the position of the central atom in the PSE on the acidity in X–H compounds

more electronegative partner, another reason for the subordinate role of silicon in life processes.

Methane   Silane

The basic conclusion for biochemistry is that element-hydrogen compounds with more electronegative elements than carbon, that is all metals and metalloids, cannot occur in nature: they would immediately and partly explosively react with water. Synthesis and storage are only possible in the absence of water (exceptions see Sect. 2.2.3).

Due to the approximately equal electronegativities of carbon and hydrogen, a situation that leads to the formation of covalent bonds, methane under biotic conditions (in water!) does not release hydride ions (H^-) or protons (H^+).

$$\overset{\oplus}{C}H_3 \;+\; H|^{\ominus} \;\longleftarrow\!\!\!\times\; CH_4 \;\times\!\!\!\longrightarrow\; |\overset{\ominus}{C}H_3 \;+\; H^{\oplus}$$
Hydride   Proton

Methane is thus inert in the absence of light or higher temperatures and can be stored in the earth as natural gas for millions of years without undergoing chemical change. This also applies to all other hydrocarbons that are components of crude oil.

Hydrogen compounds with more electronegative elements such as nitrogen (N) or oxygen (O) are chemically stable in water in contrast to those with elements that are to the left or below carbon in the PSE, since in these compounds hydrogen always carries the positive partial charge. The polarization and thus the acidity increases in the direction of the halogens (Fig. 1.38).

For example, NH_3 is not a Brønsted acid in water, i.e. ammonia does not dissociate into NH_2^- anions and protons. In contrast, the **autoprotolysis of water** is known, which is the dissociation of H_2O into hydrated protons (H_3O^+) and HO^-. However, the equilibrium lies far on the side of (undissociated) water. The ion product for this reaction is at 298 K (25 °C) about $10^{-14}\,mol^2\,l^{-2}$ and is thus very small. This is another prerequisite

Fig. 1.39 The ratio of element-hydrogen acids and corresponding bases in the 6th main group

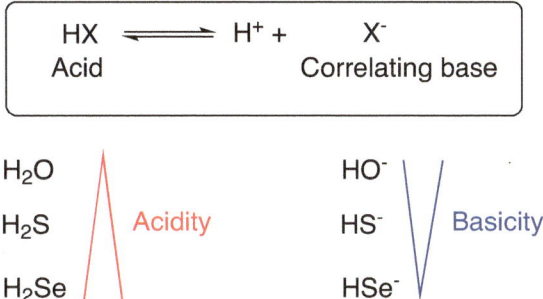

for the fact that water exclusively as a solvent for all biotic reactions on Earth is possible and in the absence of acids or bases (or the corresponding acidic or basic enzymes) is considered inert.

$$2\,H_2O \rightleftharpoons H_3O^{\oplus} + OH^{\ominus}$$

The very small concentrations of HO^- and H_3O^+ in nature are also the reason why those oxidation reactions that only become possible through strong changes in the pH value do not play a role (Sect. 2.2.1).

In the case of very large electronegativity differences between the binding partners, as in fluorohydrogen, the bond to hydrogen is particularly polarized, so that in an aqueous solution a strong acid, in this case hydrofluoric acid, results (Fig. 1.38). As already mentioned, strong acids like most mineral acids (but also bases like potassium or sodium hydroxide) only exist in the form of their salts on Earth or they are only found very rarely as an acid and then only in very extreme environments. A well-known exception is stomach acid (highly concentrated aqueous HCl) with a pH $= 1-3$. The aggressive effect is limited by mucus formation and a bicarbonate buffer, but this is not sufficient. Therefore, the acid destroys stomach cells on average every 3–5 days.

In the same main group of the PSE, the acidity of the element-hydrogen compounds increases, as the example of X–H compounds of the 6th main group illustrates (Fig. 1.39).

This means that hydrogen sulfide is more readily protonated than water. Even lower is the activation energy for the generation of HSe^- from H_2Se. This affects the base strength of the corresponding anions: In the case of an **corresponding acid-base pair**, a weak acid is in equilibrium with the corresponding strong base or a strong acid with the corresponding weak base. If this property of the PSE is combined with another, namely the size of the ionic radii, it is possible to estimate the direction of substitution reactions (Sect. 1.2.12). HS^- and HSe^- are much larger than HO^- and therefore more stable in water.

1.2.10 Molecule Geometries as an Evolutionary Criterion

The number of substituents around a central atom determines the geometry of the overall molecule. As already described, binding or free electron pairs repel each other as sites of increased electron density. As a result, compounds of the four-valent carbon with four substituents are not planar, but assume the geometric shape of a tetrahedron. Carbon dioxide is flat, while methane is three-dimensional.

2-dimensional   3-dimensional

The transition from two-dimensionality to three-dimensionality is a step towards higher complexity, which will be demonstrated later using the example of phosphoric acid esters and many organic carbon compounds.

If all four substituents at a central carbon atom are different, molecules are formed that behave like a picture to a mirror image. Because of the similarity of this phenomenon to the ratio of left to right hand, one speaks of chirality (Greek: Handiness). The central carbon atom is referred to as the chirality center, the "handy" molecules derived from it form an enantiomer pair. Enantiomers have the same energy content and therefore do not differ in physical and chemical properties.

Chiral carbon compound   Enantiomer pair

A frequently cited example concerns the amino acid L-alanine. For the assignment, according to a proposal by Emil Fischer (Fischer projection), in a two-dimensional representation, the COOH group is positioned at the top of the carbon chain and the NH_2 group to the left of it. After transfer to the actually existing tetrahedral form and the use of corresponding bond symbols (filled and interrupted wedges), different representation forms of the same compound can be graphically represented.

L-Alanine

The corresponding D-enantiomer is mirror-image constructed.

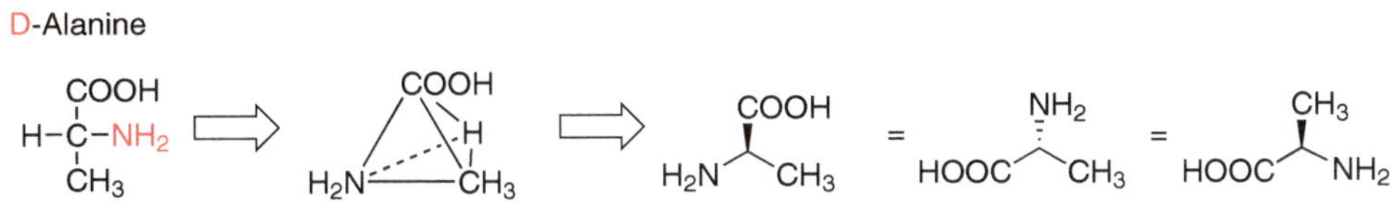

D-Alanine

If you combine several chirality centers, the number of variants increases according to the formula 2^n (where $n =$ number of chirality centers), which are called diastereomers. Diastereomers differ in their physical and chemical properties. Examples are sugar with six carbon atoms, four of which are chiral. Prominent representatives are D-glucose or D-mannose (Fig. 1.40).

In nature, only one enantiomer from a pair of enantiomers is mainly found. For example, all proteinogenic amino acids are L-configured. The majority of sugars belong to the D-series, with the orientation of the HO group at the last chiral carbon atom determining the assignment. L-Glucose does not exist in nature. This phenomenon is called homochirality. The cause lies in the selection at higher levels of complexity, where diastereomeric compounds or associates of different energy content compete with each other.

For amino acids, these are the secondary, tertiary and quaternary structures of proteins. The incorporation of a single "wrong" enantiomeric amino acid in the primary structure (sequence of amino acids) already leads to considerable disturbances in the next higher level of organization, which can lead to collapse and thus prevent the achievement of the next quality. Thus, the choice of the first enantiomeric amino acid from a pair of enantiomers opens a selection channel that subsequently allows no alternatives.

The dominance of D-glucose in nature can also be rationalized in this way. Each of the four chirality centers makes a contribution, which is reflected in the preferred chair conformation of all substituents in the energetically favorable equatorial orientation. Section 4.1.5.2 shows how the individual glucose molecules combine to form polymers.

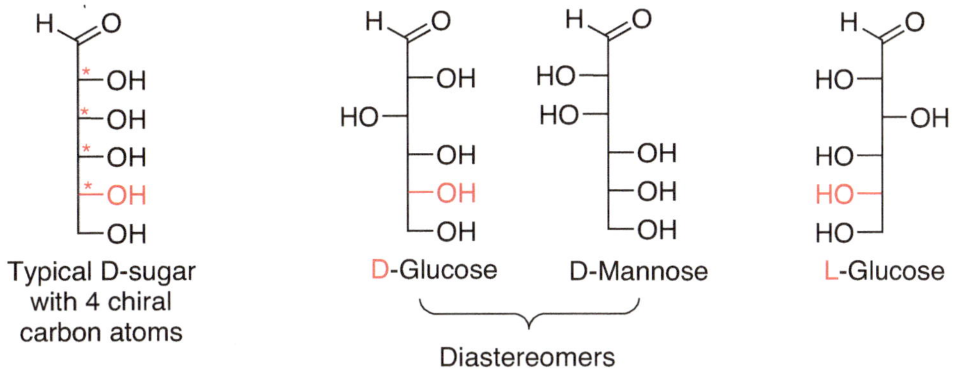

Typical D-sugar with 4 chiral carbon atoms   D-Glucose   D-Mannose   L-Glucose

Diastereomers

Fig. 1.40 Diastereomer formation using carbohydrates as an example

Only with the particularly stable glucose can the longest chains and most stable biomolecules be built compared to other six-carbon sugars.

The selection between L- and D-enantiomers, regardless of the chemical compound class, is a decision-making process. Each decision has a certain amount of information. The combination of several chirality centers multiplies this information. Thus, each biomolecule carries the information of its own formation within it, which is particularly clearly visible in the case of chirality. This applies to all current biochemical processes, but also to the evolution of life on Earth.

Deviations from the rule of homochirality are found in nature primarily at the low molecular level beyond the large compound classes with their uniform, mutually penetrating biochemical processes. They give some biological species their uniqueness. For example, Archaea and bacteria differ in the chirality of glycerol derivatives, which are the building blocks of their membranes. Different enzymes are involved in their synthesis, which transfer H_2 (in chemically bound form as NADH + H⁺) to the common reactant.

Since during the evolution of these single cells probably the present chloroplasts and mitochondria have developed by symbiosis (endosymbiont theory), the finding that these organelles still exist separately in the cells with their own DNA is an indication that this state could also have arisen through the incompatibility of these two enantiomeric basic building blocks. The fact that the same chirality is also found in eukaryotes shows the generic similarity.

Differences in homochirality and the resulting lack of compatibility can ensure the integrity of an organism against others. D-Alanine and D-glutamic acid are building blocks of the bacterial cell wall and offer protection to host organisms against enzymes that only recognize and break down proteins based on L-amino acids.

Even in higher organisms, deviations from homochirality are observed. D-Serine is one of the few amino acids that are formed by a configuration inversion from L-serine. In humans, it serves as a signaling molecule in the brain. D-Asparagine is also found in mammals. Both D-amino acids do not belong to the proteinogenic amino acids, which illustrates the evolutionary pressure exerted by hierarchically superior molecules, here the homochiral proteins, on the selective synthesis of enantiomerically pure compounds.

HO ⌇ COOH   HO ⌇ COOH   HOOC ⌇ COOH
     NH₂   NH₂   NH₂
   L-Serin   D-Serin   D-Aspartic Acid

1.2.11 Hydrogen Bonds

In X–H bonds with strongly electronegative partners X of the hydrogen, which bears the positive partial charge in such bonds, the tendency to form hydrogen bonds (Fig. 1.41) increases in the PSE. The same or different elements can be involved in a hydrogen bond.

Since the electronegativity difference between C and H is very small, methane does not form hydrogen bonds with other methane molecules. Therefore it is a gas. The strongest hydrogen bond is found in hydrofluoric acid (HF), where a hydrogen atom between two fluorine atoms mediates the bond. Even long, hydrogen-bridged chains are formed. This is the reason for the much higher boiling point of HF compared to HCl, although hydrochloric acid has the larger molar mass.

Hydrogen bonds are generally weaker than covalent bonds or ionic relationships. Hydrogen bonds, for example, are relevant in the field of physical chemistry to explain whether the compound is a gas or a liquid under normal conditions.

Strong intermolecular hydrogen bonds (Fig. 1.42) are responsible for the unusually high melting and boiling point of water and, inter alia, explain its unique position as a solvent on Earth. Due to the angled structure of the water molecule (Sect. 1.2.9),

Fig. 1.41 The tendency of X–H bonds to form hydrogen bonds in the PSE

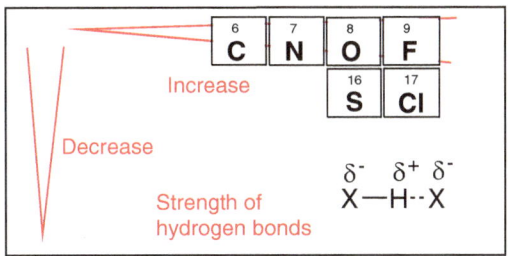

Fig. 1.42 The hydrogen bonds
in water

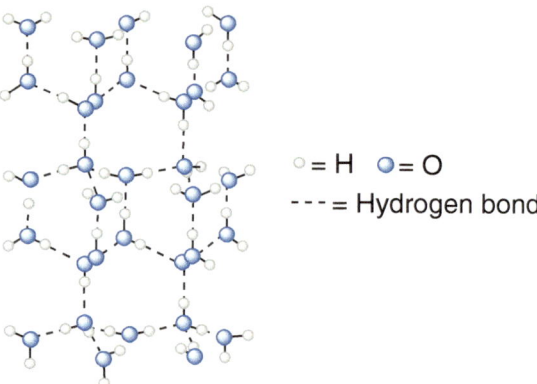

$\circ$ = H $\bullet$ = O
--- = Hydrogen bond

a three-dimensional structure is formed from dynamic hydrogen bonds in liquid water, which is even more ordered in ice crystals.

One period further down in the PSE, in the 6th main group, namely in hydrogen sulfide, hydrogen bonds are negligible weak: H_2S therefore has, in contrast to water, a very high vapour pressure and is a gas. This is also a prerequisite for being able to smell H_2S, but not H_2O. This also applies to many organic compounds containing sulfur in low oxidation states. Ammonia is also gaseous under normal conditions due to the smaller electronegativity difference between N and H and the absence of hydrogen bonds.

The formation of gaseous products in biochemical degradation mechanisms (**catabolism**) is the cause of the dynamic character of all living processes. Just as CO_2 (also a gas!) drives the degradation of carbon-containing structures according to the principle of Le Chatelier-Braun, the formation of NH_3 or H_2S (or of their lower molecular weight organic derivatives) from higher molecular weight nitrogen- or sulfur-containing natural products shifts the balance. This general principle is not changed by the partial solution of the gases in water or the formation of salts (carbonates or ammonium ions) due to the limited gas solubility.

High molecular $\longrightarrow$ CO_2, NH_3, H_2S
weight natural
products   Gases

One of the decisive requirements for this fundamental phenomenon in animate nature is thus, inter alia, the absence of hydrogen bridges in the lower molecular weight products. The cleavage of water with its high boiling point is not a driving force from this point of view.

This fact also has an effect on the properties of construction reactions. For example, the addition of water to C=C double bonds is very common in biochemical processes.

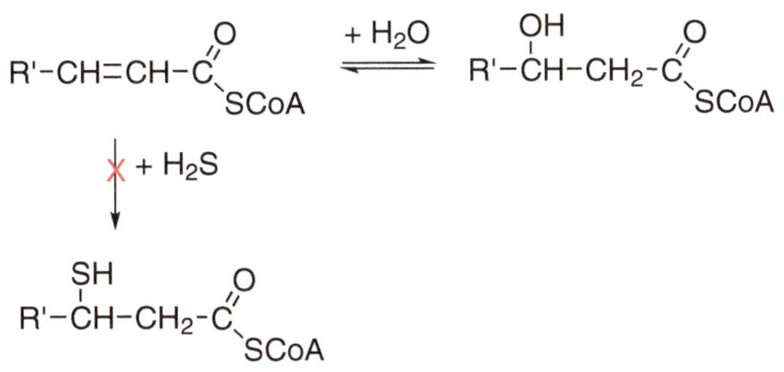

Fig. 1.43 The biochemical reaction difference between water and hydrogen sulfide

One example is the conversion of an unsaturated fatty acid ester during β-oxidation into an alcohol (Fig. 1.43).

This reaction does not take place with H_2S instead of H_2O, since H_2S would be immediately removed from the equilibrium as a gas. In such cases, the principle of "molecular weight increase" is effective in biochemistry, i.e. the volatile reagent is "incorporated" into an auxiliary compound with a high molecular weight and a low vapour pressure.

A typical case is the biosynthesis of L-cysteine from L-serine (Fig. 1.44). From L-serine, water is first split off to form an unsaturated amino acid. The addition of H_2S from any source that leads directly to L-cysteine is not possible for the above-mentioned reason. This transformation only succeeds under high pressure in an autoclave and thus in a synthesis laboratory. Under biotic conditions, therefore, a "detour" is taken: First, homocysteine, an HS-donor, is added to the double bond. Homocysteine, due to its large molar mass, is not a gas and represents a synthesis equivalent to H_2S. Under the influence of water, the formed cystathionine decomposes into ammonia and pyruvic acid and thus the

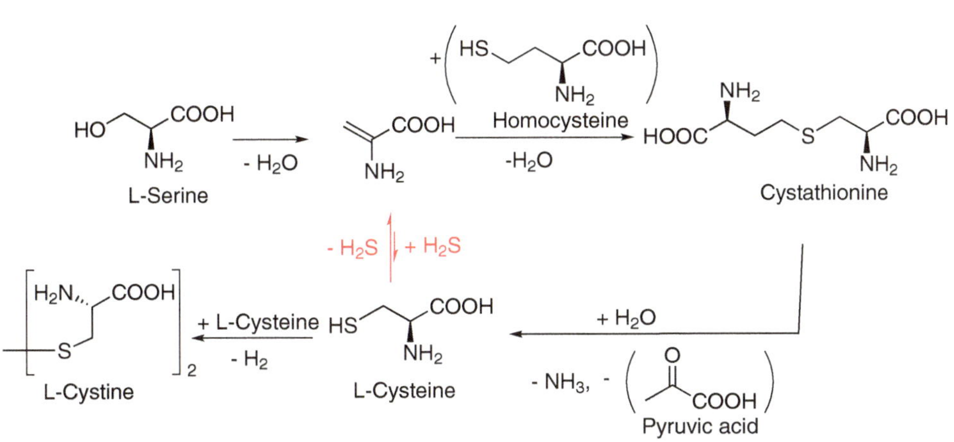

Fig. 1.44 The property-related "detour" for the biochemical synthesis of L-cysteine from L-serine

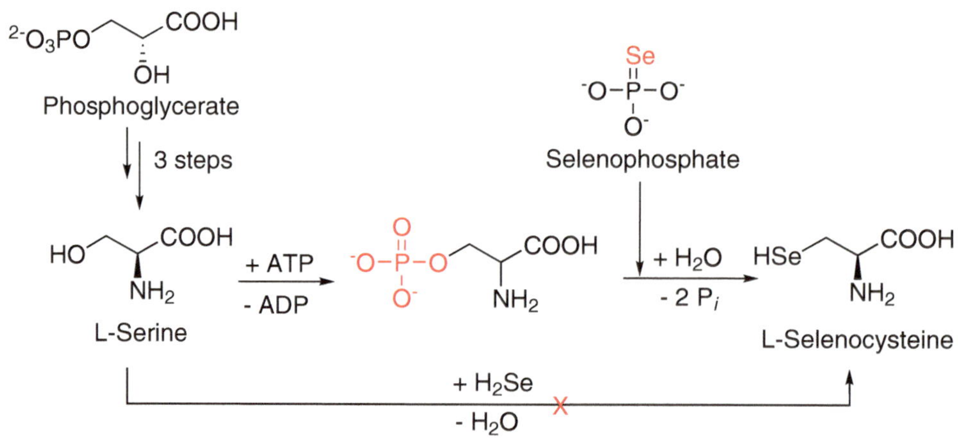

Fig. 1.45 Synthesis of L-selenocysteine from L-serine mediated by activation steps

"auxiliary reagent" for the introduction of the SH function. Cysteine is formed in parallel. The tendency to eliminate H_2S from cysteine is "slowed down" by the disulfide bridge in L-cystine.

Selenocysteine biosynthesis proceeds according to a similar principle (Fig. 1.45). In this case, the high-molecular-weight selenophosphate serves as a "selene donor" (Fig. 1.51). The properties of the poor leaving group HO^- (Sect. 1.2.12) in L-serine are improved by esterification with phosphate. After replacement of OH with SH, the two "helper reagents" that have enabled the exchange are recycled in the form of phosphate back into the organism's phosphate cycle.

In contrast to the syntheses of cysteine or selenocysteine, L-serine is produced from phosphoglycerate, a breakdown product of glycolysis. The HO group is a relic of glucose, i.e. it was already present in the molecule from the beginning. The three examples show that the three amino acids are synthesized in different ways, even though they only differ in one single heteroatom, which also belongs to the same main group. The causes of the varying mechanisms include, among other things, the different ability to form hydrogen bonds in compounds of the type H_2X (with $X = O$, S, Se), which result in different aggregate states and lead to different construction paths.

1.2.12 Nucleophilicity

The exchange of functional groups is a way to generate diversity in organic chemistry. Since all biochemically relevant functional groups are linked to the more electronegative element carbon, nucleophiles, i.e. negatively charged particles, are predestined for the exchange. The reaction is referred to as **nucleophilic substitution**. One differentiates between the attacking nucleophile (Nu), which is usually characterized by a free electron pair or carries a negative charge, and the departing leaving group Y.

The success of this reaction depends on the properties of the nucleophile and the leaving group as well as the geometry of the carbon atom attacked. In laboratory chemistry, a wide range of reagents and methods are available to enable this reaction. This includes particularly the variation of the solvent, an option that is irrelevant in biochemistry in the exclusive solvent water.

The nucleophilicity of a particle depends on its polarizability. Polarizability is a measure of the shiftability of a charge cloud in the molecule, which is induced by applying an external electric field. In a chemical reaction, nucleophile and substrate form the electric field and thus a reactive unit. Polarizability is a function of the properties of basicity and ion size of the nucleophile. In group 6, the basicity of anions of the type HX^- decreases from top to bottom and thus the tendency to take up a proton. These anions are the corresponding bases of increasingly stronger acids (Fig. 1.39). The ion size tends to increase from top to bottom and thus also the deformability of the particle. Basicity and polarizability in the PSE can thus act in opposite directions.

Especially important in the biochemical context is the exchange of hydroxy groups in alcohols by amino groups. This creates amines. This reaction should take place most efficiently by reaction of the alcohol with ammonia. Also the reverse reaction, the exchange of an amino group by a hydroxy group by means of water, would be the quickest way. In principle, ammonia is less acidic than water, which means in reverse, amide (NH_2^-) is the stronger base in comparison to hydroxide (HO^-) (Fig. 1.46). At the same time, the slightly larger ionic radius of amide compared to hydroxide contributes to a better polarizability.

The direct exchange of OH for NH_2 is not observed under biotic conditions, however. There are several reasons for this: Small particles, such as NH_3 or H_2O, are in principle only slightly polarizable and therefore poor nucleophiles. In addition, ammonia is

Fig. 1.46 Basicity and ionic size in comparison between ammonia and water

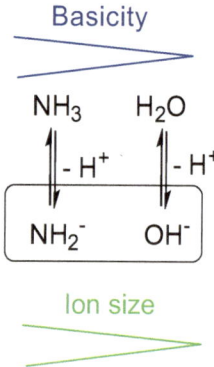

present in water primarily as an ammonium ion due to its basicity, which further reduces its nucleophilicity (1st argument).

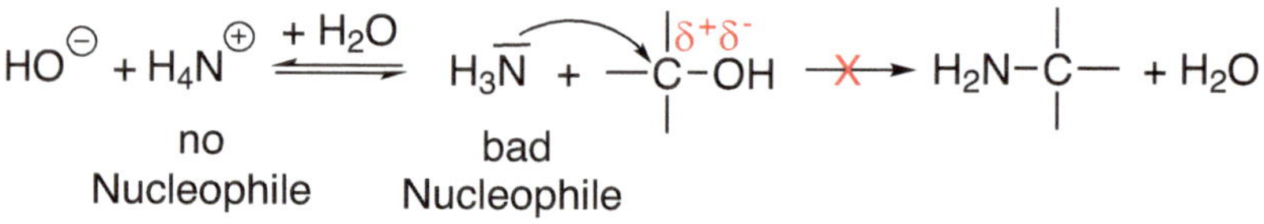

$$HO^{\ominus} + H_4N^{\oplus} \xrightleftharpoons{+ H_2O} H_3\overline{N} + {}^{\delta^+\delta^-}_{}C\text{-OH} \xrightarrow{\ \times\ } H_2N\text{-}C\text{-} + H_2O$$

no   bad
Nucleophile   Nucleophile

The exchange would be conceivable by increasing the basicity of the nucleophile, for example by converting ammonia into the stronger base amid NH_2^- (Fig. 1.47) However, under biotic conditions, amid does not exist because its generation requires an even stronger base that does not exist (2nd argument).

The hydroxide anion is a poor leaving group due to its small size and the resulting negative point charge. Protonation and the formation of water would greatly improve its leaving properties. But this would lead to the parallel formation of NH_4^+ and completely eliminate the already weak nucleophilic properties of ammonia (3rd argument, Fig. 1.48).

Similar arguments also speak against the suitability of water as a nucleophile. These relations are responsible for the fact that the substitution of HO for NH_2 and vice versa

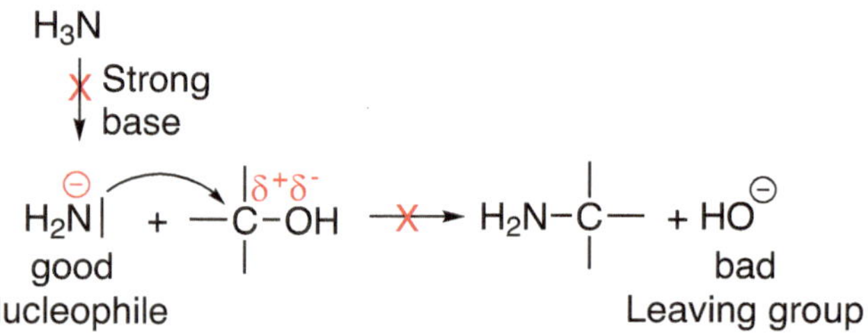

$$H_3N$$
$$\downarrow \text{Strong base}$$
$$H_2N^{\ominus}| + {}^{\delta^+\delta^-}_{}C\text{-OH} \xrightarrow{\ \times\ } H_2N\text{-}C\text{-} + HO^{\ominus}$$

good   bad
Nucleophile   Leaving group

Fig. 1.47 Why the direct exchange of HO for NH_2 does not work in biochemical systems (2nd argument)

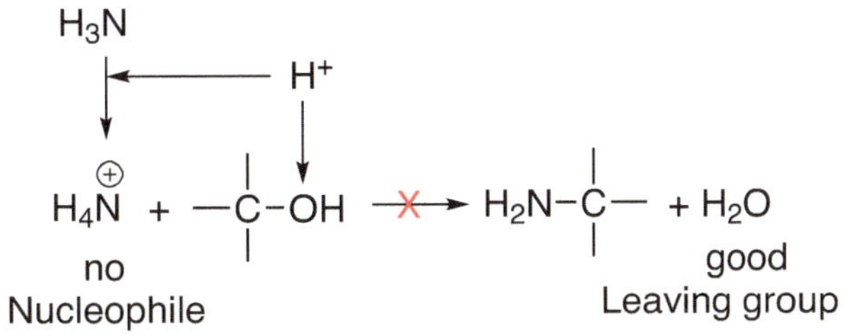

Fig. 1.48 Why the direct exchange of HO for NH$_2$ does not work in biochemical systems (3rd argument)

$$H_3N \xleftarrow{H^+}$$

$$\overset{\oplus}{H_4N} + -\overset{|}{\underset{|}{C}}-OH \xrightarrow{\times} H_2N-\overset{|}{\underset{|}{C}}- + H_2O$$

no
Nucleophile

good
Leaving group

does not take place in biochemistry. Therefore, an "indirect" route via the corresponding carbonyl compounds must be taken as part of an oxidation-reduction mechanism (Sect. 3.2).

Nucleophilic substitutions on carboxylic acid derivatives, however, also take place under biotic conditions. The stronger polarization of a C=O group compared to a C–OH group and the lower steric hindrance due to the planarity provide the conditions for this. Most of the time, the carbonyl group is additionally activated by anhydride formation with phosphate (e.g. with AMP).

An example is the replacement of the alcoholic group in esters of long-chain fatty acids by the thiol derivative CoA–SH (Fig. 1.49). Its anion is formed more easily than that of water or an alcohol, since the HS group is the stronger acid. Due to the considerable size of the anionic sulfur, CoA–S$^-$ is also easily polarized and thus a good nucleophile. The properties of AMP as a large and thus good leaving group facilitate the adjustment of the equilibrium.

The reverse reaction also takes place in living organisms. Although an alcohol, like water, is a poor nucleophile, this disadvantage is compensated by the good properties of the large leaving group CoA–S. These properties result in the fact that the thiolate anion (SH$^-$) and its organic derivatives, the alkylthiolates (SR$^-$), are well substitutable. Due to the easier cleavage of the acylthiol bond compared to the analogous ester bond (with O instead of S), the thiol bond serves as a "transporter" for all kinds of short- or long-chain

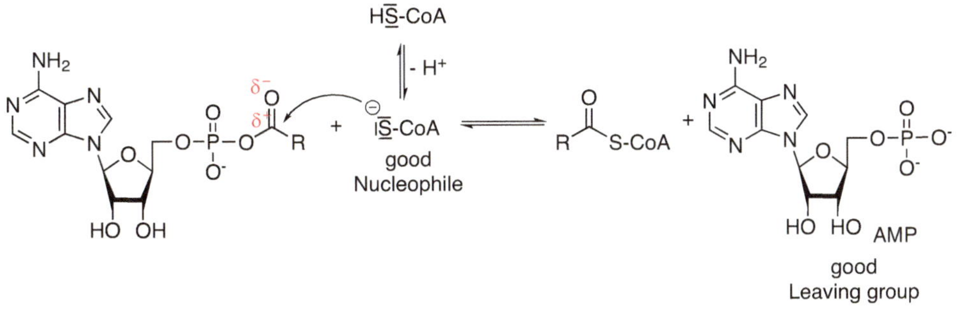

Fig. 1.49 The exchange of HO for SR as a biochemically realizable option

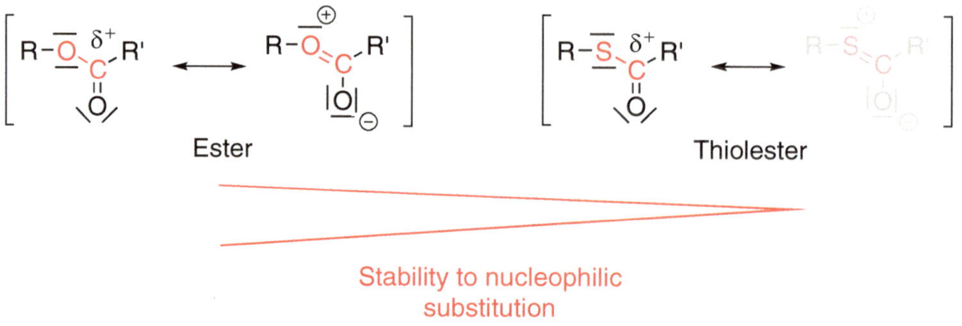

Fig. 1.50 The stability of esters and thiolesters

acyl residues, e.g. fatty acids. Esters, on the other hand, are more robust and are therefore found mainly in more permanent cell organelles such as membranes.

The higher stability of an ester compared to thiol esters can not only be explained on the basis of the poorer leaving group properties of the alcoholate, but also by the theoretical concept of mesomerism. As **mesomerism** (also resonance) one understands the formal phenomenon that only through several chemical (border) formulas the bond relationships in a molecule or ion can be adequately represented. A large number of plausible mesomeric boundary structures is an indication of the high stability of the compound or substructure. Such considerations facilitate the estimation of reactivities and the course of reactions.

The formulation of two mesomere boundary structures provides an explanation for the greater stability of the ester in comparison to the thiolester (Fig. 1.50). The structure of the ester with the charges shows that the positive partial charge on the carbon atom can also be delocalized over the doubly bonded oxygen atom. In contrast, a corresponding mesomere boundary structure with a C=S double bond and a positively charged sulfur atom describes the actual bonding state less due to the larger atomic radius of sulfur.

Esters of inorganic acids are also subject to the same trends in substitution reactions. An example of the great nucleophilicity of the HSe^- anion was already given in the transformation of L-selenocysteine from L-serine (Sect. 1.2.11). For the synthesis of this "helper reagent", hydrogen selenide HSe^- first substitutes a hydrogen phosphate group in ATP (Fig. 1.51). This reaction succeeds due to the large ionic radius of HSe^-, which results in a high polarizability. The attacked phosphorus atom is characterized not only by a larger atomic radius in comparison to carbon, but is also strongly positive due to the electron-withdrawing effect of the oxygen atoms. At the same time, the highly molecular ADP is a good leaving group. Due to these properties, the O/Se exchange takes place under biotic conditions. After a **dyadic tautomerism,** that is the migration of a proton between neighboring atoms with simultaneous displacement of a double bond, in this case from oxygen to selenium, and cleavage of the proton, selenophosphate is formed. The latter is, for example, as a higher molecular reagent for the transformation of L-serine into L-selenocysteine significant (Fig. 1.45).

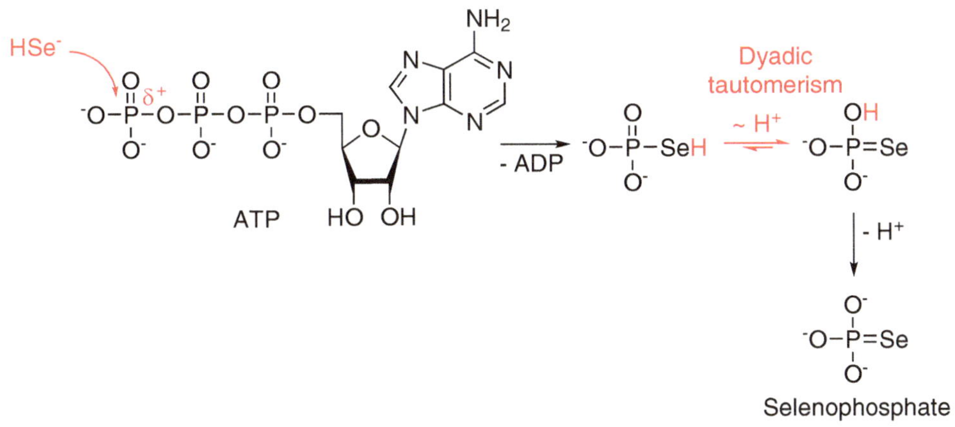

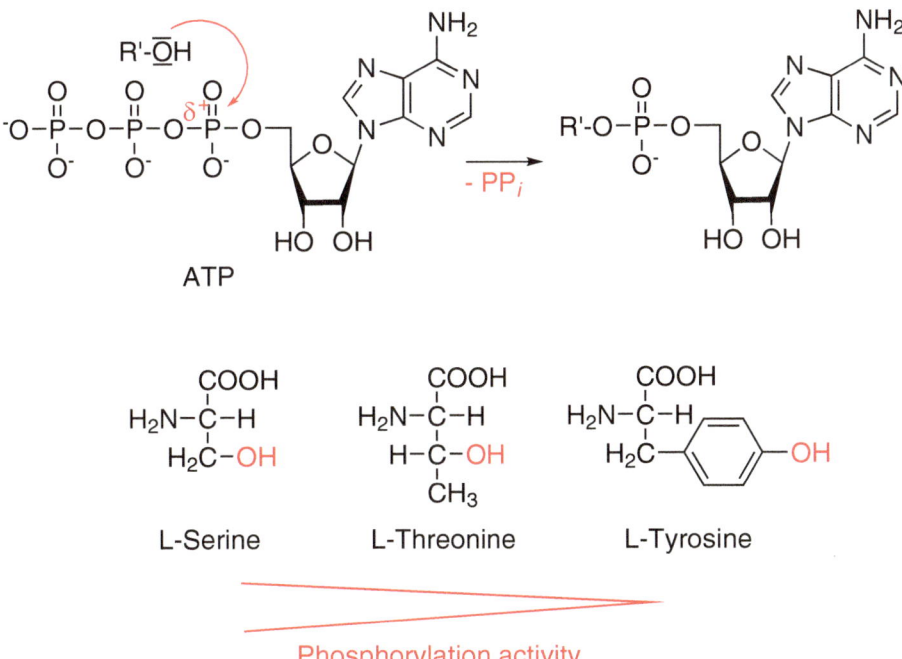

Fig. 1.51   The synthesis of selenophosphate via ATP

Pyrophosphate (PP$_i$) can also act as a leading group and thus provide a driving force for the esterification of anionic phosphoric acids with alcohols, as illustrated by the reaction of an alcohol with ATP below (Fig. 1.52). With the phosphate or hydrogen

Fig. 1.52   Phosphorylation of alcohols via ATP

phosphate anion, this reaction would only take place to a limited extent. Their reactivity is only established by integration in ATP.

In this way, a variety of functional groups in natural products, e.g. amino or hydroxy groups, are phosphorylated. The ability to phosphorylate is dependent on their properties. For example, the amino acids L-serine, L-threonine and L-tyrosine are phosphorylated in a ratio of 1800:200:1. L-Serine is a primary and L-threonine is a secondary alcohol, which causes the different accessibility to the phosphorylation reagent. L-Tyrosine is a derivative of an aromatic alcohol (Sect. 4.1.4.2), which is less nucleophilic under physiological conditions (in the absence of bases).

1.2.13 Oxidation Numbers and Oxygen Compounds

Oxidation numbers (also called oxidation states) show trends that are also reflected in the PSE. Oxidation numbers are the result of a formalistic model of the structure of molecules. They indicate the ionic charges of the atoms in a chemical compound or in a polyatomic ion if the compound or the polyatomic ion were built up from monatomic ions. In Fig. 1.53 characteristic oxidation numbers of elements of the first three periods are given. Atoms in the atomic state and elemental molecules are always assigned the oxidation number 0. This applies, for example, to elemental carbon as well as molecular nitrogen and oxygen. Also the noble gases helium (He), neon (Ne) and argon (Ar) are only given in the figure with their oxidation number of zero, because they are so unreactive

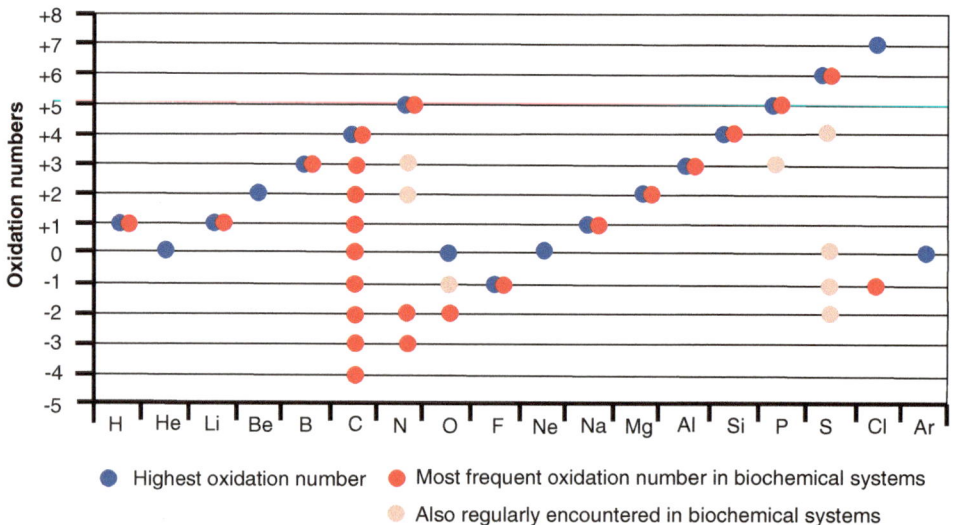

Fig. 1.53 Characteristic oxidation numbers of elements of the first three periods; highest and biologically particularly relevant oxidation numbers

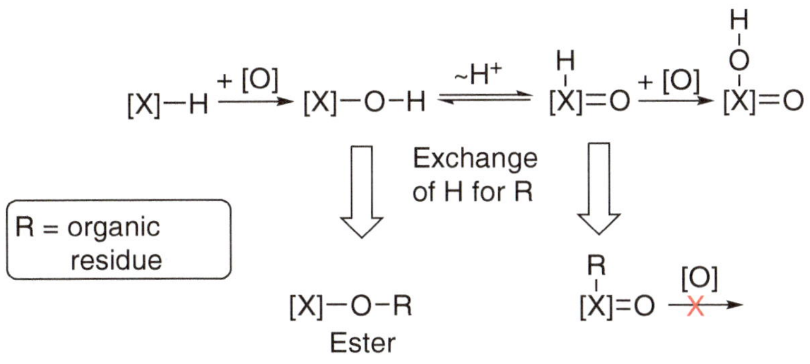

Fig. 1.54 By exchanging H for R, the oxidizability is restricted

that they are not oxidized under the conditions on Earth. The highest oxidation numbers of the other elements result from their position from the PSE, e.g. the alkali metals (1st main group) lithium and sodium with the oxidation number +1, the alkaline earth metals (2nd main group) beryllium and magnesium with the oxidation number +2, boron (3rd main group) with the oxidation number +3, etc. An exception is oxygen, for which the lowest oxidation state (−2) is dominant. In addition, there are oxygen compounds (peroxides) with an oxidation number of −1. In organic compounds, hydrogen is usually assigned the oxidation number +1 and oxygen the oxidation number −2. The carbon atom involved receives a number at the end, which leads to the neutrality of the overall or partial structure in the sum.

In most cases, the highest oxidation numbers occur in biochemically relevant elements that are predominantly found in nature, which is an indication of the element's affinity for oxygen, the primary reaction partner on Earth. In addition to these dominant oxidation numbers, there are also stable compounds with lower oxidation numbers in nature, for example some nitrogen compounds (ON = −3, −2 and +3), sulfur compounds (ON = −2, −1, 0 and +4), and phosphorus compounds (ON = +3). The large range of oxidation numbers of carbon from −4 to +4 is unique and therefore particularly noteworthy. The corresponding compounds hardly differ in their affinity for oxygen, with the exception of carbon dioxide. They all occur in nature and therefore play a central role in biochemistry.

Typical oxidation reactions proceed by formally inserting an oxygen atom into an [X]–H bond, where [X] symbolizes the molecular backbone with a central atom.

$$\text{[X]}-\text{H} \xrightarrow{+\,[\text{O}]} \text{[X]}-\text{O}-\text{H} \qquad\qquad \overline{\text{[X]}} \xrightarrow{+\,[\text{O}]} \text{[X]}^{\oplus}\text{O}^{\ominus} \text{ or } \text{[X]}=\text{O}$$

If there are several [X]–H bonds, this process will be repeated until they are all oxidized. In this way, oxoacids are formed in the reaction with nonmetals (Fig. 1.17).

Alternatively, a free electron pair on the central atom can react with oxygen to form an [X]–O bond. Depending on the properties of [X], a charged structure is formed or oxygen is bonded to [X] via a double bond. The oxidation reactions often proceed in a radical manner and via several unstable intermediate stages (for the exact chemical mechanism of carbon, see Chap. 4). For simplicity, the oxygen reactant is abbreviated below as [O], since not only the oxygen molecule O_2, but also reactive oxygen species (abbreviated ROS = *reactive oxygen species*) oxidize.

By the formal replacement of the proton of the HO group by an organic residue (R) are formed **esters** of inorganic acids (Fig. 1.54). Alternatively, the proton of the HO group can also tautomerize to the adjacent central atom and from the original [X]–O (H) single bond a [X]=O double bond is formed. In parallel, a new [X]–H bond is formed, into which oxygen is also inserted. The insertion is "blocked" if H is previously replaced by an organic residue (R). The X–C bond is not cleavable. The reaction thus remains on this oxidation level of the central atom of [X] and higher oxidation levels are no longer accessible.

In this way, the attainment of the maximum oxidation levels of nitrogen, sulfur and phosphorus is prevented in some cases. The associated natural products are often found in lower organisms, such as bacteria.

Carbon versus Silicon

Carbon reaches its highest possible oxidation state +4 in carbonic acid. Carbonic acid is formally produced by insertion of four oxygen atoms into the four C–H bonds of methane (OZ = −4) (Fig. 1.55). According to the Erlenmeyer rule, a carbon compound with multiple HO groups on the carbon atom does not exist. By successive removal of two water molecules, carbonic acid is first produced and then carbon dioxide, with the oxidation number no longer changing.

Carbonic acid is a **dibasic acid,** i.e. it is neutralized by two equivalents of a base. First, the hydrogen carbonate and subsequently the carbonate anion are produced.

Orthosilicic acid is formally produced by insertion of oxygen into the four Si–H bonds of the silane (Fig. 1.56). Since the Erlenmeyer rule does not apply to larger central atoms such as silicon, but also Si=O bonds are not stable under biotic conditions (Fig. 1.4), di- and higher molecular polysilicic acids are produced by intermolecular condensation.

The corresponding silicates are formed by neutralization. Hydrosilicates as intermediates of incomplete neutralization reactions are characterized by a varying number of anionic oxygen atoms and HO groups, which mainly depend on the pH of the environment and the structure of the mostly amorphous material.

Nitrogen

Similarly to carbon, nitrogen also stands out with several oxidation states that belong to stable compounds. However, in contrast to carbon, this diversity is realized much less frequently in biochemistry. In addition to molecular nitrogen (N_2 with OZ = 0),

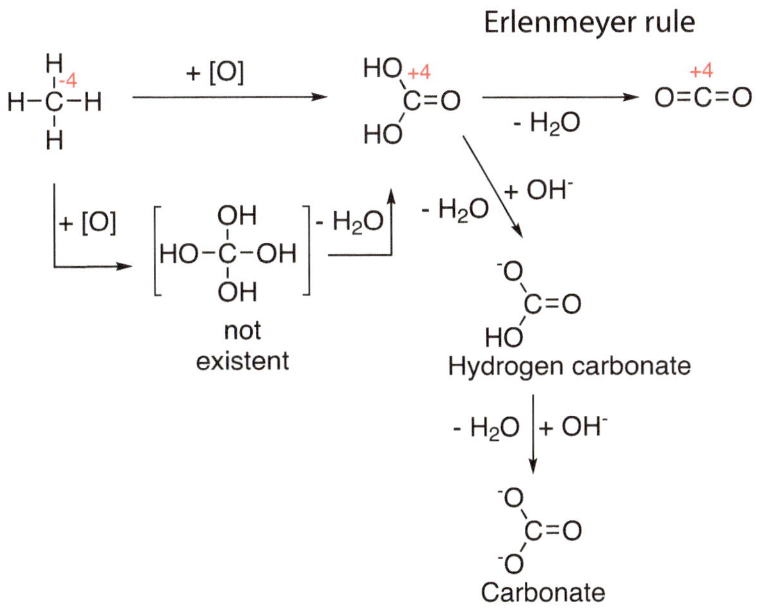

Fig. 1.55 By exhausting oxidation of methane carbonates or carbon dioxide are formed

compounds with the two extreme oxidation states −3 (ammonia) and +5 (nitric acid or its salts, the nitrates) are mainly occuring.

Molecular nitrogen is converted into ammonia by biotic atmospheric nitrogen fixation, which serves as a central starting compound for oxidative reactions. This hydrogenation provides all available nitrogen in living beings and is due to the activity of prokaryotic microorganisms. Eukaryotes do not have this ability. They are dependent on "external chemical support" by other living beings, which is often referred to as symbiosis in biology. Through industrial production (Haber-Bosch process), the biochemically available nitrogen base of plants and animals is extended by millions of tons of ammonia every year.

$$\overset{0}{N_2} + 3\,H_2 \longrightarrow 2\,\overset{+3}{NH_3}$$

Formally, all oxidation products of nitrogen can be explained by insertion of oxygen into N–H bonds of ammonia (Fig. 1.57). The structure of nitric acid can also be traced back to the effect of the Erlenmeyer rule. The insertion of oxygen into one of the two N–H bonds of hydroxylamine would theoretically lead to two HO groups on the nitrogen atom. However, due to the small atomic radius of nitrogen, such a compound, which structurally resembles the unstable hydrates of carbon, does not exist.

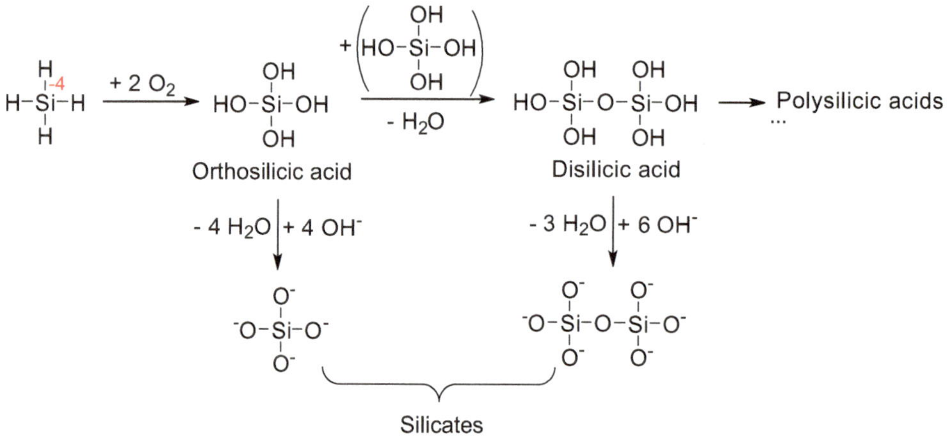

Fig. 1.56 By exhausting oxidation of silane silicic acids or silicates are formed

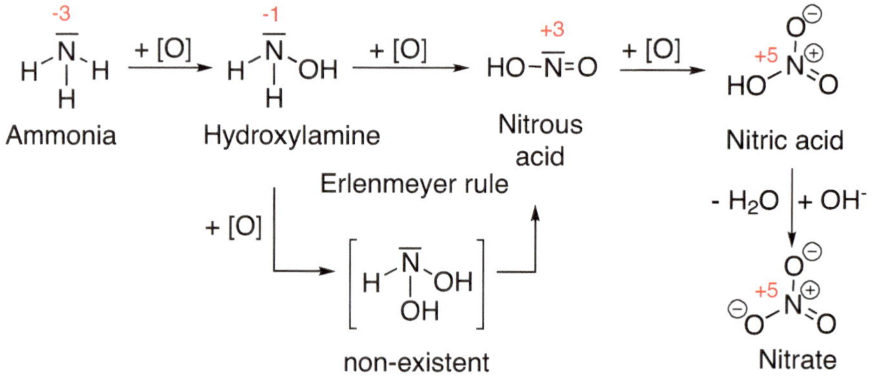

Fig. 1.57 Exhaustive oxidation of ammonia produces nitric acid or nitrate

If the free electron pair on the nitrogen atom in nitric acid is included in the oxidation process, an N–O bond is formed. Nitric acid is a **monobasic acid,** i.e. by reaction with an equivalent of a base, the nitrate anion is formed.

Natural products derived from nitric acid (hydrogen nitrite) are rare. So far, only about 200 have been detected. They are formed by oxidation of primary amines, which react via several intermediate stages to organic nitro compounds.

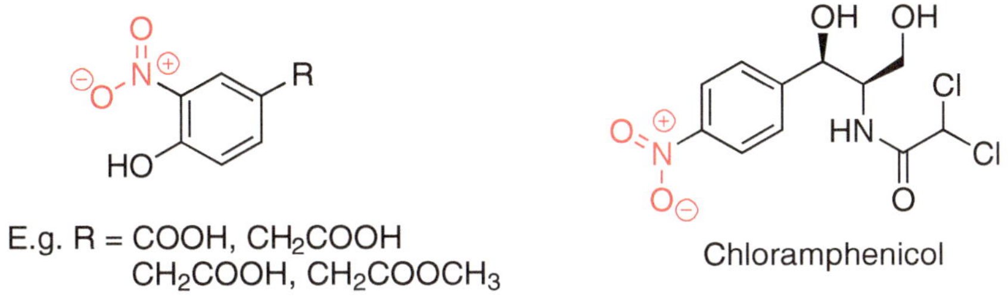

E.g. R = COOH, CH₂COOH
CH₂COOH, CH₂COOCH₃

Chloramphenicol

Fig. 1.58 Nitro-group containing natural products

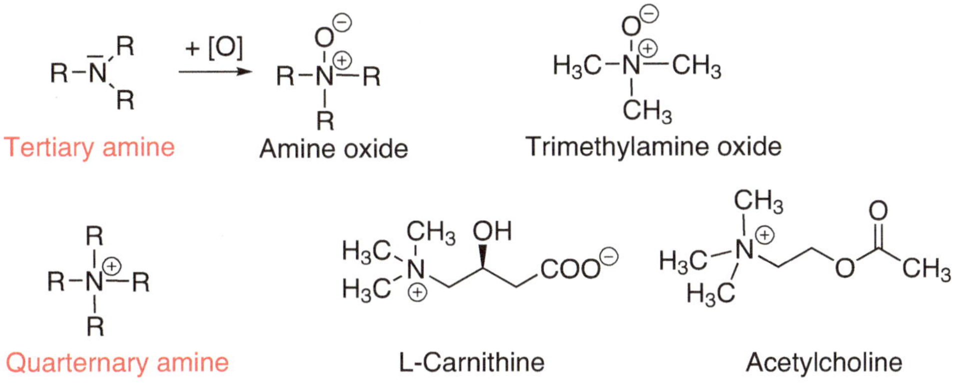

Tertiary amine   Amine oxide   Trimethylamine oxide

Quarternary amine   L-Carnithine   Acetylcholine

Fig. 1.59 Organic ammonia derivatives and examples of natural products

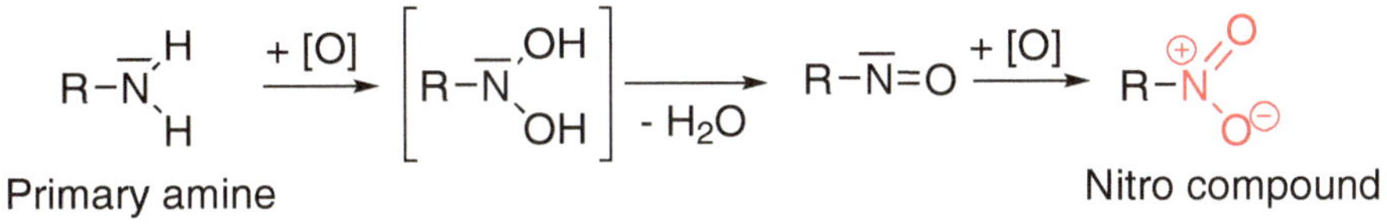

Primary amine   Nitro compound

Biogenic nitro compounds have been isolated from both plants, fungi and bacteria as well as from mammals. They are characterized by a great structural diversity. The most productive source for aromatic nitro compounds are Gram-positive bacteria *Salegentibacter* sp. T436, which occur in the Arctic sea ice. They produce more than 20 relatively simple nitro compounds on the basis of phenol (Fig. 1.58). A more complex compound with a nitro group is chloramphenicol, a compound which was first isolated from the Gram-positive bacteria *Streptomyces venezuelae* and for a long time was used as an antibiotic.

The complete exchange of all H atoms in ammonia for organic residues leads to compounds in which only the free electron pair on the nitrogen atom is left for the binding to oxygen. Aminoxides are formed, which in the following can only be oxidized further and thus degraded by means of C–H bonds in the carbon residues and not involving the nitrogen atom. A simple example is trimethylamine oxide (TMAO), a final degradation product of trimethylamine, which occurs in many organisms (Fig. 1.59). Due to the two separated and opposite charges it has a very polar character and in marine fish it serves to maintain the osmotic pressure in the body against salt water.

In a quaternary amine, even the free electron pair is missing from the nitrogen. That is why these organic structures are particularly stable. L-Carnitine (Sect. 4.1.6) benefits as a transporter of fatty acids from a quaternary ammonium function and is therefore protected against rapid oxidative degradation. Acetylcholine is an ester of acetic acid and one of the most important transmitters in the conduction of stimuli, also in humans.

Sulfur

A similar series results for the oxidation products of sulfur, with the lowest oxidation level −2 (H_2S) and the highest +6 (H_2SO_4) being of particular importance in the chemistry of life (Fig. 1.60). All organic sulfur compounds are mainly derived from sulfate, which is reduced in the course of various biochemical mechanisms. In contrast to nitrogen and phosphorus, which are treated below, organic S–S compounds, which are derived from the unstable disulfane (OZ = −1), are an important feature of biological structures (Sect. 1.2.7).

While disulfide bridges stabilize organic structures, higher oxidation products of organic sulfur compounds are usually chemical witnesses of degradation processes. A striking example is the degradation of the proteinogenic amino acid L-cysteine (Fig. 1.61). By alkylation with an allyl group, a thiolether is formed without changing the oxidation state, which is oxidized to alliin with oxygen. Alliin is a sulfoxide in which oxygen and sulfur are connected to each other by a double bond. The unsaturated amino acid residue is then eliminated and propanthial-S-oxide is formed. The acid character of the latter is made clear by the tautomeric sulfonic acid. Propanthial-S-oxide is, among other sulfoxides, the tear-inducing compound of the kitchen onion *(Allium cepa)*. Two sulfonic acids finally combine with the elimination of water to form allicin. Allicin with

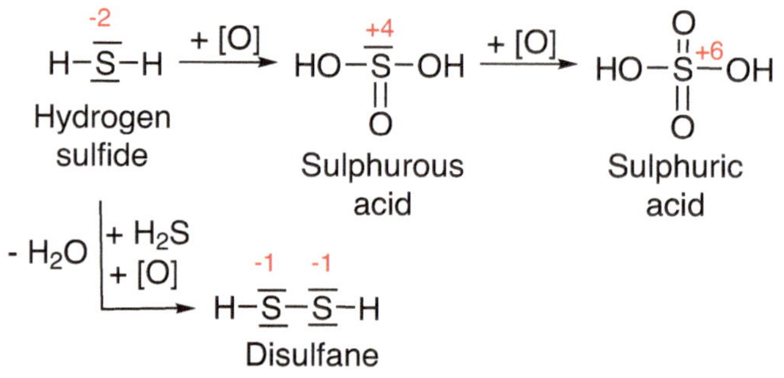

Fig. 1.60 Oxidation products of sulfur

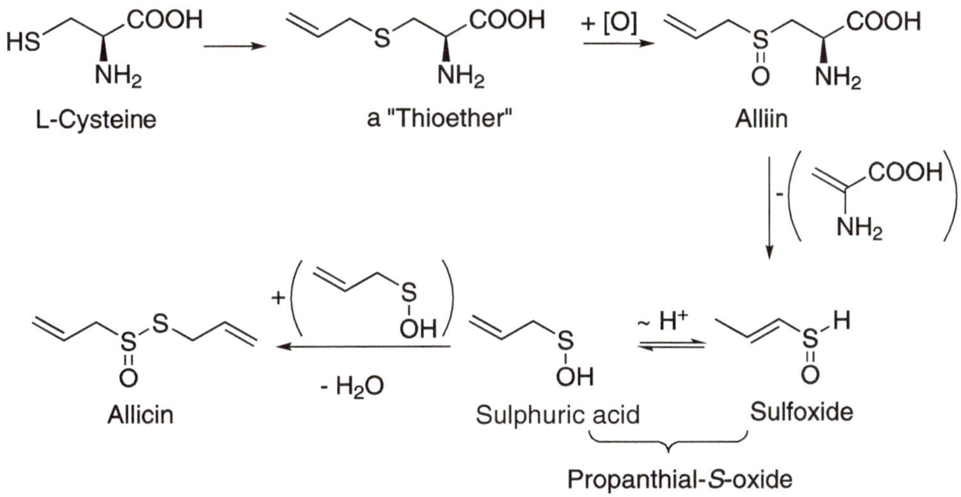

Fig. 1.61 The synthesis of allicin via sulfur compounds with different oxidation states

two sulfur atoms in different oxidation states is a partially oxidized disulfide and is the odour-determining compound of garlic *(Allium sativum)*. The cytotoxic chemical effects only occur after the onions are injured, when the catalytic enzyme and the substrate meet.

Alliin, allicin and many other oxidation products derived from L-cysteine have boiling points well above 100 °C. In comparison, H_2S is a gas. For this reason, the elimination of H_2S from L-cysteine is always in competition with these oxidation reactions, which ultimately also includes the formation of disulfide bridges as in cystine (Fig. 1.62). The degree of oxidation ultimately contributes to the individuality of the biological species.

Sulfonic acids, such as taurine, are also oxidative breakdown products of organic sulfur compounds (Fig. 1.63). The compound is formed in metabolism from the amino acid L-cysteine. Since cysteamine is an intermediate compound, taurine can also be derived from the breakdown of the central acyl group carrier CoA–SH. In an aerobic environment, L-cysteine is continuously lost in this way and must be replenished. Taurine has a variety of regulatory functions in mammals, including humans. The dog's organism is able to produce taurine itself, while cats are primarily dependent on external supply. This proves that biochemical differences can even manifest themselves between mammals in such a central class of natural products as amino acids. Such differences in chemistry ultimately contribute to their individuality.

Sulfuric acid, the compound with the highest oxidation state of sulfur, reacts successively with bases (HO^-) in two neutralization reactions first to hydrogen sulfate and then to sulfate. For this reason, it is a **dibasic acid**.

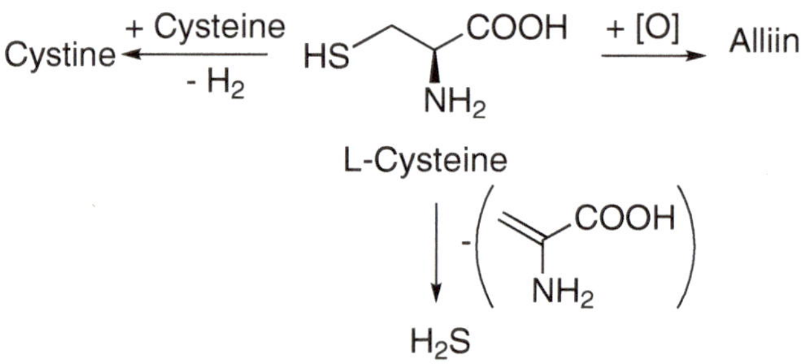

Fig. 1.62 Sulfur compounds with different oxidation states prevent the elimination of hydrogen sulfide

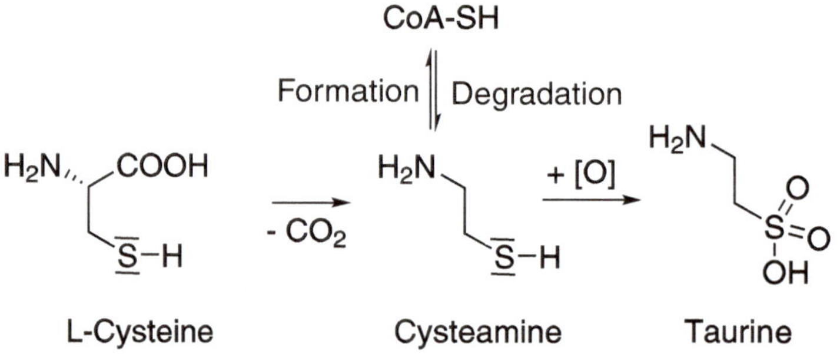

Fig. 1.63 Taurine as an example of a rare sulfonic acid in natural product chemistry

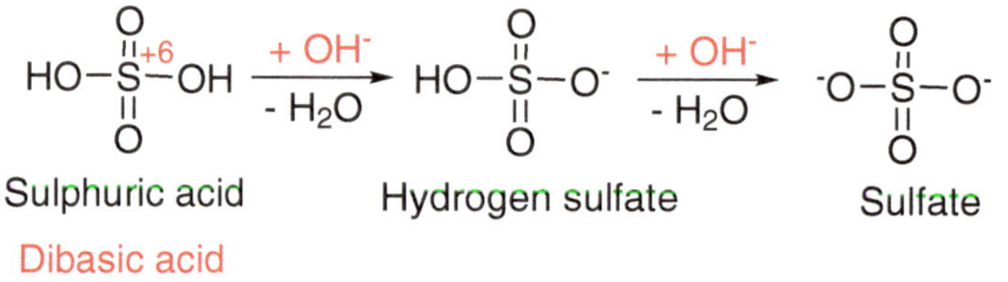

Phosphorus versus Nitrogen versus Arsenic

In contrast to ammonia NH_3, phosphane PH_3 with the most oxophilic phosphorus in the oxidation state -3 only exists in an anaerobic atmosphere, for example in swamp gas or on the neighboring planet Venus. For this reason, there are also no phosphinecarboxylic acids comparable to amino acids in living systems, such as the phosphorus-analogous glycine (Fig. 1.64). Organic PH_2 compounds can only be synthesized and stabilized under anaerobic conditions in the laboratory. For the same reason, there is also no natural product in which the NH_2 group in glycine is replaced by AsH_2. Arsenic is in the PSE

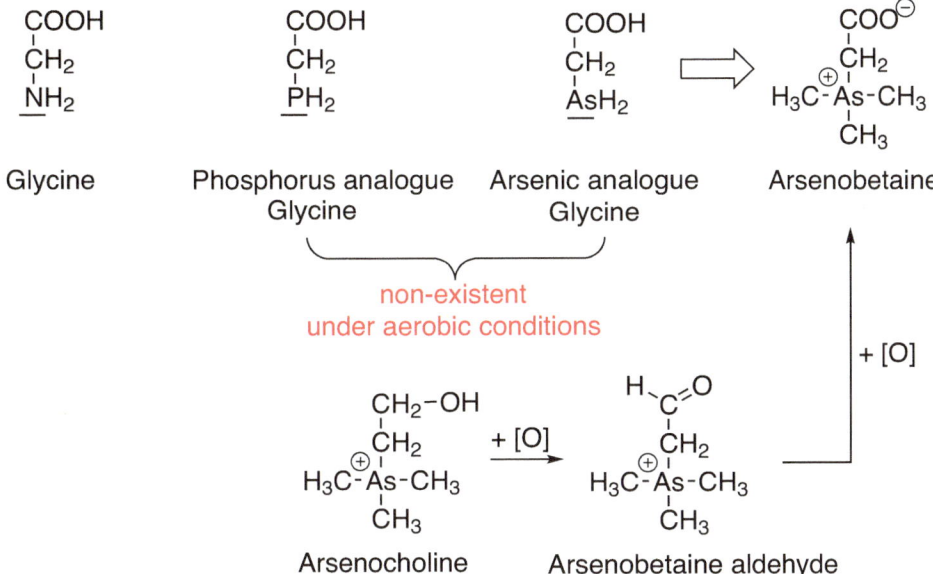

Fig. 1.64 Four alkyl groups prevent oxidation of the arsenic atom

in the 5th main group directly below phosphorus. Such an arsenic compound should be even more oxidation-prone than the comparable PH_2 compound. In fact, however, arsenobetaine occurs in marine organisms and fungi. The prerequisite for its stability is the absence of oxidizable As–H bonds and the lack of the free electron pair on the arsenic atom. This situation arises from the three CH_3 groups on the arsenic atom and internal salt formation. It is comparable to the high oxidation resistance of quaternary amines. Arsenobetaine also does not act as a poison in humans, but is excreted unchanged. Apparently there is no degradation mechanism with the corresponding enzyme equipment. Arsenobetaine can be considered a degradation product of arsenocholine, which is formed by exclusive oxidation of C–H bonds via arsenobetaldehyde as an intermediate. Arsenocholine occurs in fish (fish arsenic), with it being assumed that the primary source is algae and crustaceans.

By reaction of phosphane with oxygen, the corresponding phosphorus oxoacids with oxidation states of phosphorus between −1 and +3 are formed as intermediates. Phosphoric acid with the oxidation state +5, from which numerous biologically important salts (phosphates) are derived, is the most stable and also the most frequently occurring phosphorus species in nature. But also salts of phosphonic acid (phosphonates) represent a source of phosphorus for organisms.

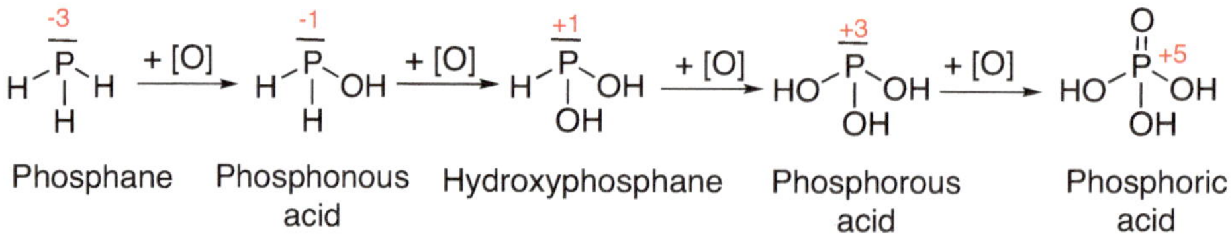

| Phosphane | Phosphonous acid | Hydroxyphosphane | Phosphorous acid | Phosphoric acid |

Organic substances containing phosphorus in an oxidation state less than +5 play a subordinate role in the biotic context. For example, organic derivatives of phosphoric acid, the phosphonic acids, as well as the derivatives of salicylic acid are formed formally by a preceding dyadic tautomerism and subsequent substitution of H by R (Fig. 1.65).

2-Aminoethylphosphonic acid occurs in the membranes of plants and many animals. Its biological role is still largely unknown.

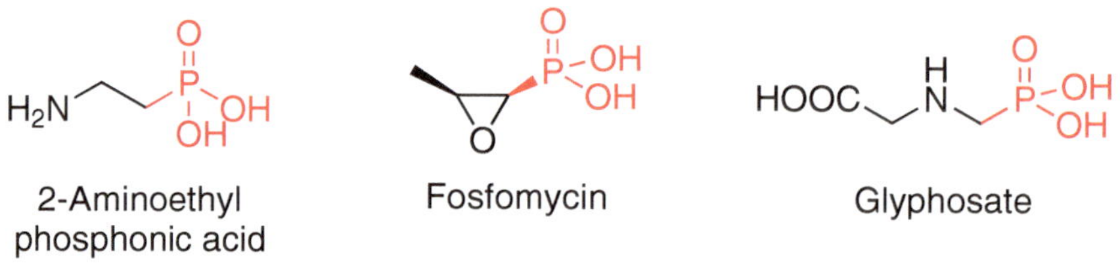

2-Aminoethyl phosphonic acid   Fosfomycin   Glyphosate

Another example is fosfomycin. It was isolated from the very diverse genus *Actinobacteria* and has been used as an antibiotic in human medicine since then. Since organic phosphonates are often used as herbicides in agriculture, some microorganisms with their rapid adaptability to changed chemical environmental conditions are now able to use them as a phosphorus source. A popular example is glyphosate, which is degraded not only by bacteria, but also by fungi. This is an indication that synthetically produced, new compounds are included in the dynamics of biochemical processes. In the end, a new biological species is created. For adaptation, it requires a transitional phase that can be of different lengths depending on the development level of the organism. Short times are a real-time demonstration of the action of biochemical evolution.

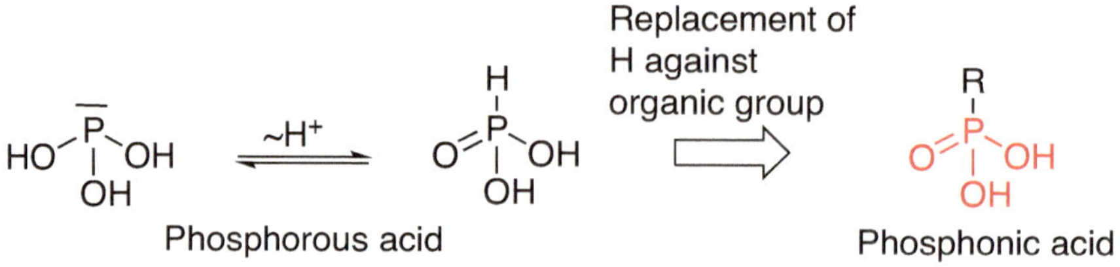

Phosphorous acid   Phosphonic acid

Fig. 1.65 Formation of phosphonic acids

Phosphoric acid, with the maximum oxidation state of phosphorus, is a **three-base acid,** i.e. it has jthree HO groups and can therefore react with three equivalents of a base in neutralization reactions. This produces successive dihydrogen phosphate, hydrogen phosphate and phosphate. The monoanionic dihydrogen phosphate and the dianionic hydrogen phosphate play the main role in biochemical mechanisms.

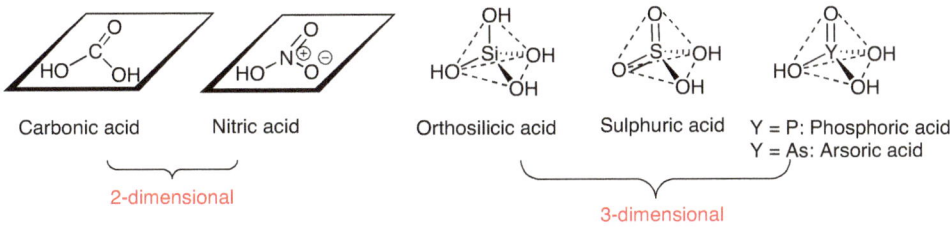

Arsorate, which is chemically similar to phosphate, is formed by the gradual neutralization of arsoric acid, also a three-base acid, via dihydro arsorate and hydro arsorate.

Comparison of the Geometries of Oxoacids

Carbonic and nitric acid molecules are flat (Fig. 1.66). In contrast, orthosilicic, sulphuric, phosphoric and arsoric acids are not planar, but adopt the geometry of a tetrahedron.

This not only applies to the acids, but also to all their associated salts and organic derivatives. This fundamental difference in geometry affects all compounds, especially where they act as bridges: they are—like all organic compounds with a four-valent carbon atom—not two-dimensional, but three-dimensional in structure. The transition from two-dimensionality to three-dimensionality leads to a higher level of evolution in

Fig. 1.66 Geometries of oxoacids

biochemistry. This is ultimately due to the atomic radius and the resulting effect of the Erlenmeyer rule.

1.2.14 Ester of Inorganic Acids and Their Evolutionary Potential

As demonstrated in the latter part of the book (see Chap. 3 and 4), the replacement of H atoms in inorganic compounds with organic groups is constitutive for organic chemistry. In the case of inorganic acids, this results in the formation of esters, which are treated separately here due to the close relationship to inorganic oxoacids. Esters are generally formed by the condensation of an acid with an alcohol (Fig. 1.67). The replacement of the HO group of the acid is initiated by the nucleophilic attack of one of the two free electron pairs of the HO group of the alcohol on the positively charged central atom of the acid. In the end, water is split off. Esterifications are equilibrium reactions. In particular, in the solvent water, the equilibrium lies on the side of the educts. The rapid establishment of equilibrium must be forced by acidic catalysts (protons). At the same time, subsequent reactions of the ester contribute to the equilibrium shift. It is important to note that the water is formed from the HO group of the acid and the proton of the alcohol and not vice versa. Thus, the presence of this HO group is essential. In salts, this HO group must first be generated before it is replaced. This can only be done by the stoichiometric amount of a stronger acid and follows the rule: The stronger acid displaces the weaker acid from its salts.

The degree of esterification of polybasic acids depends on the number of HO groups. The acidity of an HO group not only depends on the oxidation number and the electronegativity of the central atom, but also on whether a second substituent is present as an anionic oxygen or as an HO group (Fig. 1.68). In contrast to a neutral HO group, a negatively charged oxygen atom pushes electron density into the adjacent H–O bond, resulting in the negatively charged ions being less likely to give up protons than the corresponding neutral acids. The latter are therefore stronger acids.

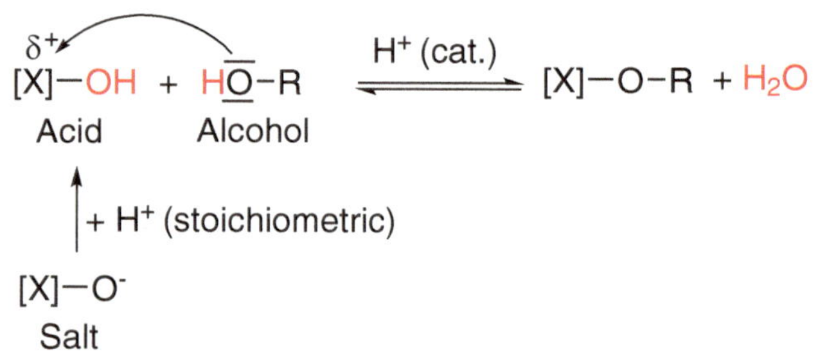

$$\delta^+ \overset{\frown}{\quad}$$
$$[X]-OH \ + \ H\underline{O}-R \ \overset{H^+ \ (cat.)}{\rightleftharpoons} \ [X]-O-R \ + H_2O$$
$$\text{Acid} \qquad \text{Alcohol}$$

$$\uparrow + H^+ \text{ (stoichiometric)}$$

$$[X]-O^-$$
$$\text{Salt}$$

Fig. 1.67 Formation of esters from salts

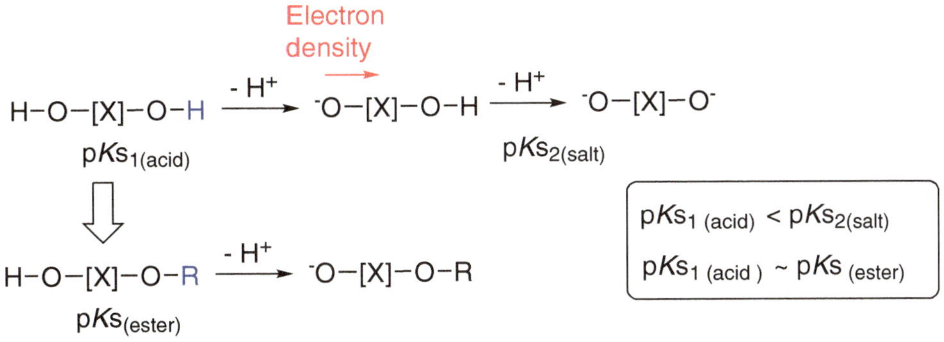

Fig. 1.68   pK_s values of polybasic oxoacids, salts and esters compared

This phenomenon can be quantified by the **pK_S-values**, which are consequently different for polybasic acids (Fig. 1.69). The pK_S-value of the 2nd (or 3rd) dissociation step is always greater than that of the previous step, as shown by the example of phosphoric acid, arsoric acid and sulfuric acid. At a pH value of 7.20, approximately equal concentrations of dihydrogen phosphate ($H_2PO_4^-$) and hydrogen phosphate (HPO_4^{2-}) are present. The concentrations of undissociated phosphoric acid and phosphates are in comparison million times smaller. Dissolved phosphate (PO_4^{3-}) only exists in a strongly basic environment, while in a strongly acidic solution the phosphoric acid (H_3PO_4) dominates. The three pK_S-values of arsoric acid are comparable to this.

$$\text{Phosphoric acid}\begin{cases} H_3PO_4 + H_2O \rightleftharpoons H_2PO_4^- + H_3O^+ & pK_{s1} = 2,1 \\ H_2PO_4^- + H_2O \rightleftharpoons HPO_4^{2-} + H_3O^+ & pK_{s2} = 7,2 \\ HPO_4^{2-} + H_2O \rightleftharpoons PO_4^{3-} + H_3O^+ & pK_{s3} = 12,4 \end{cases}$$

$$\text{Arsoric acid}\begin{cases} H_3AsO_4 + H_2O \rightleftharpoons H_2AsO_4^- + H_3O^+ & pK_{s1} = 2,1 \\ H_2AsO_4^- + H_2O \rightleftharpoons HAsO_4^{2-} + H_3O^+ & pK_{s2} = 6,9 \\ HAsO_4^{2-} + H_2O \rightleftharpoons AsO_4^{3-} + H_3O^+ & pK_{s3} = 11,5 \end{cases}$$

$$\text{Sulfuric acid}\begin{cases} H_2SO_4 + H_2O \rightleftharpoons HSO_4^- + H_3O^+ & pK_{s1} = -3 \\ HSO_4^- + H_2O \rightleftharpoons SO_4^{2-} + H_3O^+ & pK_{s2} = 1,92 \end{cases}$$

Fig. 1.69   Comparison of the pK_S-values of phosphoric acid, arsenic acid and sulfuric acid

Sulfuric acid, on the other hand, is five orders of magnitude more acidic than phosphoric acid and arsoric acid. But even in sulfuric acid, hydrogensulfate, which is present in only small concentrations under physiological conditions, is the weaker acid than sulfuric acid itself.

By replacing the proton in an HO group with an organic group R (H against C), the electronic properties hardly change. Therefore, for polybasic acids, the pK_S-values of acid and ester are approximately equal (Fig. 1.68). These relationships have fundamental effects on ester formation in an aqueous, neutral environment, which are discussed below.

Carbonic Acid and Nitric Acid Esters

It is striking that there are neither mono- nor diesters of carbonic acid in living nature, which is due to the instability of carbonic acid and the enormous stability of the two decomposition products carbon dioxide and water (Fig. 1.70). Carbonic acid diesters can only be synthesized under certain conditions in the laboratory. Even in this context, carbonic acid monoesters are highly unstable in the form of their anions. The only biogenic forms of carbonic acid are therefore carbonate and hydrogen carbonate.

Nitric acid esters have only a limited distribution in living nature. Nitric acid is only present in the form of its salts (nitrates) (Fig. 1.71). To generate free nitric acid from the nitrates, an acid even stronger than the already very strong nitric acid is necessary. Such

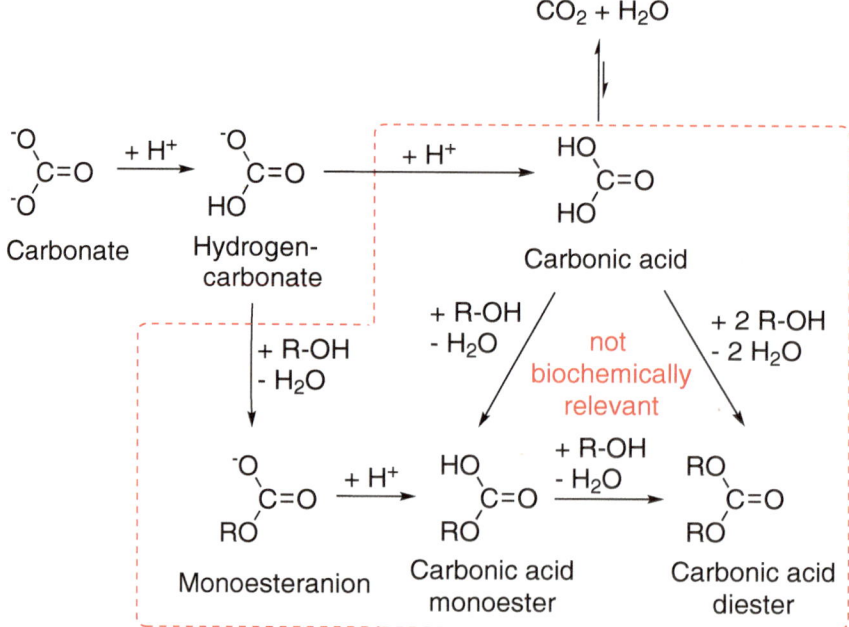

Fig. 1.70 Carbonic acid and derivatives in biochemical relevance comparison

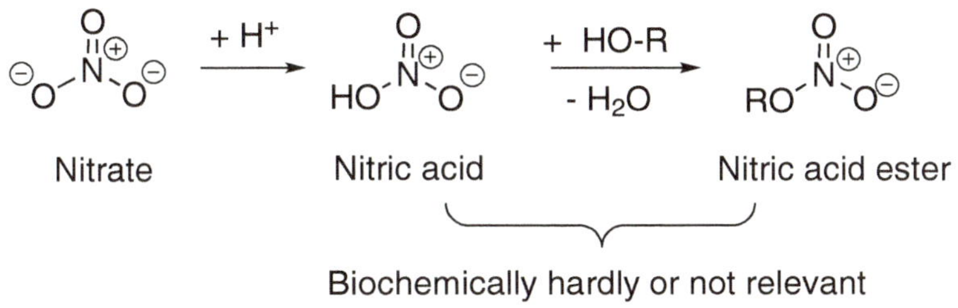

Fig. 1.71 Nitric acid and derivatives in biochemical relevance comparison

extreme conditions for the formation of nitric acid esters are completely atypical in biochemical systems. They should in principle only be synthesizable in a laboratory.

Despite this, a few esters of nitric acid have been isolated from biogenic sources. So far, there are no indications of how they are formed. For example, the subarctic coral *Alcyonium paessleri* produces a series of cyclic compounds classified as illudanes. (*Z*)-9-Tetradecenyl nitrate was identified as a component of the sex attractant of the female cotton leaf perforator moth *(Bucculatrix thurberiella)*.

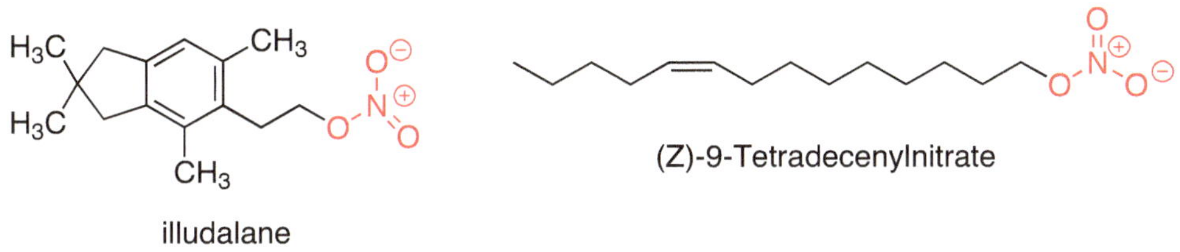

It should be noted that a second organic moiety in the nitric acid cannot be bound due to the lack of another HO group (see the effect of the Erlenmeyer rule in the total oxidation of NH_3). In other words, nitric acid esters are not existent under laboratory conditions. The logical consequence of this is that nitric acid cannot act as a bridge between two organic moieties.

Sulfuric Acid, Phosphoric Acid and Arsoric Esters

Both nitric acid and sulfuric acid are very strong acids. This statement can be derived from an **diagonal relationship in the PSE** (Fig. 1.72). diagonal relationships are the similarities in properties between those elements and their compounds that are diagonally opposite each other in the PSE. In the present case, this refers to the acid properties of the two oxoacids.

However, in contrast to nitric acid, sulfuric acid has two esterifiable HO groups and can therefore be transformed into a diester (Fig. 1.73). The main occurrence of sulfuric acid in nature is sulfate. In the approximately neutral milieu of the living nature,

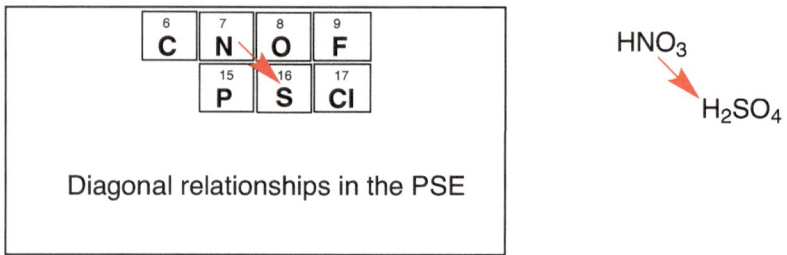

Fig. 1.72 Diagonal relationships in the PSE, application to the estimation of acid strengths

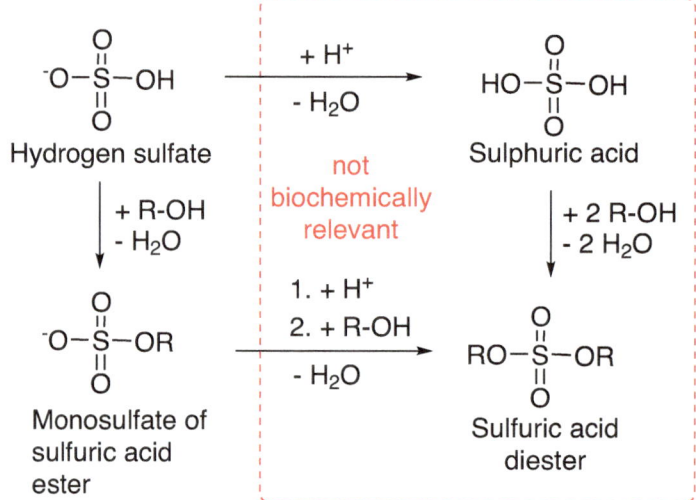

Fig. 1.73 Sulfuric acid and derivatives in biochemically relevant comparison

only small concentrations of the mostly water-soluble salts of hydrogen sulfate exist. Hydrogen sulfate has only one HO group. If this is replaced with an equivalent of an alcohol, the salt of a sulfuric acid monoester is formed. Another HO group can only be generated by the action of a very strong acid—comparable to the equilibrium between nitrate and nitric acid—which, however, is not given under biotic conditions. For this reason, only sulfuric acid monoesters, but no diesters, are important as natural substances.

Salts of sulfuric acid monoesters are widespread in living nature. An example are the heparins, which act as blood clotting agents (Fig. 1.74). The negative charge on the sulfate oxygen atom further increases the intrinsic polarity of the sulfate group and attractive interactions with other polar or even charged compounds are possible.

Sulfatides (Fig. 1.75), a group of glycosphingolipids consisting of a sugar and fatty acids that occur as components of brain matter, are salts of monoesters of sulfuric acid.

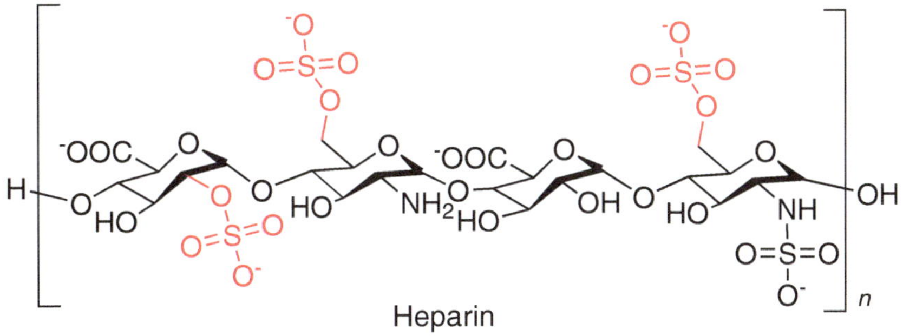

Fig. 1.74 Heparin as an example of the salt of a sulfuric acid monoester

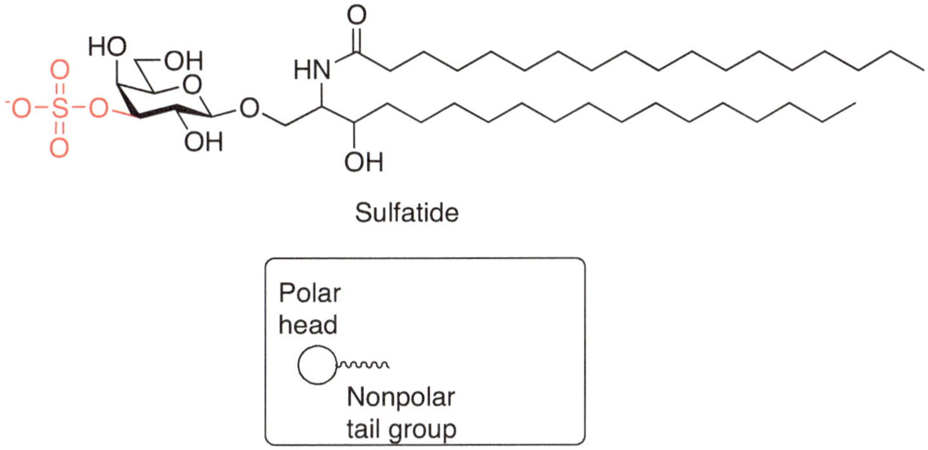

Fig. 1.75 A long-chain sulfatide as a typical example of an amphiphile

The sulfate-based sugar moiety and the long unpolar chains confer polar properties (head) and unpolar properties (tail) on the molecule. Such compounds are referred to as **amphiphiles**. They are the structural prerequisite for the formation of ordered aggregates that serve as substructures of biological membranes together with fatty acid esters and phospholipids (Fig. 1.82).

Phosphoric acid is weaker than sulfuric acid in comparison. This is due to the position of phosphorus in 5th main group and that of sulfur in the 6th main group and is attributable to the different oxidation states of the central atoms.

$$\overset{+5}{H_3PO_4} \quad < \quad \overset{+6}{H_2SO_4}$$

Acid strength

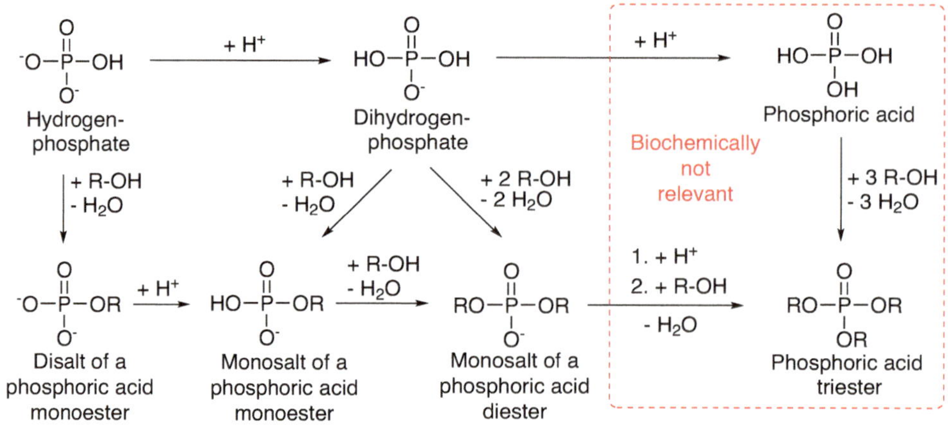

Fig. 1.76 Phosphoric acid and derivatives in biochemically relevant comparison

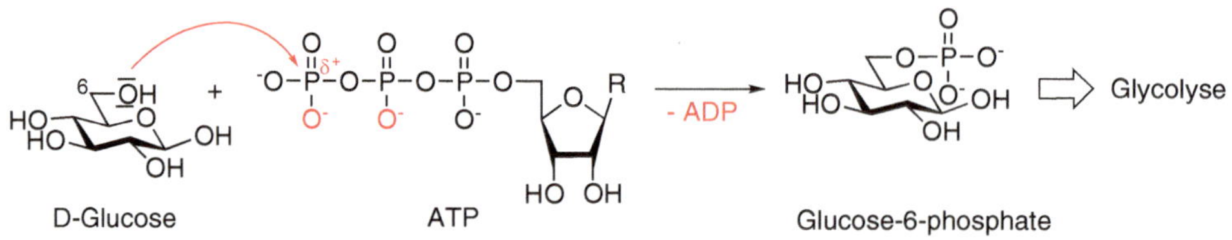

Fig. 1.77 Formation of a biochemically important phosphoric acid monoester via ATP

As already shown by the pK_S values (Fig. 1.69), hydrogen phosphate and dihydrogen phosphate are present in approximately equal concentrations in the neutral medium and predominate over phosphate and phosphoric acid. This means that they are converted into the corresponding mono- or diesters in the presence of alcohols under physiological conditions (Fig. 1.76), although often in the presence of activated phosphoric anhydrides (e.g. ATP). The non-esterified HO groups are usually neutralized, i.e. they are present in anionic form. Triesters of phosphoric acid play no role.

The salts of phosphoric acid mono- and diesters are among the most outstanding classes of natural products derived from an inorganic basic structure. An example of a monoester is glucose-6-phosphate (Fig. 1.77). The compound is formed by esterification with the outer phosphate group of ATP. In particular, the repulsive effect of the anionic oxygen atoms in pyrophosphate and the size of ADP as a leaving group favor substitution under biotic conditions. Glucose-6-phosphate is at the entrance of glycolysis, the central mechanism for the breakdown of glucose.

Phosphoric acid esters are in principle stable in the absence of catalysts. However, the C–O–P cleavage can be supported by properties of the organic residue, with energy left over for another reaction. An example is the energy-consuming synthesis of ATP

Fig. 1.78 Cleavage of a phosphoric acid ester for the coupled generation of ATP

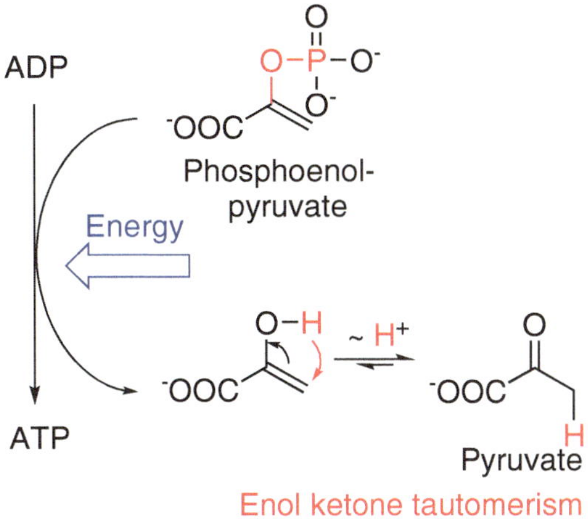

Phosphoenol-
pyruvate

Enol ketone tautomerism

from ADP at the end of glycolysis from phosphoenolpyruvate (Fig. 1.78). The energy for the hydrolysis of the phosphoric acid ester and for the simultaneous construction of the energy-rich phosphoric anhydride structure in ATP is used up by the parallel, irreversible enol-keto tautomerism to pyruvate.

Phosphoric acid monoesters can react reversibly under physiological conditions to the corresponding diesters. An example is the conversion of AMP to the phosphoric acid diester *c*AMP (cyclic AMP) and back (Fig. 1.79). For esterification, the HO group attacks the $C^{3'}$ of the sugar at the phosphorus atom and formally replaces one of the two anionic oxygen atoms. However, their replacement is only possible by the preceding protonation, which generates the leaving HO group. In the end, it leaves the reaction in the form of water.

The retroreaction and thus the cleavage of one of the two ester groups is made possible by the special structure of the *c*AMP. The cyclic AMP is in contrast to its precursor, the AMP, composed of two annulated rings (six-membered and five-membered ring are linked to each other via a common bond). An "anomeric effect" (Fig. 4.56), caused by

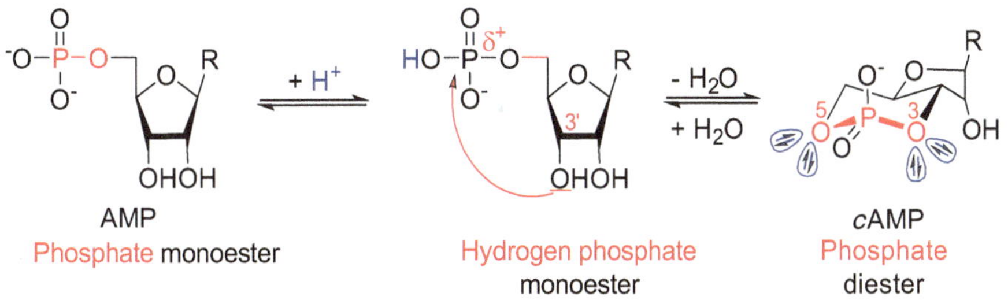

AMP
Phosphate monoester

Hydrogen phosphate
monoester

*c*AMP
Phosphate
diester

Fig. 1.79   pH-Dependent equilibrium between phosphoric acid monoester and phosphoric acid diester

repulsive free electron pairs on the two oxygen atoms in the vicinity of the phosphorus atom, is responsible for the fact that the compound is more tense and thus more energy-rich than AMP. The diester cAMP thus easily reverts to the monoester AMP, with the sterically more heavily loaded O^3–P bond taking precedence over the O^5–P bond (secondary versus primary alcohol, Sect. 4.1.4.2).

The position of the chemical equilibrium with ATP to cAMP via AMP as an intermediate product is determined by the concentration of phosphate according to the principle of Le Chatelier-Braun. High concentrations of phosphate (energy-poor) shift it in dependence on the pH value in favor of ATP (energy-rich). On the other hand, high concentrations of ATP lead to increased formation of cAMP via AMP. Since the phosphorylation of biochemical substrates represents an energy-consuming step, numerous biochemical transformations are initiated by a high concentration of cAMP or P_i. In this respect, cAMP reacts to the lack of glucose and acts as a "hunger signal" not only in humans, but even down to bacteria.

The effect can be illustrated using the example of the cleavage of D-glucose from a glycogen chain with an excess of hydrogen phosphate (Fig. 1.80). Glycogen (Fig. 4.60) serves in animal organisms as a storage form of glucose. Phosphorylation is thus the beginning of glycolysis and subsequent to the citrate cycle, the most important catabolic processes for energy generation.

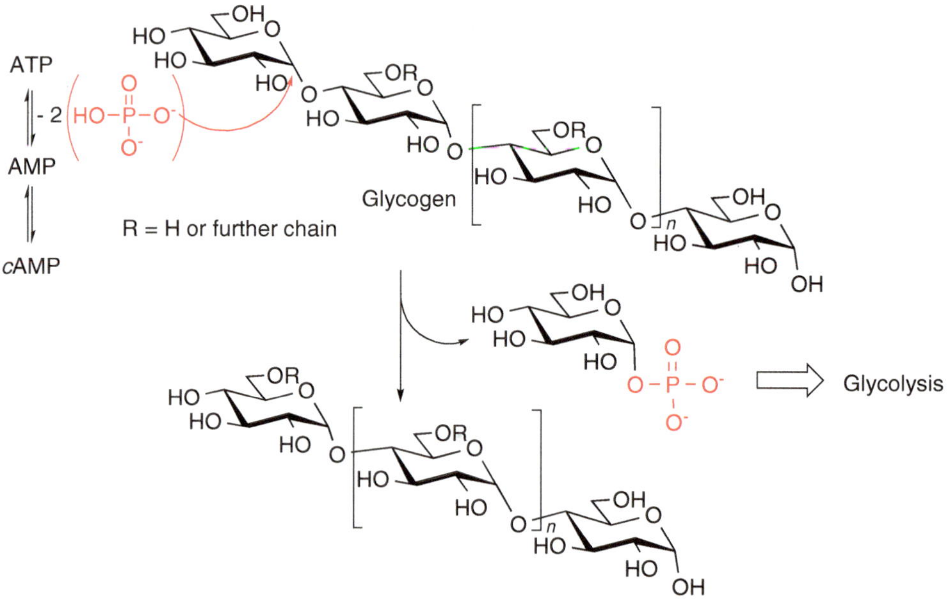

Fig. 1.80 Cleavage of a glucose monomer from glycogen by phosphorylation

Due to the central role of ATP in energy-consuming reactions in all organisms, the degradation products ADP and AMP are also present in all cells. Their transfer to compounds with different functional groups (alcohols, carboxylic acids, carboxylic acid amides, Fig. 4.108) is driven by the energy released during the hydrolysis of ATP to ADP or during the hydrolysis of pyrophosphate PP_i to two phosphates P_i in parallel. In biochemistry, this is referred to as a high "group transfer potential".

Phosphoric acid esters are of great biochemical importance for the formation of intramolecular bridges between complex organic structures. For example, if dihydrogen phosphate is linked to a glycerol and a charged choline unit, such an ester is formed (Fig. 1.81). By linking the two remaining HO groups on the glycerol to long-chain fatty acids, amphiphilic phospholipids are formed.

Like sulfatides, phospholipids form ordered aggregates of different organizational quality in water. The simplest structure is that of a bilayer, where the non-polar chains face each other and the polar head groups are oriented towards the polar water phase (Fig. 1.82).

Higher aggregates are micelles and liposomes. All structures have in common that they form biological membranes for organelles and cells in water and thus create the prerequisite for reactions in delimited spaces. By structuring amphiphiles in polar head and non-polar tail groups, not only ordered aggregates are formed, but membranes are also characterized by an asymmetrical structure. One can distinguish between an outer and an inner side, which, in cooperation with proteins and rare carbohydrates, ensure a selective permeability.

Another very prominent example of a phosphoric acid ester is the bridging of nucleosides in the chains of the polynucleic acids RNA and DNA (Fig. 1.83). The chain

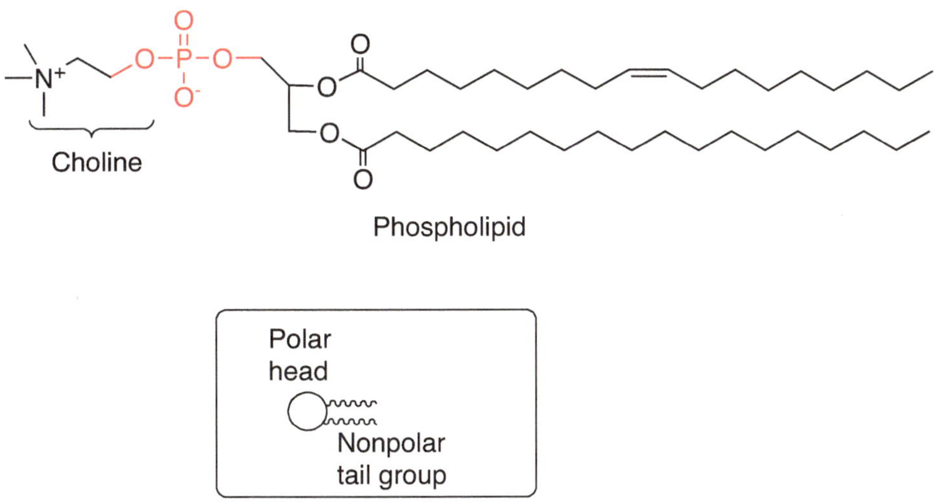

Fig. 1.81   A phospholipid as an example of an amphiphile

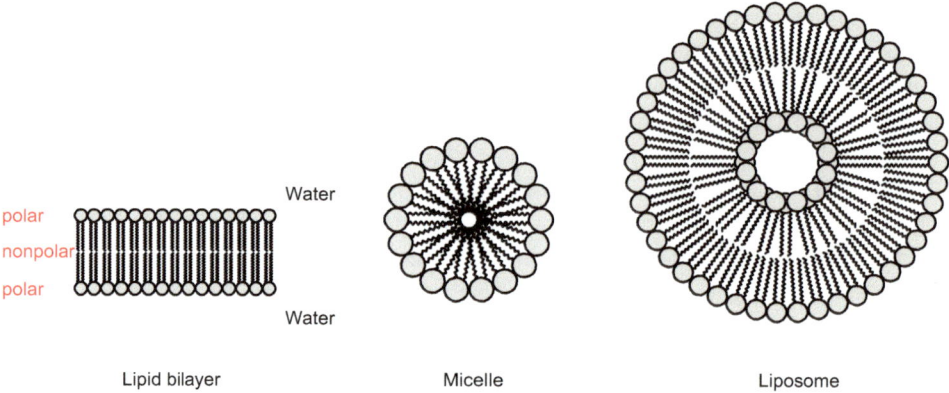

Fig. 1.82 Concentration-dependent formation of ordered aggregates from amphiphiles

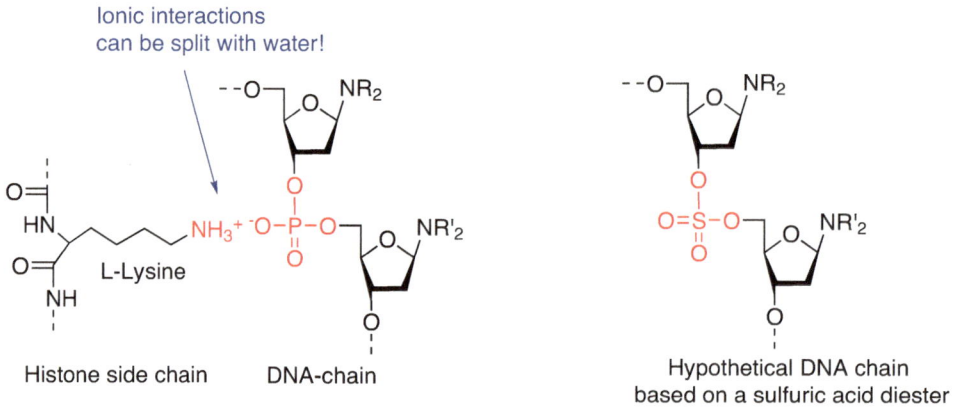

Fig. 1.83 Phosphoric acid or sulfuric acid as a bridge builder in DNA—a comparison

consisting of nucleoside monomers is formed by the two ester groups. At the same time, the third HO group is present at the phosphorus atom as an anionic oxygen. This is the prerequisite for ionic interactions with charged side chains of proteins (e.g. with the ammonium group of the amino acid L-lysine). In this way, an associate is formed between a genetic basic structure and a protein. Ionic bonds can be easily cleaved by water. In fact, the genetic information is not only transferred from the DNA to the proteins during protein synthesis, but special proteins, called histones, designate the sections to be read. Functionally, a higher level of evolution is achieved, which even founded a new area of research in biology: genetics is supplemented by the **epigenetics**.

Aside from the pH-dependent non-existence of corresponding diesters in a biotic milieu, this evolutionary pathway would not present itself for sulfuric acid. The S=O

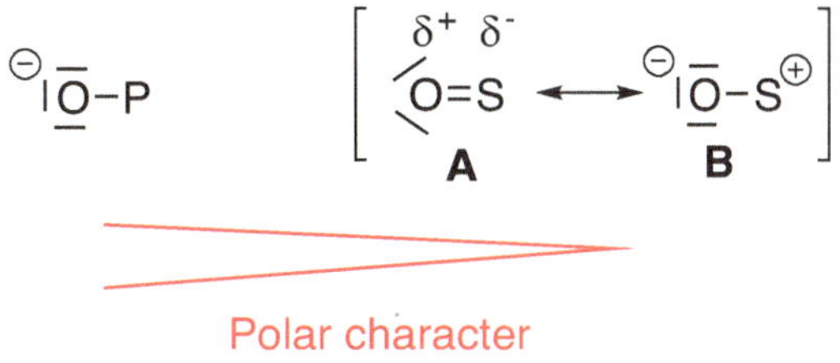

Fig. 1.84 Anionic P-O versus semipolar S=O bond

bond **A** does indeed contribute partially to the double bond rule single bond character **("semipolar bond"")**, as is illustrated by the corresponding **mesomere boundary structure B** (Fig. 1.84). In comparison to the negative charge in the anionic phosphate, however, the formation of ionically based attractive interactions is not possible; sulfuric acid diesters are neutral.

ADP and AMP occur as substructures in a variety of natural products, including coenzymes, in which the energy or information aspect takes a back seat. In this context, the anionic phosphate oxygen atoms increase water solubility or are responsible for attractive electrostatic interactions with surrounding proteins and embed the catalytically active structures in the final enzyme. Well-known examples are the hydrogen acceptors NAD^+ and FAD, where the terminal ends are formed by anhydride-bonded ADP (Fig. 1.85). The ADP substructure is also found in the acyl group carrier CoA–SH (Sect. 4.1.6), which is of great importance in the degradation of carbohydrates as well as of fats.

AMP as a basic structure and derivative of (weak!) phosphoric acid is an impressive demonstration of the multiple and thus economic use of certain chemical basic structures in different biochemical processes. At the same time, it becomes clear that the chemistry of life is subject to stronger limitations in terms of the number and diversity of natural products compared to synthetic chemistry in research laboratories and the chemical industry.

Due to its chemical similarity to phosphoric acid, arsoric acid can also form mono-, di-, and triesters. However, these are extremely unstable in aqueous solution in comparison to phosphoric acid esters. The half-life of arsoric esters and diesters in water is less than 0.02 s. One reason is the approximately 10 % larger length of the As–O bond in relation to the P–O bond, which is due to the larger atomic radius of arsenic (3rd period of the PSE!). At the same time, the oxygen in a P–O bond is more negatively charged in comparison to arsenic in an As–O bond, making it predestined for nucleophilic substitutions. In contrast to phosphorus(V), arsenic(V) is easily reduced to arsenic(III). In particular, thiols play a role here, which subsequently brings the toxic properties of arsenic into effect. For these reasons, phosphorus and its oxidation products have a much higher evolutionary potential than arsenic, although both are adjacent in group 5 of the PSE and arsenic (in small concentrations) occurs almost everywhere in the soil.

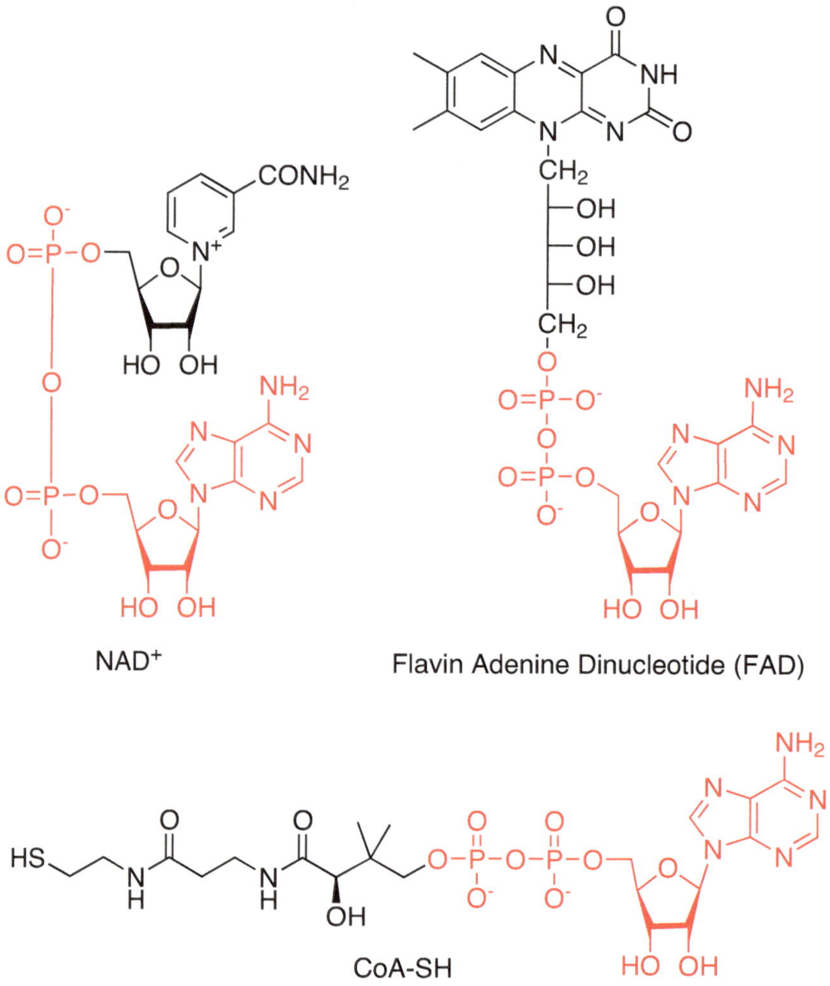

Fig. 1.85 Natural products with AMP or ADP substructures

On the Dynamics of Anionic Phosphoric Acid Diesters

Despite the unique structural advantages, the potential trivalency of phosphoric acid destabilizes the RNA chain structure. As a result, a new chemical selection channel is opened, which has revolutionary consequences for biochemistry and biology. Two effects are to be discussed (Fig. 1.86):

1. By an intramolecular **transesterification reaction,** in which the neighboring hydroxy group of the sugar participates at $C^{2'}$, a five-membered ring is formed on the basis of a phosphoric acid diester. The reaction proceeds via a nucleophilic substitution at the phosphorus atom, with the formation of the required HO-leaving group from the

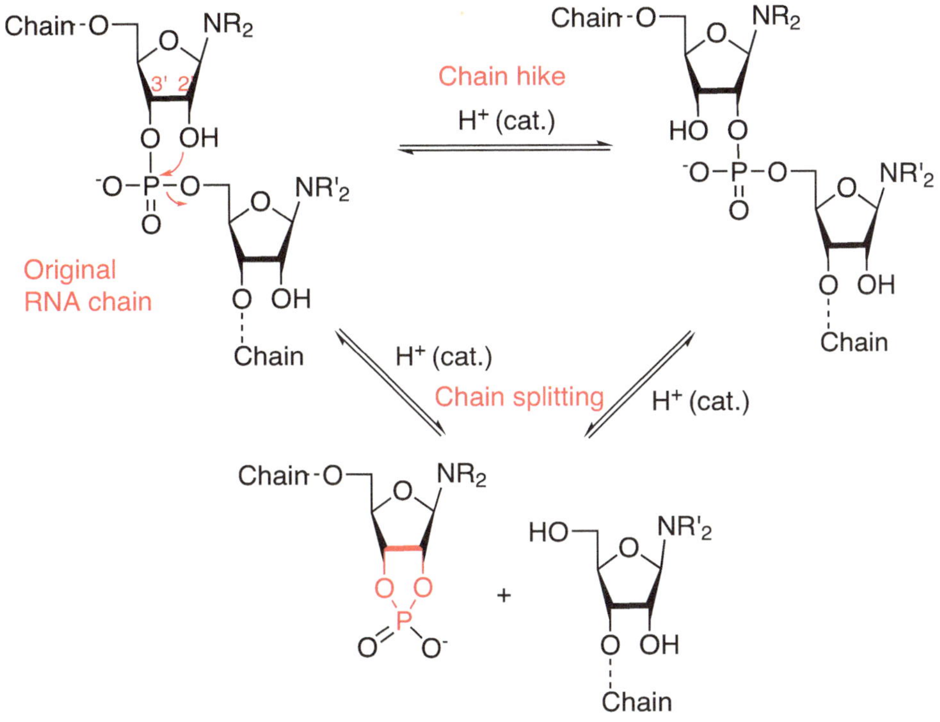

Fig. 1.86 Dynamics of phosphoric acid diesters in the presence of another alcohol

anionic oxygen being prepared by catalytic protonation. Through this reaction, the RNA chain is cleaved.

2. In a subsequent step, the chain migrates from $C^{3'}$ to $C^{2'}$. This reaction is also made possible by previous catalytic protonation of the anionic oxygen.

In both transesterification reactions, which are independent of the change in pH, the structure of the original RNA is disturbed; it loses its function as an informational molecule. Such rearrangements are responsible for the fact that RNA viruses, which carry their genetic information exclusively in RNA, quickly change their genetic blueprint and thus their infectivity. But the relatively short chain lengths of all RNAs in higher organisms are also a consequence of this.

The "removal" of the hydroxy group at $C^{2'}$ of the sugar (by reduction) prevents the chain break of the RNA or the migration of the chain from $C^{3'}$ to $C^{2'}$ (Fig. 1.87): RNA becomes DNA. The latter is thus already much more robust as a single-stranded molecule. By double-stranded formation via the nucleobases by means of hydrogen bonds, a further gain in stability arises (Fig. 4.128). Ultimately, the transition from RNA-based organisms, especially viruses, to DNA-based and thus higher organisms, such as

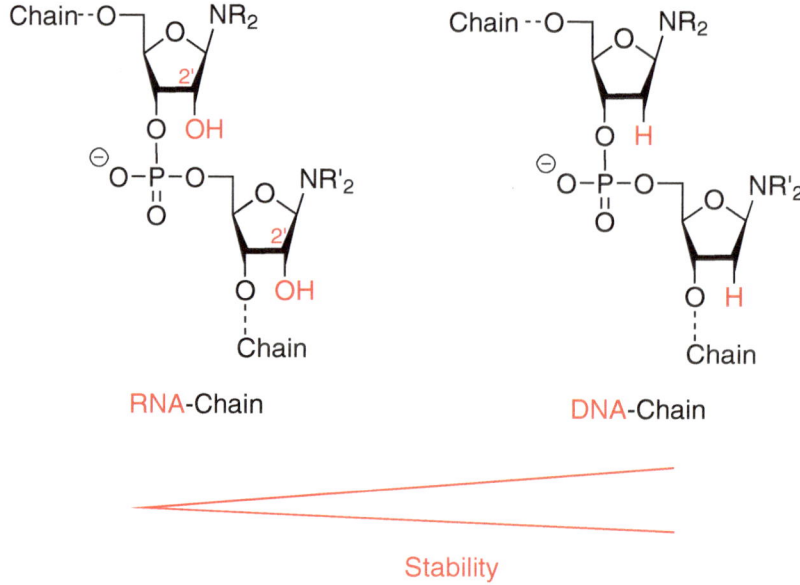

Fig. 1.87   The stability of a DNA chain compared to an RNA chain

bacteria, plants and animals, takes its origin in the difference of a single HO group (other causes see Chap. 4).

Conclusion of the Biochemical Evolutionary Potential of Oxoacids

Oxygen is inserted into X–H bonds (X = C, Si, N, S, P, As, Cl) to form oxoacids. By neutralization, the corresponding salts are formed. The condensation of the X–OH groups with alcohols leads to esters. The occurrence and importance of esters of oxoacids in the biotic context depend on two influencing variables:

1. the potential for the synthesis of esters under biotic conditions.
2. the evolution of complex superstructure, which is a consequence of the number of HO groups in oxoacids and their geometry.

In terms of evolutionary potential, the following picture emerges: At the lowest level, carbonic acid falls out of the selection because of its instability. Although nitric acid is much more stable, it only exists in the form of its salts, which do not react with esters under biotic conditions. In addition, it is not suitable as a bridge builder as a monobasic acid. Orthosilicic acid, sulfuric acid, phosphoric acid and arsoric acid are potentially capable of bridge building. Orthosilicic acid condense into polysilicic acids under biotic conditions before esterification can take place. The synthesis of defined silicic acid esters is only possible under laboratory conditions. Monoesters of sulfuric acid esters are biologically relevant as organic derivatives of hydrogen sulfate. However, the corresponding diesters are excluded from the selection as derivatives of a very strong acid and thus as bridge builders. Arsoric acid esters are too unstable to play a role in the biotic context. Only the properties of phosphoric acid as a weak tribasic acid with the extremely high concentrations of dihydrogen phosphate and hydrogen phosphate under physiological conditions predestine it for the construction of multiple reusable and stable structures in the living world.

By the three-dimensional structure of phosphoric acid derivatives, a new chemical evolutionary level is reached, which culminates in genetics. In fact, phosphoric acid in the backbone of RNA or DNA causes these polynucleic acids to be not flat, but to form a helix. Although the individual **nucleotides** and their associates mediated by hydrogen bonds are approximately planar themselves, a three-dimensional structure develops from these two-dimensional structures by linking them as phosphoric acid diesters, thereby increasing the quantity and quality of information storage and transmission.

Halogen oxyacids do not play a role in nature. For example, the best known, perchloric acid (Sect. 1.2.5), with the chlorine oxidation state of +7 is an extremely strong oxidizing agent, i.e. the compound transfers oxygen partially explosively to other compounds and, in the process, passes into chlorine compounds with lower oxidation states. In addition, it is an extremely strong acid. Even without this property, perchloric acid, as a monobasic acid comparable to to nitric acid, would not be suitable as a bridge building block. The biochemical evolutionary potential of perchloric acid is thus already exhausted at very low levels.

Extrinsic Properties that Favor the Occurrence of Elements in Biological Context

2

Essentially, two extrinsic conditions can be distinguished that are relevant to the access of an element to biochemical mechanisms to varying degrees:

- The frequency of occurrence of the element concerned and its distribution on Earth
- The reactivity under the conditions on Earth, taking into account the graded reactivity to water and oxygen.

2.1 The Frequency of Occurrence of the Element Concerned and Its Distribution on Earth

The frequent occurrence of an element on Earth is an important prerequisite for it to play a role in living organisms at all. Life-relevant gases such as nitrogen, oxygen or carbon dioxide are ubiquitous in the atmosphere. Molecular nitrogen N_2 is the main component of air with 78%, followed by molecular oxygen O_2 with a current share on Earth of 21%. The share of CO_2 in the Earth's atmosphere has increased since the beginning of industrialization from about 280 ppm (parts per million) to about 415 ppm at present, mainly due to the combustion of fossil fuels. Nitrogen and oxygen are taken up by the organisms directly from the air. Either they are chemically processed further in microorganisms (N_2) or in special organelles of higher organisms that have developed from single cells, the mitochondria (O_2). Carbon dioxide serves as the basis for the synthesis of organic compounds (sugar) by means of photosynthesis. It is taken up through the stomata of the plants and fixed differently depending on the type of plant, C3 or C4 plants.

Two fundamental differences in the availability of vital gases arise from whether the organism lives in water or on land. In water, the concentration of gases is much lower than in air, which is due to the limited gas solubility. The solubility of O_2 and CO_2 in water decreases with increasing temperature. In contrast to synthetic chemistry, gas pressure does not play a role in solubility in biochemical systems, it is approximately constant. However, dissolved solids reduce gas solubility. Therefore, less oxygen is dissolved in salt water than in fresh water. The transition of life from water to land during the course of biological evolution could have been a major driving force to exploit these resources for energy gain and growth to a greater extent. Due to the low concentration of oxygen in water compared to air, fish have evolved other mechanisms for energy generation. For example, glycolysis, an anaerobic mechanism, is dominant. As a result, the typically red oxygen carriers (especially myoglobin) play a subordinate role and the flesh of many fish is white.

Solid compounds are more unevenly distributed than gases on Earth. For biological relevance, it is decisive where the element occurs: on the entire Earth or only in the Earth's mantle, crust or dissolved or undissolved in the water of the oceans. In Fig. 2.1 the mass fractions of the elements in different spheres of the Earth are shown.

As in the entire universe, the frequency of distribution on Earth decreases with increasing atomic mass of the elements, which is also reflected in the declining importance in living nature. For example, the so-called **rare earths** already form no basis for life due to their rarity, as the collective name suggests. Also iridium and other platinum metals (rhodium, platinum, palladium, ruthenium), which probably arrived on Earth through the impact of a meteorite and are therefore only found in a very thin geological layer on Earth (**iridium anomaly**), are not found in living organisms. In addition, they are not converted into reactive species on Earth by oxygen or water under the conditions on Earth (Sect. 2.2.1).

Hydrogen, carbon, oxygen, nitrogen, phosphorus and sulfur, on the other hand, were already present in large quantities either elemental or later in chemically bound form since the beginning of the Earth's history and are therefore logically found in numerous natural substances. Possibly a primitive "iron-sulfur world" existed before the advent of free oxygen. Sulfur-containing organic compounds, such as a number of thiolesters (e.g. the acyl group transferrer CoA–SH), in today's biochemical mechanisms could be a legacy of this prebiological development.

Non-precious metals and semi-precious metals, which were already converted into water-soluble compounds by oxidation in the ur-oceans, still serve today as central metals for a variety of enzymes that are either strongly (including the transition metal ions Fe^{2+}, Fe^{3+}, Cu^{2+}, Zn^{2+}, Mn^{2+} or Co^{2+}) or weakly (including the ions of alkali and alkaline earth metals such as Na^+, K^+, Mg^{2+} or Ca^{2+}) bonded to proteins.

Not only the frequencies of elements on Earth are decisive for their biological relevance, but also the actual availability on site. Water plays a central role as a solvent and transport medium from the environment into the organism (Fig. 2.2).

In water, sparingly soluble compounds, e.g. many oxides and some salts, are not taken up by living organisms and therefore play no role in biochemistry. In principle, a distinction can be made between concentrated and less concentrated solutions. A mathematical

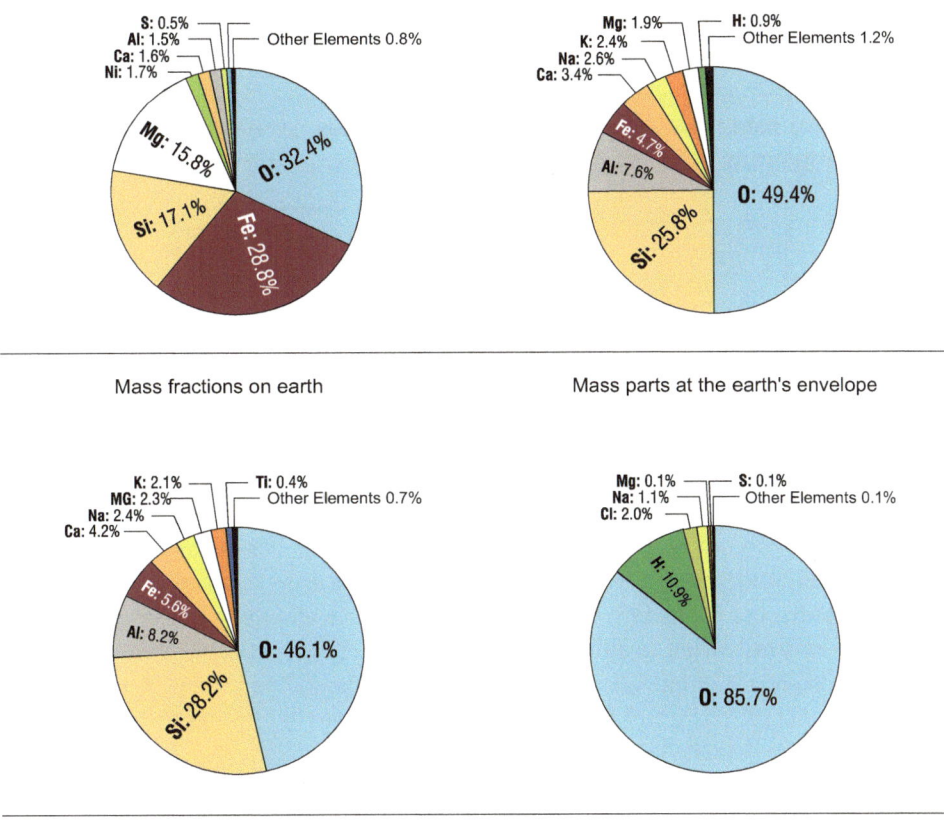

Mass fractions on earth

Mass parts at the earth's envelope

Mass fractions of the elements of the earth's crust

Volume fractions of the elements in the oceans

Fig. 2.1 Mass fractions of the elements in different spheres of the Earth, adapted from https://de.wikipedia.org/wiki/Liste_der_H%C3%A4ufigkeiten_chemischer_Elemente

description of solvation processes is carried out using the **solubility product**. In a saturated solution, the dissolved components A and B are in chemical equilibrium with the solid body AB, which is expressed by the **law of mass action**. Since the concentration of the solid is independent and thus constant, the equilibrium constant K is combined into a new constant K_L, the solubility product. The solubility of solids depends on the bonding relationships in AB and the individual interactions of the dissociated particles A and B with the solvent, here water.

$$AB \rightleftharpoons A^+ + B^- \qquad K = \frac{[A^+] \cdot [B^-]}{[AB]} \qquad \Rightarrow \qquad K_L = [A^+] \cdot [B^-]$$

Solubility product

Environment   **Organism**

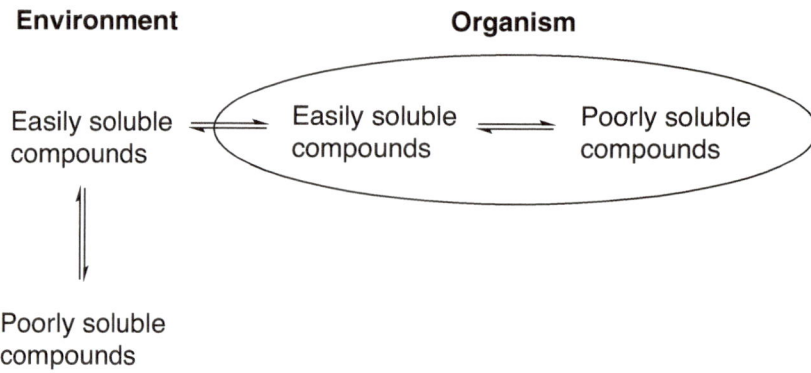

Fig. 2.2 Transfer of compounds from the environment into an organism

In biotic systems, all conceivable combinations occur, from saturated to highly diluted solutions. The solubility of a solid in water is temperature-dependent. Therefore, local climatic conditions are also of importance. Seas and oceans can differ in temperature by up to 30 degrees depending on the region of the earth and the water layer. Therefore, temporary **concentration gradients** also play a role. In this context, weather phenomena must also be taken into account: It is known that the iodine content in marine fish is higher than that of freshwater fish in lakes, which is due to the stronger mixing of oceans by storms that lead to the stirring up of sediments.

The enrichment of insoluble compounds in living organisms by **biomineralisation** is based on a mechanism in which initially soluble compounds are converted into insoluble compounds in the organism itself and in this form fulfil biological functions. Typical examples relate to the entire calcium metabolism in plants and animals. With the exception of alkali carbonates (Li^+, Na^+, K^+), inorganic carbonates are only slightly soluble, so that most metal ions are precipitated in aqueous solution in the reaction with alkali carbonates. This also applies to the biologically relevant carbonates of the alkaline earth metals (Mg^{2+}, Ca^{2+}). Their solubility depends on the pH value.

The insoluble calcium carbonate (aragonite), for example, is in equilibrium with calcium hydrogen carbonate in corals (Fig. 2.3). The consumption of CO_2 by photosynthesis to form glucose leads, according to the principle of Le Chatelier-Braun, to the formation of the insoluble calcium carbonate. The latter protects the corals with a limestone skeleton and anchors them to the seabed. As always, a chemical process is the basis for a biological phenomenon. At the same time, this keeps the concentration of $Ca(HCO_3)_2$ at a low level and $CaCO_3$ acts as a CO_2 depot. Due to increased concentrations of CO_2 in the air, which is currently mainly due to anthropogenic causes, calcium carbonate dissolves in favour of the soluble calcium hydrogen carbonate and thus the corals.

Organic calcium salts can also be insoluble and then offer a biological evolutionary advantage to the relevant organism. Calcium oxalate occurs in some plants (shield ferns,

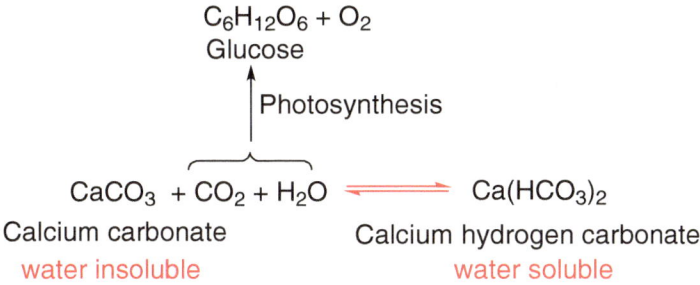

Fig. 2.3 The carbonate/hydrogencarbonate equilibrium in corals

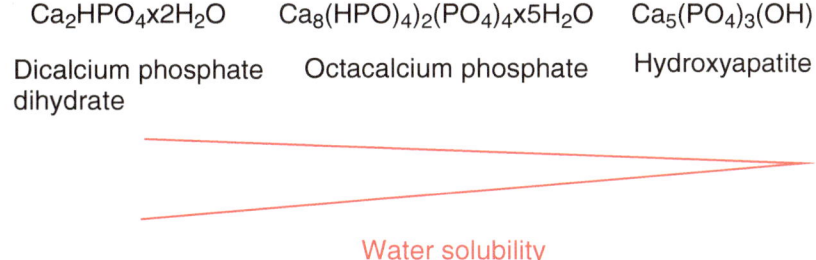

Fig. 2.4 Water solubility of calcium phosphates

parsnips, dieffenbachias) and serves as protection against predators. On the other hand, pathogenic kidney and bladder stones in humans are also mainly made up of calcium oxalate, an indication that the same chemical compounds can have divergent biological effects in different organisms. This not only applies to different species, but concentration differences can be advantageous or disadvantageous in the same organism.

Phosphates also represent limited resources for the growth of organisms. Insoluble calcium phosphates form the hard substance (bones and teeth) of all vertebrates. At the same time, such phosphates represent phosphate stores for readily soluble organic phosphate derivatives (ATP, AMP, FAD, NAD^+, etc.), which changes the biochemical application area (Fig. 5.1) and at the same time creates competition in the context of various, usually only temporarily established, balances. Insoluble calcium phosphates are formed by the mineralization of calcium-deficient hydroxylapatite with calcium. The smaller the calcium content, the higher the water solubility. In other words, the solubility product of the associated phosphates decreases with the increasing incorporation of calcium ions. Of the biotically relevant calcium phosphates, dicalcium phosphate dihydrate has the greatest water solubility (Fig. 2.4). It is followed by octacalcium phosphate. Hydroxylapatite is the most stable and therefore least soluble salt in this series.

Hydroxylapatite is involved, for example, in the construction of the human skeleton and teeth with a share of more than 50% and makes up about 90% of the mineral substance of the body.

Pathological calcifications in humans, such as the dissolution of dental calculus by caries, chondrocalcinosis (pseudogout) and bladder stones, are attributed to more water-soluble precursors of hydroxylapatite.

All biogenic hydroxyapatites are not stoichiometrically built, which opens up space for evolutionary changes. The often quoted formula $Ca_5(PO_4)_3(OH)$ [sometimes also formulated as $Ca_{10}(PO_4)_6(OH)_2$] is therefore an approximation. By contact with carbonate ions, for example from the blood, a replacement of PO_4^{3-} by CO_3^{2-} takes place. But also organic compounds such as citrate or proteins can replace the anions, which increases the strength of the material. Under biotic conditions, the calcium ions are partly replaced by cations of potassium, sodium, magnesium or zinc. But also foreign ions are integrated into the crystal lattice of the hydroxylapatite. This is a prerequisite, for example, that fluoride ions can be incorporated into tooth enamel by means of toothpaste or fluoridated drinking water, which prolongs the life of the teeth.

$$Ca_5(PO_4)_3(OH) + F^- \longrightarrow Ca_5(PO_4)_3F + OH^-$$

$$\text{Fluorapatite}$$

The resulting fluorapatite has a much lower solubility product at the same pH value. Far fewer fluorapatite molecules dissociate in an aqueous solution than hydroxylapatite molecules. Therefore, fluorapatite is more resistant than the body's own hydroxylapatite. The particularly insoluble calcium sulfate (gypsum) does not play a role in biological systems, unlike calcium carbonate.

2.2   Reactivity Under Earth Conditions

Chemical compounds must not only be sufficiently water-soluble to be introduced into biochemical cycles, they must also react in stages. The most important reaction partners on Earth are oxygen and water. Both are, with the exception of a few places on Earth, such as volcanoes (absence of oxygen and water), deep-sea areas (absence of oxygen) or deserts (absence of water), omnipresent. The reaction with these two partners not only converts numerous elements into reactive compounds, but is the first step towards water solubility.

The reactivity of elements is independent of their occurrence on Earth. For example, the noble gas argon, with a volume fraction of just under 1%, is present in large quantities in the atmosphere, but plays no role due to its lack of reactivity. This also applies to other noble gases such as helium or neon, which occur in much smaller quantities. The reasons for the lack of biocompatibility of silicon, although it is the second most common element in the Earth's crust after oxygen, were already discussed in Chap. 1.

2.2.1 Reactions with Water and the Electrochemical Voltage Series

Water serves as a solvent in all biochemical reactions. Therefore, biotically relevant compounds or their precursors must have sufficient water solubility. The prerequisite is created by reaction with oxygen and/or water. The oxidation of non-metals and their conversion into oxoacids were already discussed in Sect. 1.2.5, below the availability of metals is in focus. The always insoluble metals are first converted into water-soluble compounds, which is achieved by the action of water (Fig. 2.5). This results in metal hydroxides. This reaction also produces molecular hydrogen. Metal hydroxides are also formed by the direct oxidation of metals with oxygen and the subsequent reaction of the resulting metal oxides with water. By the reverse reaction, the dehydration (water removal) of metal hydroxides, metal oxides are formed again. For this reason, they are also referred to as (basic) **anhydrides**.

In principle, metal hydroxides are more water-soluble than pure metals. However, in comparison to many metal salts, their water solubility is often only low.

Whether metals are oxidized by water can be seen in the **electrochemical series**, where their standard redox potentials are listed against the standard potential of hydrogen. The latter is assigned a standard potential of zero volts.

$$H_2 \rightleftharpoons 2\,H^+ + 2\,e^- \quad \text{Standard potential} = 0\ eV$$

In general: The more negative the redox potential, the stronger the reducing power, the easier the element is oxidized. In biological systems, the standard redox potential refers to a standard hydrogen electrode at a pH of 7.0 and a partial pressure of hydrogen of 1 bar. All metals that have a negative standard redox potential in comparison are oxidized by water or aqueous acids, which has led to the designation **base metals**. This includes the biologically important metals potassium, sodium, magnesium, zinc and

Fig. 2.5 How metals become water-soluble

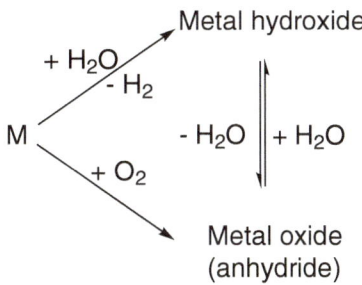

calcium. "**Semi-precious metals**" such as copper, bismuth or antimony have a positive redox potential in comparison. Even more so for the **precious metals** such as gold, platinum, iridium, palladium, osmium, silver, mercury, rhodium or ruthenium. The latter in particular are not oxidized under these conditions, which has direct consequences for their biochemical availability.

The electrochemical series is not limited to metals and allows non-metallic redox systems (e.g. ubiqinone/ubihydroqinone, $FAD/FADH_2$, $NAD^+/NADH+H^+$, ascorbic acid/dehydroascorbic acid) to be included in the consideration. This knowledge is particularly important in electron transport chains in which both metal complexes and purely organic redox systems are involved. By the participation of several redox systems with only minimal differences in the redox potentials, fast oxidations, in particular the "Knallgas reaction", which normally proceeds explosively, are slowed down. The best-known example can be found in the electron transport chain that runs parallel to the citrate cycle and as a result phosphorylates AMP to ATP.

The redox potential is not only dependent on the position in the voltage series, but also on the concentrations of the participating partners in the redox equilibrium, consisting of **oxidizing agent** (Ox) and **reducing agent** (Red). The oxidizing agent is reduced and the reducing agent oxidized. This process is based on the transfer of electrons. The number of these (z) depends on the chemical properties of the participating reaction partners.

$$Ox + z \text{ Electrons} \rightleftharpoons Red$$

The concentration-dependent potential (E) of a redox system can be described by the modified Nernst equation, in which the limited temperature range in which biochemical processes take place has already been taken into account to a large extent (Fig. 2.6). Since the concentration of the oxidized form [Ox] also includes the concentration of protons (or alternatively the concentration of HO^- ions), it becomes clear that with increasing concentrations of acids (or bases), the potential of the redox system (E) in comparison to the standard potential (E^0) changes.

Fig. 2.6 Nernst equation for estimating the oxidizability

$$E = E^0 + \frac{0{,}059 \text{ V}}{z} \cdot \lg \frac{[Ox]}{[Red]}$$

E : Potential of the redox system
E^0 : Standard potential
z : Number of electrons absorbed or released
[Ox] : Concentration of the oxidized form
[Red] : Concentration of the reduced form

The synthesis chemistry benefits from the acid-induced increase of the standard potential when the oxidation of noble metals is intended, since these can only be converted into the corresponding salts by very strong acids. The concentration of protons then enters the Nernst equation as a power, which explains their decisive effect. A well-known example is aqua regia, a mixture of concentrated (37%) hydrochloric acid and concentrated (65%) nitric acid in a ratio of 3:1, with which it is even possible to oxidize gold and convert it into a water-soluble metal complex ($HAuCl_4$).

$$Au + 4\ HCl + HNO_3 \longrightarrow HAuCl_4 + NO + 2\ H_2O$$
$$\text{Water-soluble}$$

The reaction is also successful with metallic platinum. But such conditions do not exist in nature. Therefore, noble metals have not played any role in the chemical-biological evolution since the beginning of the earth until today, which, in addition to their rarity, is the most important exclusion factor. Synthetically produced noble metal compounds often act toxic in organisms because there are no evolutionary chemically established utilization or degradation pathways. This is exploited in pharmaceuticals, for example, to specifically combat tumors with platinum compounds, such as *cis*-platinum $[Pt(NH_3)_2Cl_2]$.

In large quantities, the semi-noble metals bismuth, technetium, rhenium or antimony and their oxidation products are also toxic. In contrast to the noble metals, most of the semi-noble metals are attacked not only by dilute acids, but also by weak bases. Their salts are found on earth and have thus been integrated into biochemical cycles in small concentrations during evolution.

2.2.2   The Role of Oxygen

Oxygen plays the central role as a reactant in biochemistry. At the beginning of the Earth's history, free oxygen played no role. Organic compounds were also not yet present, with the exception of methane. The early atmosphere was a reducing atmosphere, i.e. the few existing compounds not only had a very small molar mass, but also contained elements in low oxidation states. The atmosphere at that time probably consisted of ammonia (NH_3), hydrogen (H_2), water (H_2O), hydrogen sulfide (H_2S), phosphane (PH_3) and hydrogen cyanide (HCN). Carbon monoxide (CO), carbon dioxide (CO_2) and nitrogen (N_2) may also have been part of it. Oxygen was only present in bound form, i.e. primarily in water, in some metal oxides (e.g. silicon dioxide) or in carbonates (e.g. calcium carbonate). From these basic building blocks, the first chemical building blocks of life were formed under the influence of solar radiation, radioactivity and volcanic activity, from which the first living beings, the prokaryotes, evolved.

Some of these primitive organisms were able to generate free oxygen (O_2) for example by water splitting. From them, the chloroplasts of today's green plants later developed, which transferred the hydrogen obtained in a parallel reaction to carbon dioxide and ultimately produced glucose ($C_6H_{12}O_6$) powered by solar energy. Both processes form the gross reaction of photosynthesis.

$$6\ H_2O \longrightarrow 3\ O_2 + 6\ H_2 \qquad \text{(Water splitting)}$$

$$6\ H_2 + 6\ CO_2 \longrightarrow C_6H_{12}O_6 + 3\ O_2 \quad \text{(Hydrogenation of carbon dioxide)}$$

$$6\ H_2O + 6\ CO_2 \longrightarrow C_6H_{12}O_6 + 6\ O_2 \quad \text{(Photosynthesis)}$$

The oxygen content in the atmosphere stagnated for a long time during the development of the Earth at a low level (Fig. 2.7). The rapid transition from a reducing to an oxidizing atmosphere was "braked" by numerous elements and compounds that were initially almost completely oxidized by the oxygen produced. This led to the formation of large reserves of metal oxides, phosphates and sulfates, which characterised many geological strata. With the advent of oxygen-producing single cells, the number of minerals on Earth increased from 250 to 5000. This is evidence that not only do fundamental changes in the environment take place with the appearance of humans, but that organic life itself already represents a dramatic intervention in the global chemistry of the Earth. Parallel to the rise of free oxygen, a stable, versatile and highly functional organic compound,

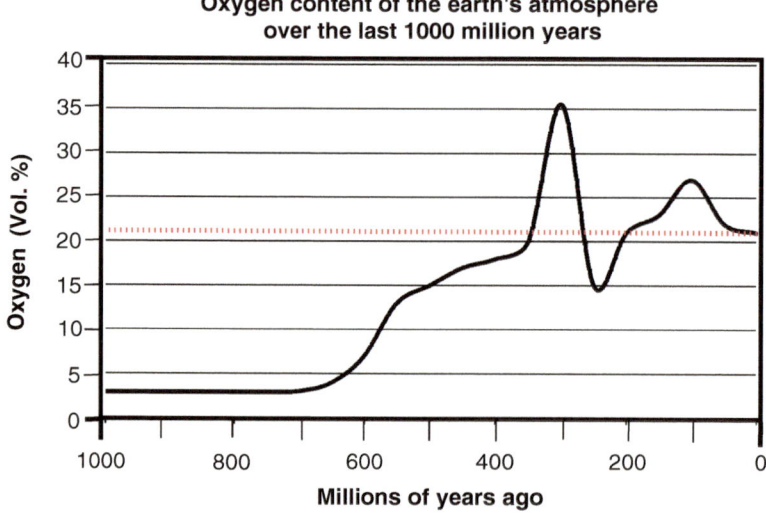

Fig. 2.7 Development of the oxygen content over the last 1000 million years. *Source* A. Börner A (2019) Chemistry—Compounds for Life, WBG, Darmstadt

glucose, was produced, which other single cells used for their own energy production. With the increase in atmospheric oxygen, more and more oxygen was also incorporated into organic compounds with biochemical relevance. In addition to water, metal and non-metal oxides, organic oxidation products of carbon also developed into "oxygen sinks", which, however, are considerably less stable and react with each other in comparison to inorganic oxygen compounds. This led to the formation of compounds with very high molecular weights such as cellulose, chitin or keratin, which served as scaffold components for more complex forms of life. The higher the oxygen content, the larger biological organisms could become. Witnesses of this geological development are giant sequoia trees, giant insects and dinosaurs.

Molecular oxygen acts as an oxidizing agent and accepts electrons. It can be reduced to oxidic oxygen in water. The prototype of this oxidation reaction takes place during respiration in mitochondria. Inversely, as explained above, the oxygen in water is oxidized back to molecular oxygen during photosynthesis.

$$O_2 + 2\,H_2 \underset{\text{Photosynthesis}}{\overset{\text{Respiration}}{\rightleftharpoons}} 2\,H_2O$$

The reduction of molecular oxygen proceeds over several one-electron transfer steps, in which oxygen oxidation numbers between 0 (molecular oxygen) and -2 (in water) are accepted (Fig. 2.8).

Even molecular oxygen has a radical character, which is even more pronounced in the three intermediate compounds superoxide radical (hyperoxide anion), hydrogen peroxide and hydroxyl radical. Such radicals play an outstanding role in cells by giving life processes their typical dynamic character. The same species can also occur in the oxidation of water to molecular oxygen (reverse reaction). In order to limit the unspecific effect of such radicals, redox reactions with oxygen usually take place in special organelles in the cell, the chloroplasts or mitochondria. If the radicals leave the place of their formation, the term "free radicals" is used in nutrition science. The separation of radical-generating processes in special organelles, which also serves to protect the particularly sensitive

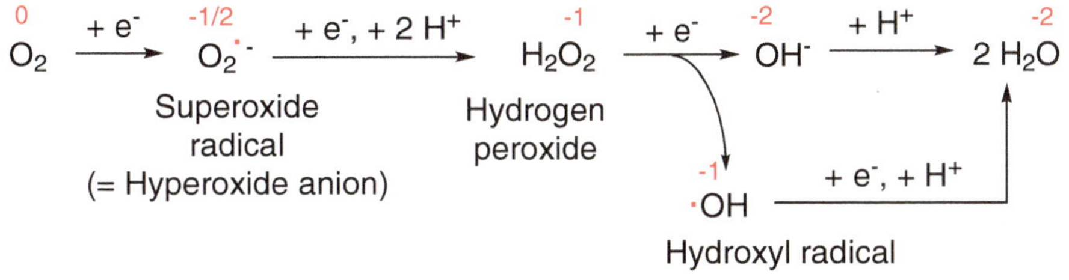

Fig. 2.8 Oxidation states of oxygen

DNA, which is located particularly carefully in the cell nucleus, is a beautiful proof of how chemical natural laws are reflected in the general cell structure.

2.2.3 Solubility in Water: Oxides, Hydroxides, Salts and Complex Compounds

In principle, all non-precious metals and many transition metals are oxidized in an aerobic atmosphere, that is, by molecular oxygen, to the corresponding oxides. By reaction of the oxides with water (hydration), basic metal hydroxides are formed. They are soluble in water depending on the solubility product and dissociate into metal cations and HO^--ions. By neutralization of these bases with acids, salts are formed. This reaction sequence is illustrated using the example of lithium (Fig. 2.9).

Similar reactions can be formulated for the alkali metals sodium and potassium or for the alkaline earth metals magnesium, calcium, strontium and barium, with the cations Na^+, K^+, Mg^{2+}, Ca^{2+}, Sr^{2+} and Ba^{2+} being formed. Their salts are often well soluble in water. Therefore, some of the "naked" metal ions play a central role in fast conduction systems in living organisms, which are mediated by electrochemical processes.

Most of the biologically relevant metals are classified as so-called **trace elements**, including iron, copper, zinc, cobalt, manganese, molybdenum and nickel. Meanwhile, vanadium and chromium have also been added. The term trace element refers in this context to their low occurrence in organisms and not to their occurrence on Earth. In fact, they can be found on Earth in very different concentrations. For example, iron is the fourth most common element in the Earth's crust, while cobalt is only 0.004% present. In this respect, living organisms not only represent delimited places where new compounds are produced that do not occur in their environment, but also certain elements or compounds are concentrated from the environment. They can be referred to as "chemotopes" in analogy to the common term biotope.

Metallions in varying oxidation states play a central role in biochemistry, as they act as oxygen or electron acceptors or donors. Important examples are the **corresponding redox pairs** of iron: $Fe^{2+}/Fe^{3+}/Fe^{4+}$, of copper: Cu^+/Cu^{2+}, or of cobalt: $Co^+/Co^{2+}/Co^{3+}$.

$$\text{4 Li} \xrightarrow[\text{Oxidation}]{+\ O_2} \text{2 Li}_2\text{O} \xrightarrow[\text{Hydration}]{+\ 2\ H_2O} \text{4 LiOH} \underset{\text{Dissociation}}{\rightleftharpoons} \text{4 Li}^+ + \text{4 OH}^-$$

$$\text{2 LiOH} + \text{H}_2\text{CO}_3 \xrightarrow{\text{Neutralization}} \text{Li}_2\text{CO}_3 + \text{2 H}_2\text{O}$$

Fig. 2.9 The transformation of metallic lithium to water-soluble compounds

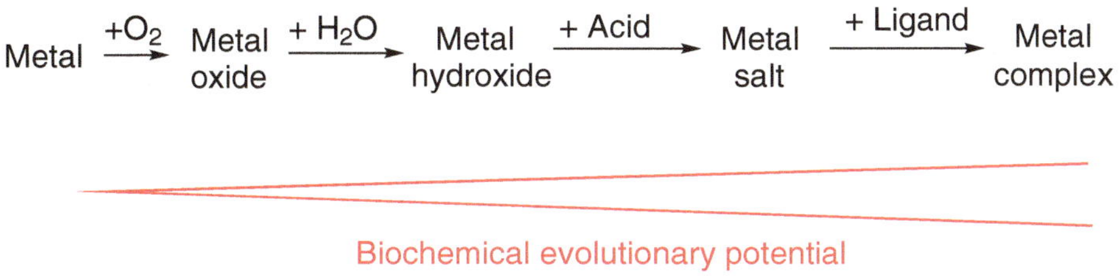

Metal $\xrightarrow{+O_2}$ Metal oxide $\xrightarrow{+ H_2O}$ Metal hydroxide $\xrightarrow{+ Acid}$ Metal salt $\xrightarrow{+ Ligand}$ Metal complex

Biochemical evolutionary potential

Fig. 2.10 The biochemically evolutionary potential of different metal compounds

The intrinsic chemical properties of metallions are changed by their binding, above all, to organic molecules. This results in the formation of **complex compounds**, which have a much higher biochemical evolutionary potential than simple metal salts (Fig. 2.10). In comparison to the corresponding salts, most complex compounds are characterized by larger molecular masses and greater complexity in structure. Through complex formation, the metal is held in the aqueous solution and does not precipitate as an oxide or hydroxide, which would make the metal's mobility and reactivity impossible. The formation of complex compounds represents a quantum leap in chemical evolution, on which central properties of biochemical reactions are based.

Complex compounds are generally characterized by the binding of simple inorganic or higher molecular organic **ligands**. Such ligands can carry charges or be neutral. The binding to the central metal atom is via hapt atoms. In nature, these are usually heteroatoms such as oxygen, nitrogen or sulfur, which have free electron pairs (Lewis bases) that interact attractively with the **central metal**. The simplest type of complex compound is the aqua complex, in which water molecules are the ligands of salts dissolved in water.

$$M(L)_x \qquad \begin{array}{l} M = \text{Central metal} \\ L = \text{Ligand} \\ x = \text{Number of ligands} \end{array}$$

Inorganic ligands such as water, but also oxygen or carbon dioxide, often coordinate only temporarily on the metal. They are activated by this binding and can be converted more easily in the coordinated state than in the uncoordinated state. Organic ligands are often proteins in living things. They can be permanently bound and hardly change their structure after binding to the metal. In particular, high-molecular organic ligands stabilize the central metal atom for a longer period of time. By coordination, they change both the electronic properties of the metal and the steric properties around it. The resulting molecular ensemble has new and biochemically advantageous electronic and steric properties in comparison to the "naked" metal ion, which make it the prototype of many coenzymes. In particular, the electron distribution in the metal center is modified, which affects its redox potential and reactivity.

Ligands can be coordinated with one, two or more hapto (binding) atoms on the metal. The latter surround the central atom like a crab claw, ancient Greek χηλή, *chēlé* for "claw" or "crab claw", which is why the term **chelating ligand** has become established. One distinguishes between one-, two-, three-toothed etc. ligands. The hapto atoms do not have to be the same. Multi-toothed ligands allow the coordination of several metal atoms. The corresponding complex compound is referred to as a di- (or also trinuclear) complex, in which the same or different metals can be bound. The metals influence each other's reactivity or participate in common reactions in concert.

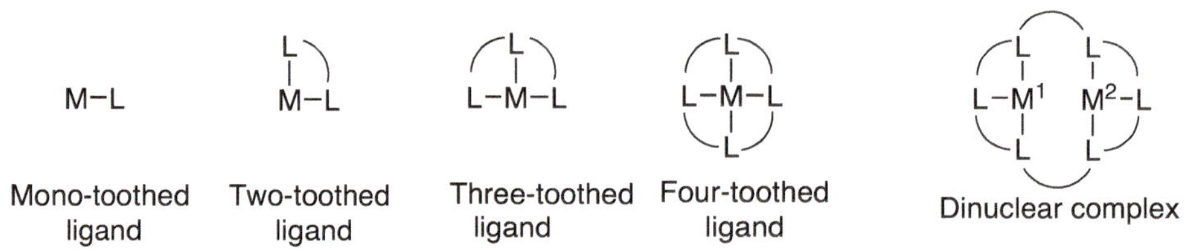

| Mono-toothed ligand | Two-toothed ligand | Three-toothed ligand | Four-toothed ligand | Dinuclear complex |

One of the best-known complex compounds with a four-toothed ligand is the red blood pigment hemoglobin, which is responsible for oxygen transport in the blood (Fig. 2.11). In hemoglobin, an iron (II) ion is "framed" by four *N* ligands. The ligand prevents the iron from converting to the water-insoluble iron oxide during oxygen transport, thus preventing it from fulfilling its transport function repeatedly. The surrounding protein, globin, contains four such hemoglobin units and synergistically controls the activation and transfer of molecular oxygen. In the two forms of chlorophyll (*c1* and *c2*), the green pigment of leaves, a similar structure is found. It is noteworthy that the two complexes occur in completely different organisms, i.e. animals and plants, which indicates biochemically determined synthesis pathways and connection uses.

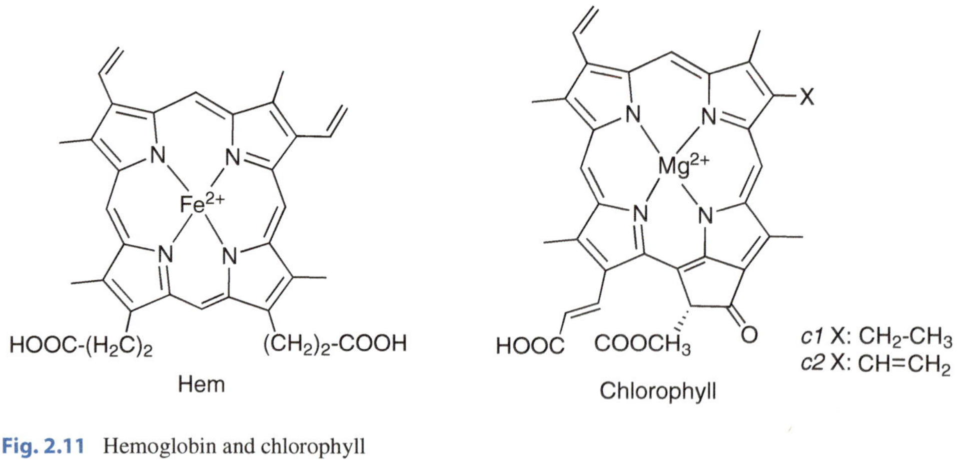

Fig. 2.11 Hemoglobin and chlorophyll

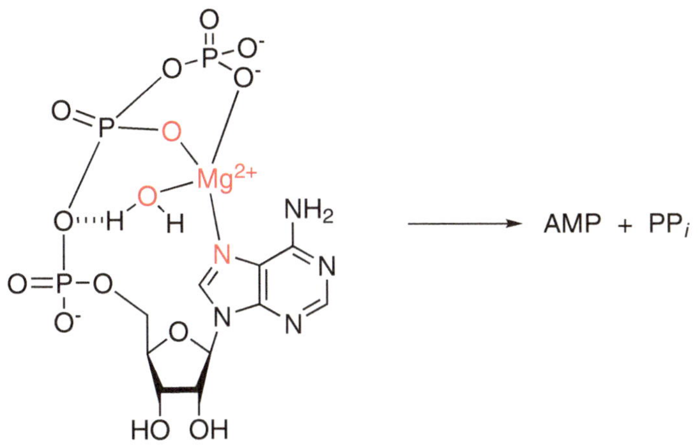

Fig. 2.12 Magnesium complex of ATP

The ubiquitous ATP is also stabilized as a metal complex, which prevents the rapid hydrolysis, i.e. the reaction with an excess of water (Fig. 2.12). The entire triphosphate group is coordinated in the form of a three-toothed *N,O,O* ligand to a Mg^{2+} central ion. This also inhibits the coordination at the metal together with the esterification with ribose, the formation of inorganic and water-insoluble polyphosphates with more than three phosphate units (Sect. 1.2.1). Remarkably, the fourth coordination position at the metal is occupied by a water ligand, which also establishes the contact to a bridging oxygen atom in the anhydride via a hydrogen bond. The complex thus represents the "frozen" reaction of the hydrolysis of ATP to AMP and pyrophosphate (PP_i), which only takes place under the influence of an enzyme.

Organometallic compounds, in which a ligand is bonded to a metal via a carbon atom (Fig. 2.13), are rarely found in nature. The situation is comparable to that in metal hydrides and their typical charge distribution in a metal-hydrogen bond (Fig. 1.37): The metal atom bears the positive charge and the more electronegative hydrogen atom

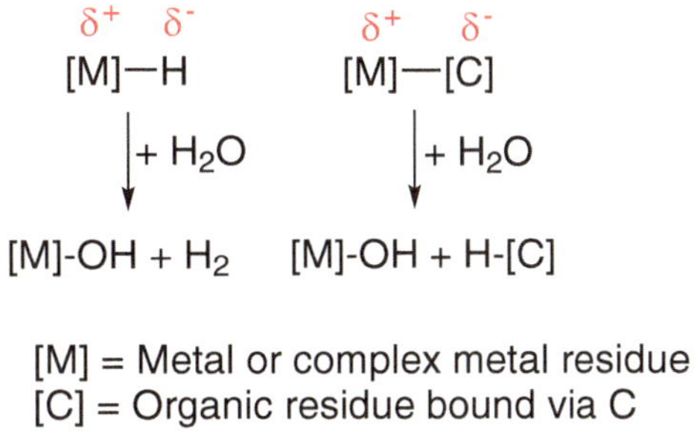

Fig. 2.13 Analogy of the reaction of M–H- to M–C-bonds with water

the negative partial charge. In water, the corresponding metal hydroxide and molecular hydrogen are formed immediately.

Also in organometallic compounds the metal atom is more electropositive than the carbon atom. Organometallic compounds usually react with water by cleavage. The carbon atom at the metal atom is replaced by an HO-ligand, and in the original carbon ligand a C–H bond is formed, which immediately suspends coordination. For this reason, the vast majority of organometallic compounds can only be synthesized and stabilized in the laboratory under water exclusion.

The few exceptions for metal-carbon bonds in living organisms are, for example, mixed organometallic complexes of nickel and iron, in which low-molecular-weight ligands such as carbon monoxide or cyanide coordinate to the metal atom. Through special binding relationships (**π-back binding**), to which not only the adhesive atom, but also other parts of the ligand can contribute, the charge distribution between the metal and carbon atom described above is canceled. For example, compounds of this type act as hydrogen-transferring enzymes (hydrogenases) in archaea, which under anaerobic conditions produce methane from a variety of organic compounds in different oxidation states (biopolymers, fats, short-chain carboxylic acids, etc.) (Fig. 2.14).

Some organometallic complexes of cobalt are also found in higher organisms and even in human biochemistry. Cobalt belongs to the electron-rich transition metals, which explains the singular properties of some of its metal complexes. Thus, a very simple hydrogen-containing complex such as $HCo(CO)_4$ is a very strong acid (release of H^+) in water and has no hydride properties, which would mean the release of H^-.

Biologically relevant organometallic compounds are often high molecular (Fig. 2.15). An example is coenzyme B_{12}, commonly known as vitamin B_{12}, in which the steric shielding of the Co–C bond by parts of the large organic ligand plays a role in the selective catalytic reaction. In the related methylcobalamin, cobalt is bonded to a CH_3 unit, which can be transferred.

The effect of coenzyme B_{12} in the **isomerization** is well documented (rearrangement) of methylmalonyl-CoA, a branched C4 unit and a degradation product of fatty acids with an odd number of carbon atoms, to succinyl-CoA (Fig. 2.16). For malonate bound to CoA there is no connection mechanism, while succinate is unbranched and can be utilized in the citrate cycle. The rearrangement is realized by migration of the CoA–C=O

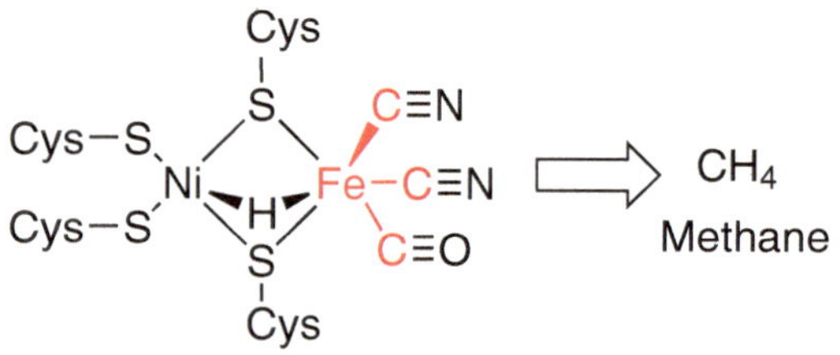

Fig. 2.14 A biochemically rare case of an organometallic complex

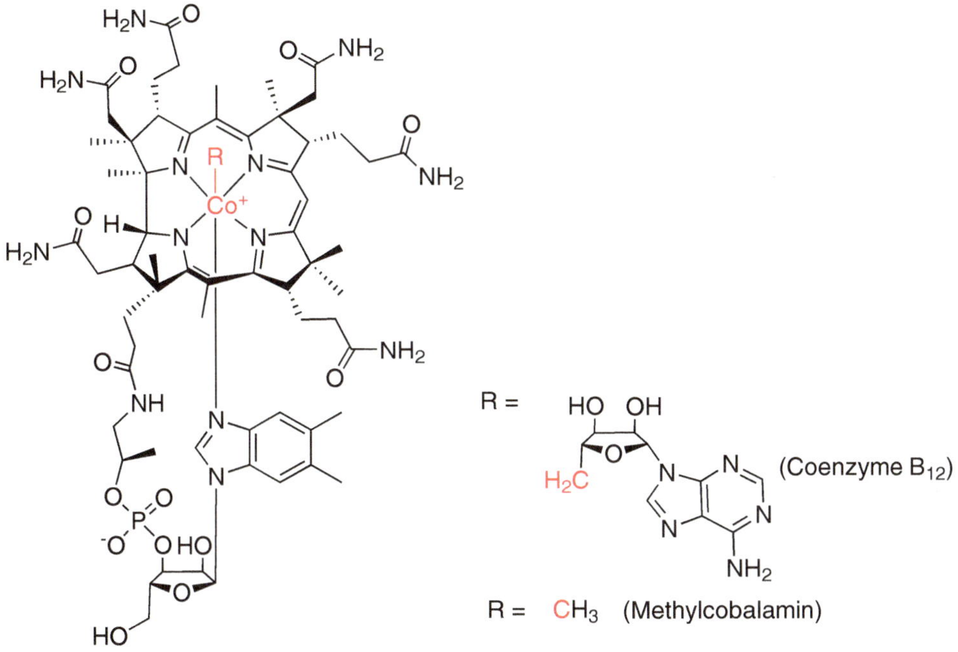

Fig. 2.15 Biochemically relevant organometallic complexes of cobalt

unit from C^1 to C^2. The carbon intermediates, which are generated by alternation from cobalt (II) to cobalt (III) and back to coenzyme B_{12}, are central. The bond cleavage between cobalt and carbon atom takes place due to the low polarity in this bond in a radical mechanism, i.e. there remains one electron on the original binding partner.

Methylcobalamin also transfers a methyl group to L-homocysteine in a radical mechanism, resulting in the proteinogenic amino acid L-methionine.

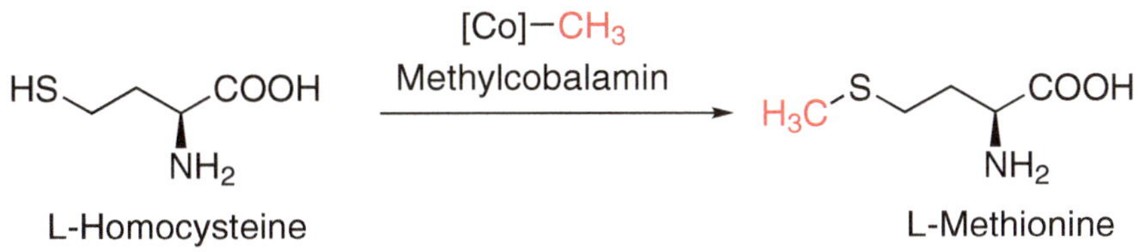

It must be emphasized that radical C–C bonds are rare in biochemistry. C–C Coupling reactions are mainly carried out via polar mechanisms (Sect. 4.1.1).

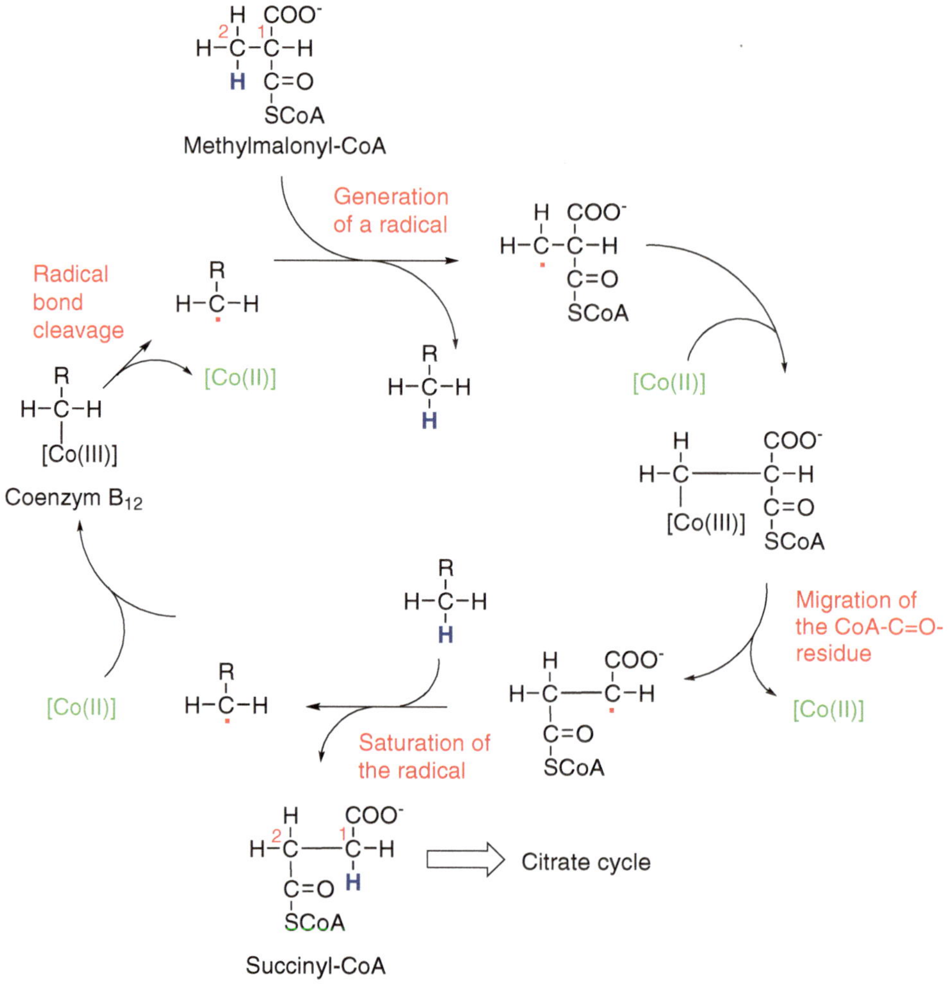

Fig. 2.16 The role of an organometallic complex as a catalyst in radical isomerization

2.2.4 Biocatalysis by Enzymes

Many complex compounds, like the cobalt complexes described above, represent chemical **catalysts** in that they accelerate the establishment of chemical equilibrium by lowering the activation energy and are not consumed themselves. In the presence of multiple competing reaction paths, they select a specific one. In biochemistry, catalysts are referred to as **enzymes**. Most biochemical reactions are only possible because of them. Enzymes gain their functional existence by converting biochemical substrates into products. The substrates are, in most cases, when they coordinate with the metal, also ligands,

but only have a very short residence time on the metal. Such "substrate ligands" are frequently structurally modified during the reaction with "reagents", which also coordinate
briefly as ligands, and at the end of the catalytic cycle are cleaved off as no longer coor-
dinatively capable product. A typical catalytic cycle is illustrated schematically below,
with $M(L)_x$ representing the prototype of a catalyst (Fig. 2.17). The number x,y and z of
each ligand can remain constant or change during the catalytic process. This makes it
clear that each metal complex has new properties during the catalytic cycle, which makes
the analysis of catalytic reactions more difficult than stoichiometric reactions.

The only short-term binding of substrates and reagents to the enzyme and the "theoretical" intactness of the enzyme at the end of the reaction characterize the nature of
any catalysis and fundamentally distinguish catalytic processes from stoichiometric
reactions. Catalytic reactions are in principle reversible reactions. Whether catalytic
processes have arisen from stoichiometric reactions during biochemical evolution is
probably, but must remain speculative at this point. There are countless examples from
laboratory chemistry where a significant acceleration of the original reaction occurs by
the addition of additives and where the additive is not consumed. On the other hand, the
lifetime of each catalyst is also limited, contrary to the theory. In particular, enzymes
whose protein components are constantly subject to degradation reactions are continuously synthesized anew, which is an indication that catalysis may have arisen from stoichiometric reactions and that enzymes have undergone an optimization process during
evolution.

Catalysts are then judged by how often and how quickly they catalyze a chemical
reaction. The number per unit of time is described in the **turnover frequency** (abbreviated TOF) how often the catalyst has transformed a certain substrate in a certain time.
The **catalytic productivity** (*turnover number,* abbreviated TON) on the other hand indicates how much product is formed under certain reaction conditions per catalyst. So
actually every reaction with a TOF > 1 could be provided with the term catalysis, which
in turn refers to the generic kinship of stoichiometric and catalytic reactions.

The products of chemical reactions can in turn influence their own formation. In this case it is called autocatalysis. If the products are proteinogenic α-amino acids, which are subsequently incorporated into enzymes, the feedback takes place much later, but it does.

It is important to note that enzymes do not support reactions that are chemically impossible. Enzymes are adapted to specific substrates or reaction types. As a result, they are divided into numerous classes and subclasses. Their structure and mode of action are of particular interest in biochemistry. However, since they all have the basic function of reducing the activation energy for certain reactions, this book, which has the basic chemical principles in view, does not go into this any further.

The Derivatization of Inorganic Compounds with Carbon Residues

3

Organic chemistry can be seen as an extreme extension of inorganic chemistry, where new structures with new properties arise. Organic chemistry was before the "man-made" synthetic chemistry biochemistry! Organic compounds were for hundreds of millions of years both a prerequisite and a product of life. Only for about one hundred years have people emancipated themselves from the narrow framework conditions of living nature with academic and industrial synthetic chemistry. Nevertheless, the laws of nature of chemistry continue to act.

As was already mentioned in connection with the treatment of oxoacids with regard to esterification (Sect. 1.2.14), in inorganic compounds in principle all hydrogen atoms can be replaced by carbon residues. This process almost extends the property- and numerically strongly limited range of inorganic compounds to infinity, which is a prerequisite for evolution. The reduction of organic chemistry to inorganic chemistry also has a valuable cognitive advantage: Many of the properties of the inorganic parent compound from the PSE can be transferred to the organic derivatives. At the same time, however, properties also change. Below, characteristic effects are to be analyzed in more detail and at the same time differences between inorganic compounds and their organic derivatives are to be worked out by way of example.

3.1 Water, Alcohols and Ethers

Alcohols and ethers can be considered derivatives of water (Fig. 3.1). If one of the hydrogen atoms in H_2O is replaced by a carbon residue, an alcohol is formed. When the second hydrogen atom is replaced, an ether is formed.

Representatives of the connection classes of alcohols and ethers have comparable geometries to water, i.e. the substituents on oxygen are not at a 180° angle to each other,

A. Börner and J. Zeidler, *The Chemistry of Biology*,
https://doi.org/10.1007/978-3-662-66521-3_3

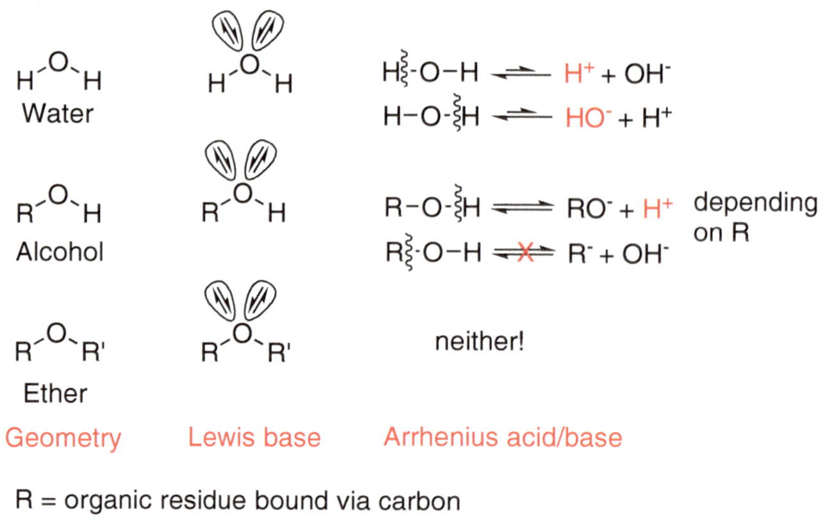

Fig. 3.1 Comparison of some properties of water with its organic derivatives

but the molecules are angled. Alcohols also have a hydrogen atom, which in principle can be given off as a proton. Since water is only a very weak Arrhenius acid, it can be assumed that this also applies to alcohols. In contrast, the C–O bond is so strong that alcohols are not Arrhenius bases. Ethers have lost the potential to be an Arrhenius acid by the loss of both hydrogen atoms. However, due to the free electron pairs on oxygen, they, like water and alcohols, are Lewis bases.

Atom groups that always contain the same heteroelements in the same bonding ratios, as shown here for alcohols or ethers, are referred to as **functional groups**. In addition to heteroelements such as oxygen, nitrogen, sulfur or halogens, these groups can also contain carbon and hydrogen. The corresponding compounds are grouped together in a substance class.

The loss of potential acid-base properties of the inorganic parent compound is more than compensated for by the enormous number of new compounds and the gain in variability of organic derivatives. Only then is the basis for life created. Organic residues transfer their electronic and steric properties to the adjacent functional groups, which thus experience an extension of their original properties. This phenomenon will be illustrated below using the example of alcohols in comparison to water.

Water is subject to a constant dissociation equilibrium on Earth, which is determined by the (unchanging) electronegativities of oxygen and hydrogen. As already mentioned, this equilibrium lies far on the side of the undissociated water. The replacement of a hydrogen atom by organic parts influences the equilibrium. The cause lies in the different polarization of the O–H bond by the organic part: electron-pushing parts (illustrated by

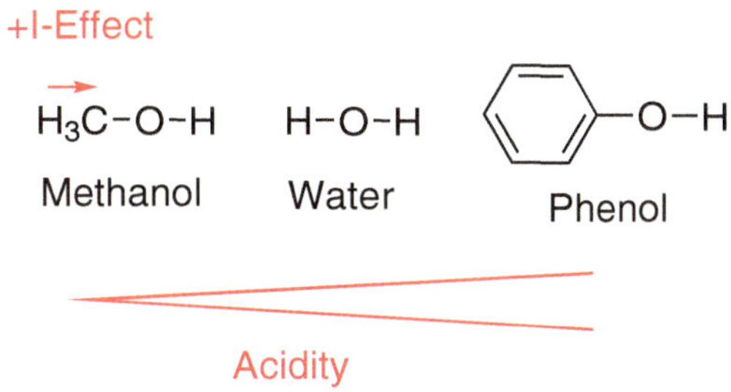

Fig. 3.2 Examples of the
influence of organic residues on
acidity

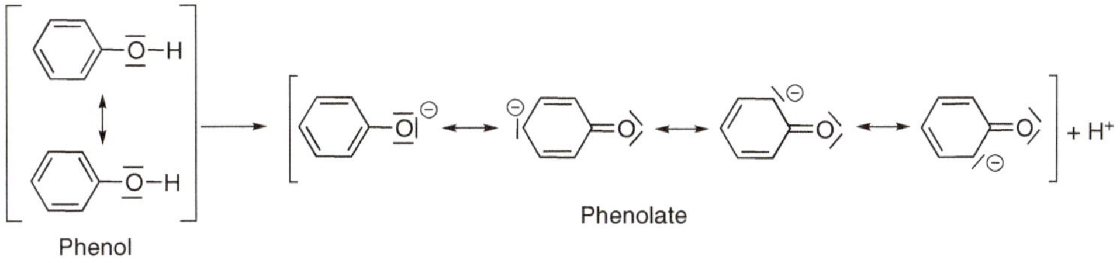

Fig. 3.3 Mesomerism as an explanation for the increased acidity of phenol

a red arrow in the direction of the electron shift) weaken the polarization of this bond, thus the H^+-donor properties of the alcohol are further weakened in comparison to water. Electron-withdrawing parts, on the other hand, increase the polarization of the HO bond. As a result, a stronger acid results in comparison to water.

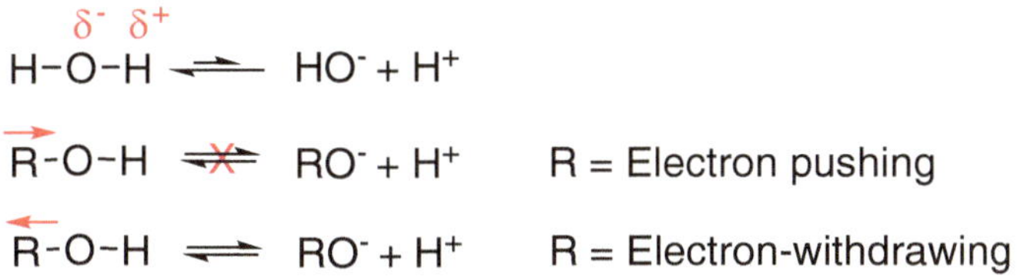

The properties of the functional group, which are intrinsic and not changeable in the inorganic compound, are now dominated and modified by those of the organic part. In general, alkyl groups, for example a CH_3 group, have a **+I-effect** (positive inductive effect), which causes the electron density to be shifted to the adjacent groups (Fig. 3.2). Methanol is thus even less acidic than water.

The experimentally found increased acidity of phenol compared to methanol is described by mesomerism of the phenolate (Fig. 3.3). In comparison to the reactant phenol, four mesomeric boundary structures can be constructed for the product phenolate. This is not possible for methanol.

In the biochemical context, for example, this phenomenon finds expression in the existence of two proteinogenic α-amino acids. An amino acid with an aliphatic hydroxy group, such as L-serine, always comes into action when no acidic properties are necessary or when acidic properties would be counterproductive. The aliphatic HO group is the preferred site of phosphorylation reactions (Fig. 1.52). In contrast, the phenolic side chain of L-tyrosine is less frequently phosphorylated. It is present as (weak) acid instead.

L-Serine

L-Tyrosine

Protons are the structurally simplest form of acidic catalysts. The effect of a phenolic hydroxy group can be illustrated by the desamination of amino acids, a central mechanism for the degradation of α-amino acids, using vitamin B_6 (pyridoxal) (Fig. 3.4). In the

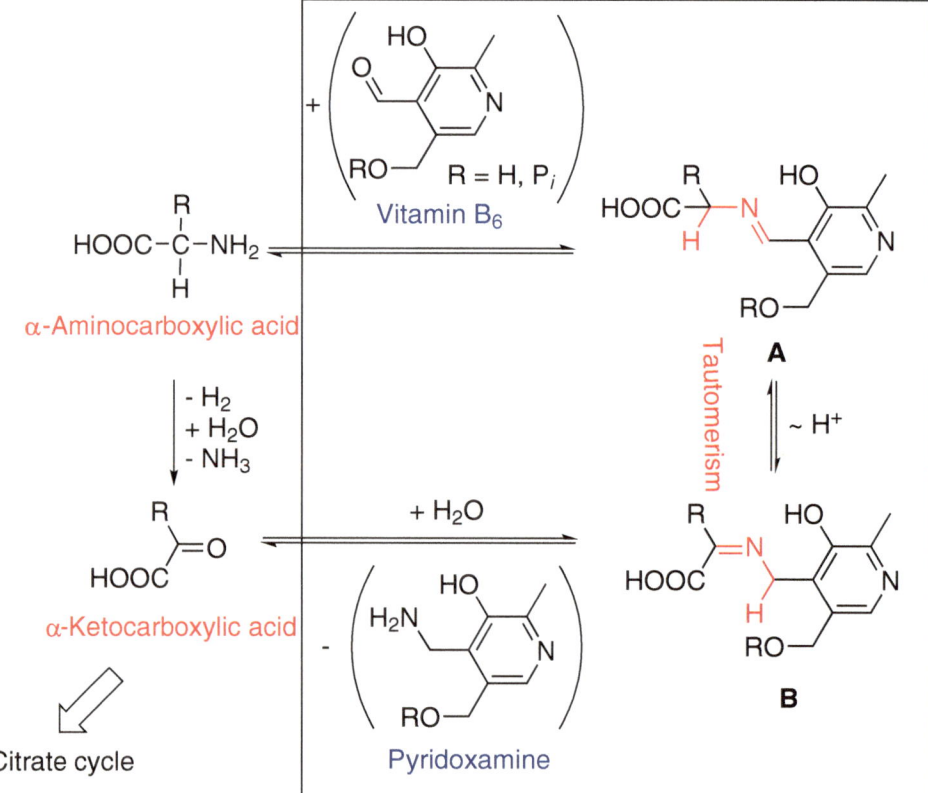

Fig. 3.4 Desamination of amino acids

overall reaction of this process, α-ketocarboxylic acids are formed. The relevant part of the mechanism is shown on the right in the box.

The reaction of the intermediate A to B is the special case of a tautomerism. In the structure of A it becomes clear why vitamin B$_6$ must have this special structure. An essential structural element is the aromatic hydroxy group. Is is placed in vitamin B6 next to the original aldehyde group for a certain reason. Only in this position can a hydrogen bond with the basic imino nitrogen atom be formed, as the scheme shows in part (Fig. 3.5). It should also be noted that the H-bond forms an energetically favorable ring, a six-membered ring. The imino nitrogen atom is protonated by this acid-base reaction. The resulting iminium cation (positive charge!) then exerts an electron pull on the adjacent C–H bond. The hydrogen atom concerned is thus acid and migrates as a proton in the course of tautomerization towards the aromatic. At the same time, the original C=N double bond shifts. The newly formed C=N bond is cleaved with water and the original amino acid has lost its NH$_2$ group. This product leaves as an α-ketocarboxylic acid and is further transformed in the subsequent citric acid cycle. Pyridoxamine is oxidized to vitamin B$_6$ in a parallel mechanism, which is not discussed here, and thus regenerates the coenzyme.

In summary, the central role of an aromatic hydroxy group in a certain geometric position as an "internal" acidic catalyst becomes clear; without it, the reaction sequence would not take place under the typical moderate conditions of biochemistry.

This mechanism is used to desaminize all α-amino acids. As an example, the **desamination** of L-alanine is shown (Fig. 3.6). First, acetic acid (or its salt, pyruvate) is formed. The final hydrogenation leads to L-lactic acid. Acetic acid is converted into the citrate cycle. Lactic acid is the product of homolactic fermentation under anaerobic conditions,

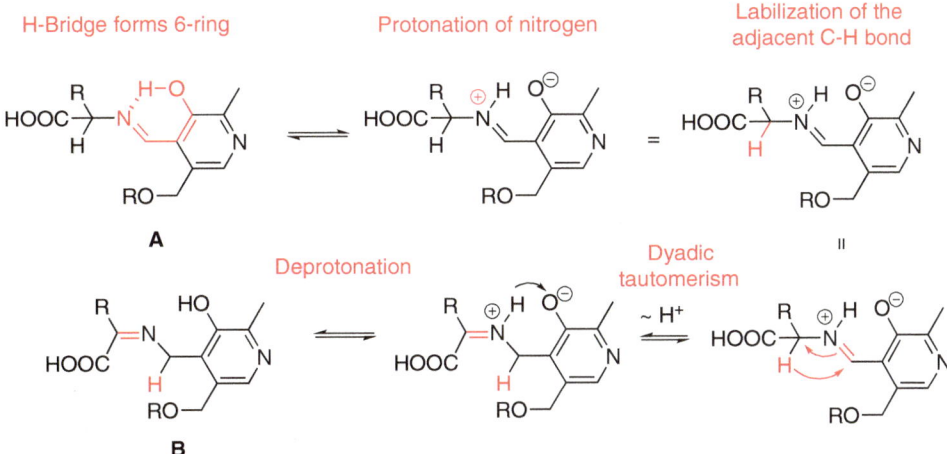

Fig. 3.5 Importance of position and acidity of a hydroxy group during the deamination of amino acids

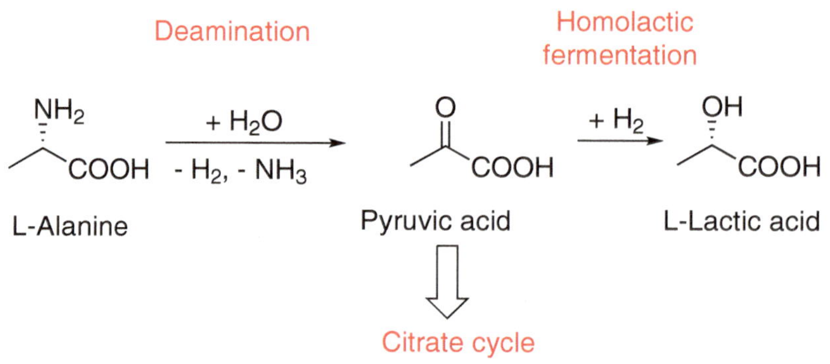
Deamination
Homolactic fermentation
NH2
+ H2O
- H2, - NH3
COOH
O
COOH
+ H2
OH
COOH
L-Alanine
Pyruvic acid
L-Lactic acid
Citrate cycle

Fig. 3.6 The deamination of L-alanine and follow-up transformations

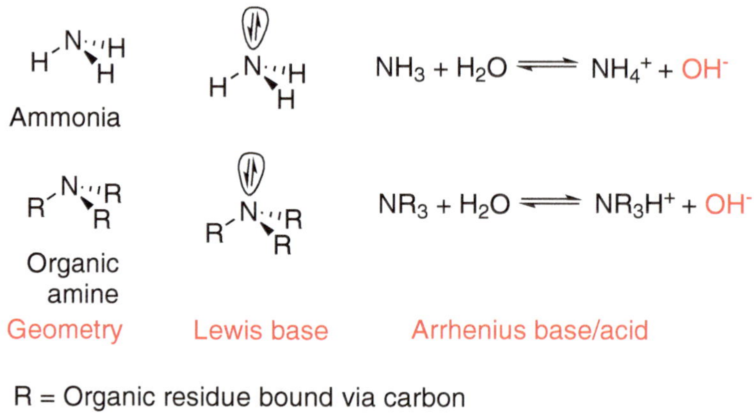
Ammonia
NH3 + H2O ⇌ NH4+ + OH-
NR3 + H2O ⇌ NR3H+ + OH-
Organic amine
Geometry
Lewis base
Arrhenius base/acid
R = Organic residue bound via carbon

Fig. 3.7 The influence of organic radicals on the properties of amines compared to ammonia

e.g. in yeast. But it also occurs in mammals, which once again proves the uniformity of many compounds and processes in organisms of different evolutionary stages.

3.2 Ammonia and Amines

A similar consideration to the relationship between water, alcohols and ethers can be applied to the organic derivatives of ammonia, the amines (Fig. 3.7). The latter are formally formed by successive replacement of the three hydrogen atoms in ammonia with organic radicals. Like the inorganic parent compound, ammonia, they are Lewis bases due to the free electron pair. The free electron pair leads to the formation of a tetrahedral structure. Amine also generate hydroxide ions with water. They are thus also bases

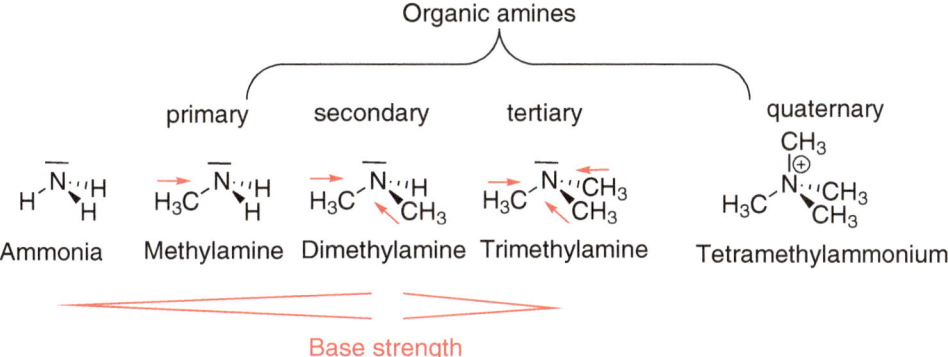

Fig. 3.8   The acidity of amines as a function of the number of methyl groups

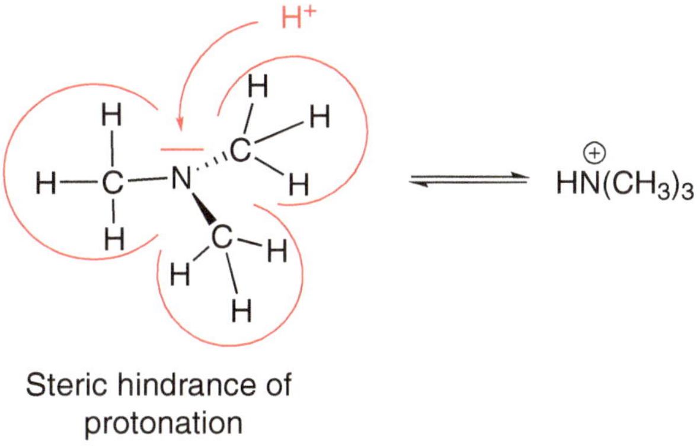

Fig. 3.9   Explanation for the
reduced acidity of tertiary
amines

Steric hindrance of
protonation

according to the Arrhenius theory. The corresponding ammonium ions are the corresponding acids.

As already seen with the organic derivatives of water, the number of possible derivatives is increased by the replacement of H with organic residues. Up to four alkyl (or aryl) groups can be bonded to the nitrogen atom. Consequently, a distinction is made between primary, secondary, tertiary and quaternary amines. Similar to the alcohols, the electronic properties of these residues also influence the properties of the functional group. Alkyl groups increase the Lewis acidity at the nitrogen atom or, in other words, the "availability" of the free electron pair at the nitrogen atom for binding to a proton is positively influenced. Consequently, the acidity increases in the series ammonia, methylamine and dimethylamine (Fig. 3.8).

The effect of three methyl groups as in trimethylamine should, in principle, produce an even stronger amine, since a triple +I effect occurs. However, this trend is not experimentally observed. The decrease in basicity when moving from secondary to tertiary amine is explained by the increased space demand of the organic residues compared to

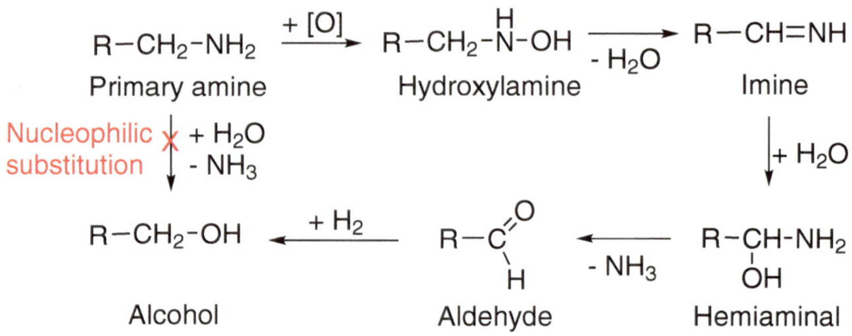

$$R-CH_2-NH_2 \xrightarrow{+ [O]} R-CH_2-\overset{H}{\underset{}{N}}-OH \xrightarrow[- H_2O]{} R-CH=NH$$

Primary amine   Hydroxylamine   Imine

Nucleophilic substitution $\downarrow$ + H$_2$O, - NH$_3$ $\downarrow$ + H$_2$O

$$R-CH_2-OH \xleftarrow{+ H_2} R-\overset{O}{\underset{H}{C}} \xleftarrow[- NH_3]{} R-\overset{}{\underset{OH}{CH}}-NH_2$$

Alcohol   Aldehyde   Hemiaminal

Fig. 3.10 Indirect transformation of amines into alcohols in the biochemical context

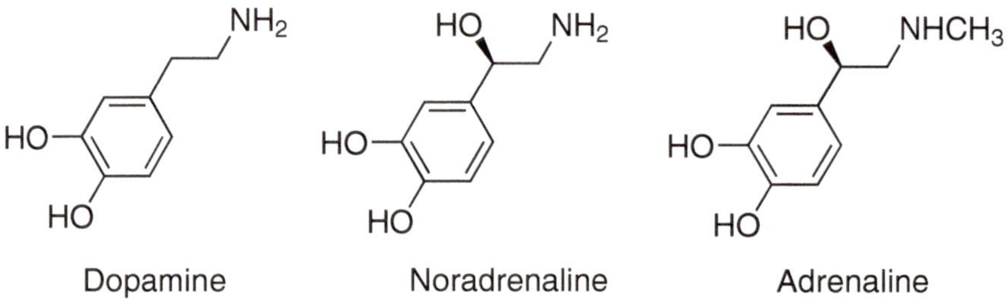

Dopamine   Noradrenaline   Adrenaline

Fig. 3.11 Examples of biogenic amines

the "small" hydrogen atoms. This hinders the formation of a bond between the proton and the free electron pair at the nitrogen atom (Fig. 3.9). This is the typical case of a **steric hindrance**. At the same time, the very large trimethylammonium ion is less effectively surrounded by a water shell, which negatively influences the equilibrium shift in favor of the product.

As carried out in Chap. 1, oxygen can be inserted into all three N–H bonds in ammonia, resulting in the formation of inorganic nitrogenous acids after the water has been split off due to the effect of the Erlenmeyer rule. In a similar way, most organic amines are degraded in living organisms. In the first step, the oxidation of an N–H bond (Fig. 3.10) takes place, analogous to the reaction with ammonia.

Hydroxylamine, the first oxidation product, splits off water, and an imine is formed. The reattachment of water leads to a hemiaminal, which in turn reacts under elimination of ammonia to form an aldehyde. In eukaryotes, this oxidation is catalyzed by monoamine oxidases (MAOs). In many biochemical systems, a hydrogenation follows, resulting in the formation of an alcohol in this case. The conclusion of the entire reaction sequence, in which oxidation and reduction alternate, is: The basic and therefore basically hostile amino group has been removed.

This is the alternative to nucleophilic substitution, which does not take place under biotic conditions (Sect. 1.2.12). Therefore, the "detour" via an oxidation-reduction sequence is taken.

In this way, the biogenic amines dopamine, noradrenaline and adrenaline are deaminated in the postsynaptic gap (Fig. 3.11). The biogenic amines are formed from canonical α-amino acids by decarboxylation and have central physiological tasks as transmitters by means of their amino function. The duration of action of the amines is limited by deamination, which expresses their important but only temporary effect in stimulus conduction.

It should be remembered that the free electron pair on the nitrogen atom can also be used for a bond with the oxygen atom, as demonstrated by the oxidation of ammonia to nitric acid (Fig. 1.57). The replaceability of hydrogen atoms in ammonia by organic residues limits the oxidizability. This situation is particularly striking in the reaction of tertiary amines with oxygen. For example, the reaction of trimethylamine with oxygen only leads to trimethylamine oxide (TMAO), since no oxidizable N–H bond is available (Fig. 3.12, reaction a). Tertiary amines are therefore degraded in biochemical processes by the competing insertion of oxygen into adjacent C–H bonds (reaction b).

First, an unstable hemiaminal is formed (Fig. 4.69), which reacts with water under elimination of formaldehyde to give dimethylamine. The latter represents a secondary amine and is deaminated by the usual oxidation mechanism. At the end of the entire reaction sequence, ammonia is formed, which as a gas shifts the equilibrium of all degradation reactions of N-containing compounds in organisms in favor of the end products according to the principle of Le Chatelier-Braun and thus provokes the constant replenishment of amino acids and other nitrogen-containing compounds.

Tertiary amines are also stable end products in the biochemical degradation of nucleobases, as they occur in RNA or DNA. A well-known example is the alkaloids theobromine, theophylline and caffeine, which are formed from adenine (Fig. 3.13). Adenine is first hydrolyzed by water to hypoxanthine. Then the only C–H bond in the six-membered

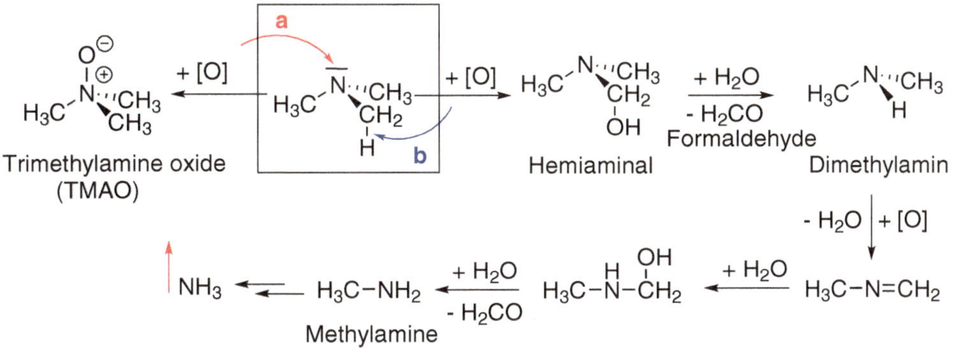

Fig. 3.12 Oxidation reactions on a tertiary amine

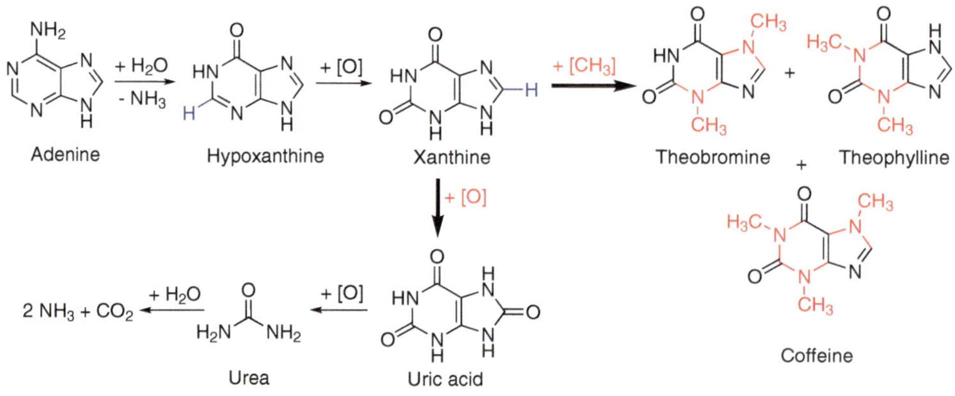

Fig. 3.13 Oxidative degradation products of adenine

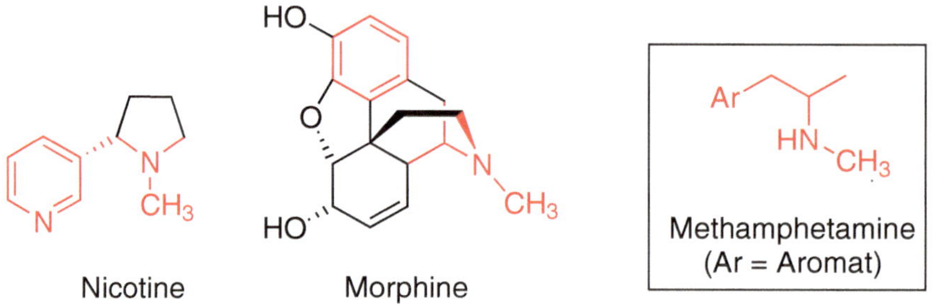

Fig. 3.14 Tertiary amines with psychoactive effect

ring is oxidized and xanthine is formed. At this stage, the further degradation pathway is decided. Either the C–H group in the adjacent five-membered ring is also oxidized, which leads to uric acid. Uric acid is then further decomposed in the presence of oxygen depending on the animal species into urea (frogs, mammals) or ammonia and carbon dioxide (fish, tadpoles). However, if the methylation at the N-atoms is dominant, the stable plant alkaloids are formed, which have a long-lasting biological effect. They can be mutagenic to bacteria, fungi and algae. Furthermore, a defense effect on snails, various larvae and insects is discussed, which in any case suggests an advantage for the producing plants. These examples show that evolutionary fitness effects, which only become effective at a higher level, determine the direction of the oxidation chemistry in the sense of a feedback.

Some biogenic drugs also represent tertiary amines, including nicotine and morphine (Fig. 3.14). Both are derived from the methamphetamine base structure (Ar = phenyl), a secondary amine. Unlike the latter, they are degraded in the first step by the insertion of oxygen into a C–H bond adjacent to nitrogen. This altered degradation mechanism can be a cause of the often stronger effect of drugs on psychologically active (biogenic) amines in the human body compared to endogenous ones.

3.3   **Carboxylic Acids**

In contrast to alkyl groups with their electron-pushing effect **acyl groups** have an electron-withdrawing effect on a neighboring N–H group. This has significant consequences: In carboxylic acid amides, the Lewis basicity of the nitrogen atom is greatly reduced. The theoretical proof can be based on a mesomeric boundary structure in which the free electron pair is involved in the bond to the adjacent carbon atom. There is a **single bond with partial double bond character**. Double bonds are characterized by hindered rotation of the substituents. Two geometric **isomers** can be distinguished: *cis* (Z = together) and *trans* (E = opposite), in which the two residues R^1 and R^2 are on different sides. Only at higher temperatures (ΔT), which are usually not reached under biotic conditions, are they in equilibrium.

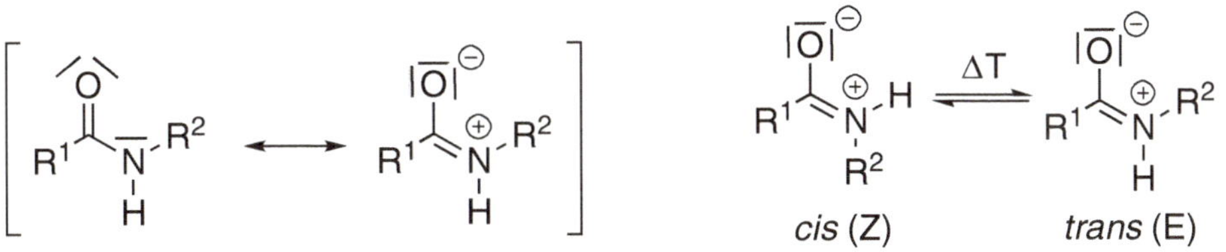

Ammonia is (like primary and secondary amines) not an acid in the sense of the Arrhenius definition. It also does not form hydrogen bonds, which is the reason why ammonia is a gas under normal conditions. However, the electron-withdrawing effect of the acyl group in carboxamides (in short: amides) leads to a stronger polarization of the N–H bond (Fig. 3.15). The proton becomes acidic. This is the prerequisite for a hydrogen bond to an electronegative element (X).

Such hydrogen bridges between acetyl amide groups are, for example, responsible for the softness and flexibility of chitin. The compound consists of long chains of D-glucose monomers, in which the HO group at C^2 is replaced by an acetamide group (Fig. 3.16). Remarkably, the gluco configuration is retained during the exchange and thus also the orientation of the acetamide group in the equatorial position at the six-membered ring (Sect. 4.1.5.2). This is another proof of the outstanding stability of structures derived from the geometry of glucose. The chitin chains are interconnected by hydrogen bridges

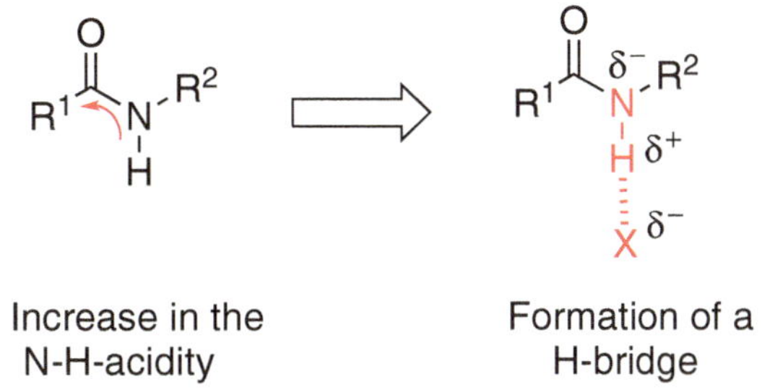

Fig. 3.15 Acyl groups increase the acidity of N–H bonds and are the prerequisite for hydrogen bonds

Increase in the
N-H-acidity

Formation of a
H-bridge

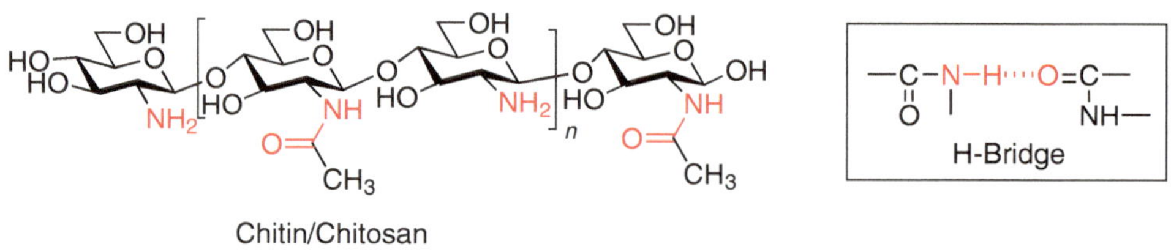

Chitin/Chitosan

Fig. 3.16 Hydrogen bridges emanating from chitin

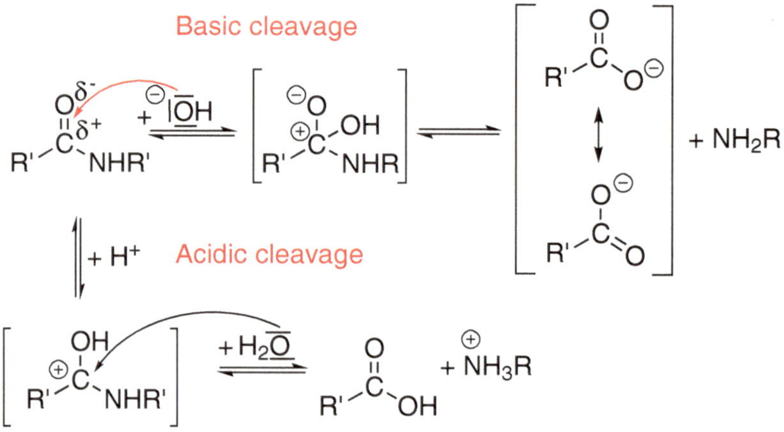

Fig. 3.17 Why an acidic or basic environment is hostile to life

between N–H and C=O, which are even stronger than hydrogen bridges between hydroxy groups, for example in cellulose. The chain length of chitin depends on the acetylation degree of the amino group. If it is lower than 50%, it is called chitosan. Chitin occurs in many lower eukaryotes such as algae, fungi, arthropods and molluscs, but not in vertebrates. In the latter, keratin, a biopolymer based exclusively on proteins, predominates. In arthropods, chitin is the main component of the exoskeleton, which only becomes hard and stable through the interaction with the structural protein sclerotin. Chitin is the second most common biopolymer after cellulose with an estimated bioproduction of 10^9–10^{11} t per year.

Carboxamides are split by strong bases, resulting in a mesomerically stabilized carboxylate anion at the end (Fig. 3.17). In water as a solvent, hydroxide ions are the simplest base. Hydrolysis in the presence of acids is also known, with ammonium ions and free carboxylic acids being formed in addition to the ammonium ion. This is the reason why both a basic and an acidic environment are counterproductive for the chemistry of life. Therefore, it is self-evident that life preferably takes place in a neutral environment.

Further biochemical consequences are presented in connection with the organization of proteins and genetic structures in Sect. 4.1.6.

The Singular Properties of Carbon as the Basis for the Emergence of Life

4

Biochemistry is essentially chemistry based on carbon and hydrogen, extended by the chemistry of some other elements—mainly oxygen, nitrogen, sulfur, phosphorus—of some non precious metals and seminobel-metals and a few halogens. This fact results from the singular variability of carbon compounds. If you only combine a few vital elements with each other and limit the molar mass to 500 g/mol, 10^{62}–10^{63} variants are possible. In this calculation, only those compounds were taken into account which are relatively stable towards water and oxygen. The limitation to 500 g/mol excludes all macromolecules and polymers which also play an important role in living nature, i.e. the possibilities of variation are really much greater. Some causes of the unique evolutionary potential of carbon have already been discussed in the first parts of this book (points 1–4). Others, which are discussed in detail below, result from the consequences of the reaction to oxygen (points 5 and 6).

1. Due to the relatively small atomic radius of carbon, the formation of single bonds, double bonds and triple bonds with itself or small heteroatoms is possible.
2. Stable chains based on C–C bonds can contain up to one million carbon atoms and probably even more. Ring formation and annulation are also possible and unique.
3. The electronegativity of carbon is comparable to that of hydrogen. C–H Bonds are therefore hardly polarized in the absence of adjacent electron-withdrawing groups. They are particularly stable against water.
4. Carbon is present on Earth in considerable quantities.
5. By radical insertion of oxygen into C–H bonds, oxygen-containing functional groups are formed, which represent the starting point for functionalization with other heteroelements such as N, S, P or halogens.
6. As an element of the first 8 periods of the PSE, carbon is able to form stable C=C or C=X bonds (X = O, N, S), which are sufficiently labile to serve as central "switches" in dynamic reaction networks.

A. Börner and J. Zeidler, *The Chemistry of Biology*,
https://doi.org/10.1007/978-3-662-66521-3_4

If one applies the principle of biological evolution originating from Darwin to the chemistry of life as well, it becomes clear that only on the basis of the element carbon a sufficiently large number of compounds is possible which form the basis for biochemical processes of enormous complexity and mutual penetration. Evolution in the context dealt with here means that in parallel selection processes chemical structures arise which interact with others and thus condition each other. At the same time, they are subject to continuous modification and further development by internal and external influences. The new structures and mechanisms are selected by feedback effects to structures and mechanisms on lower levels of complexity down to the elements. In this way, chemical stability islands arise. The dynamics and stability of these processes determine the phenotype, behaviour and life span of biological organism classes and their individuals.

4.1   The Redox Chemistry of Carbon

The "opening reaction" in the chemistry of life and thus the access to biochemically relevant compounds consists in the insertion of oxygen into C–H bonds. This is to be demonstrated by the stepwise oxidation of methane, the simplest hydrocarbon compound, to carbon dioxide (Fig. 4.1). The formal designation [O] for an oxygen donor has been chosen for simplification, the precise chemical mechanism will be explained below.

First, methanol is formed by inserting an "oxygen atom" into one of the four C–H bonds of methane. The repetition of this process leads to a new class of compounds, here generally referred to as hydrates. Already during the discussion of properties of carbon compounds on the one hand and of silicon, phosphorus and sulphur compounds on the other hand (Sect. 1.2.1) it was pointed out that compounds with more than one HO group

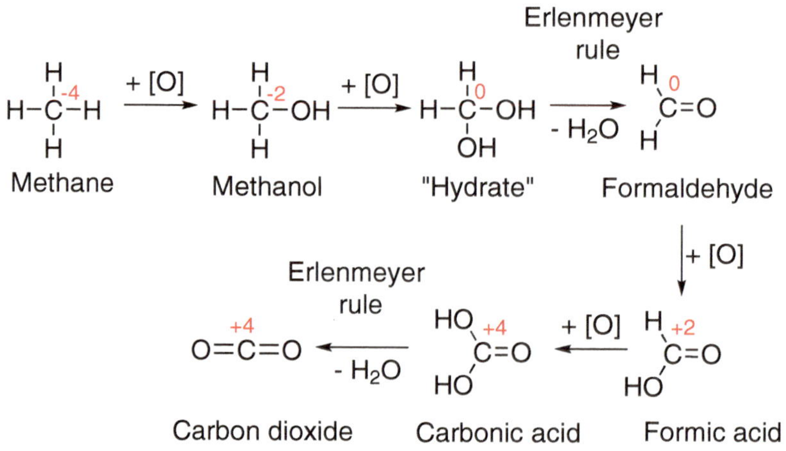

Fig. 4.1 The stepwise oxidation of methane

show a different stability depending on the radius of the central atom. Hydrates of (small) carbon are not stable and split water according to the Erlenmeyer rule. Formaldehyde is formed as a product. In formaldehyde, two C–H bonds are still left. Their oxidation first leads to formic acid and ultimately to carbonic acid. The latter decomposes again according to the Erlenmeyer rule to water and carbon dioxide.

When comparing the oxidation numbers, it becomes apparent that after each insertion of oxygen into a C–H bond, the oxidation level of the carbon atom increases by two units. Starting with -4 in methane, compounds with the even oxidation numbers -2, 0, $+2$ and $+4$ are generated. Methane, methanol, formaldehyde and acetic acid represent the first representatives of the homologous series of the **alkanes**, **alcohols**, **carbonyl compounds** and **carboxylic acids**, which are mostly dealt with in the didactics of organic chemistry in the form of compound classes.

A methyl group at the end of a hydrocarbon chain can thus also be degraded by insertion of oxygen into C–H bonds up to carbon dioxide (Fig. 4.2). It should be noted that with the exception of acetic acid ($R = H$), all other carboxylic acids no longer contain any C–H bonds. CO_2 is generated in this case by decarboxylation and the alkane chain is shortened by one CH_2 unit. In this way, "alkane 1" becomes "alkane 2".

The oxygen-containing intermediates are the starting point for the formation of numerous derivatives in which oxygen is replaced by other heteroatoms. The totality of the substructures thus formed leads to natural products and represents the chemical basis of biology (see Fig. 4.3). The return of the substitution products to oxygen-containing basic structures requires numerous and partly very complex biochemical reaction sequences and thus time. From this formal context, which represents one of the basic theses of this book, it can be concluded that a large part of biochemistry can be considered as **slowed-down total oxidation of the electron-rich carbon**. Processes in which

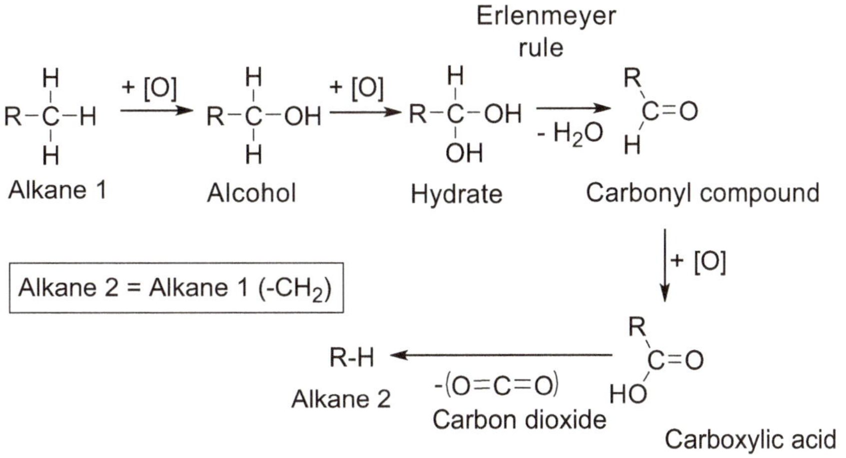

Fig. 4.2 Shortening an alkane chain by oxidation and decarboxylation

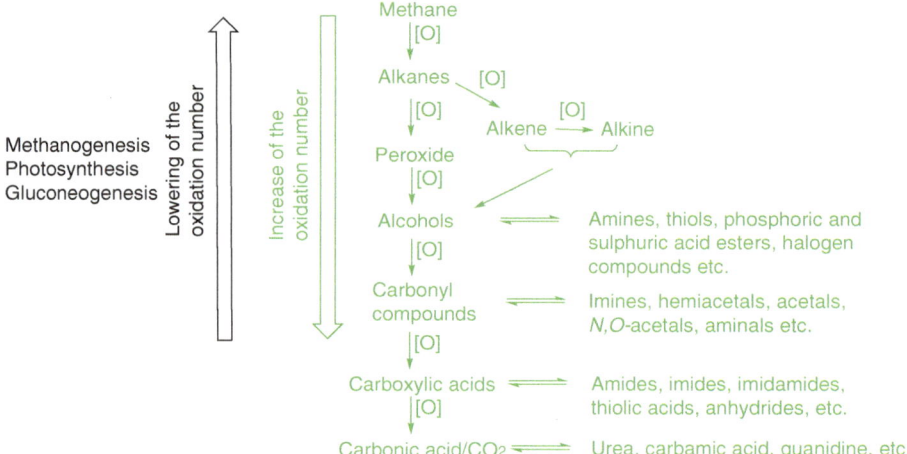

Fig. 4.3 Life as slowed-down total oxidation of the electron-rich carbon

intermediarily compounds with carbon in lower oxidation states (e.g. by hydrogenation) are formed do not change this general phenomenon of any organic life. Exceptions to this general phenomenon of any organic life are methanogenesis and photosynthesis, in the course of which carbon dioxide is converted into products with low oxidation numbers such as methane or glucose, which requires the supply of energy (e.g. sunlight). Also, fatty acid synthesis in algae and gluconeogenesis, in which small amounts of D-glucose are formed from non-carbohydrates and which occurs in all organisms, belong to these exceptions.

It is remarkable that only few natural products and biochemical mechanisms are associated with the construction of such energy-rich compounds. Their origin often goes back to processes in selected single-celled organisms. Cyanobacteria and methanogenic bacteria are exemplary. Even if these single cells were integrated into symbiotic processes by higher host organisms, such as green plants (chloroplasts) or ungulates (methanogenic bacteria) during the course of biological evolution, the basic mechanisms remained the same. This means that diversity and individuality of biological organisms are expressed mainly by oxidative degradation processes, which differ in scope and direction. The approach chosen in this book, to attribute the totality of biological phenomena to delayed oxidative degradation processes, finds its justification in this.

It should be noted that in many exchange reactions, the respective oxidation number of the carbon atom in question does not change. This important conclusion is that there are, in principle, two types of compounds or substructures in biochemistry:

1. those that have the same oxidation number on the carbon atom, and
2. those that differ in their oxidation numbers on the carbon atom.

The determination of oxidation numbers in compounds or substructures and types of reactions is therefore important in order to recognize similarities or differences. For many structures, e.g. *N*-heterocycles, which do not contain oxygen or hydrogen atoms, this is not immediately possible, since according to the rules only the latter provide the basis for the calculation of oxidation numbers. However, by formally transforming to oxygen-hydrogen structures, the oxidation numbers can be easily determined, which considerably simplifies the classification of reactions (substitution, elimination, addition, redox reaction) in biochemical events and provides a red thread for the entire biochemistry.

4.1.1   C–C Bond Formation Reactions

There are not only compounds of carbon with even oxidation states (Fig. 4.1), but also those with odd oxidation numbers. The prototype results from the coupling of two methane molecules to ethane (Fig. 4.4).

As a result, both carbon atoms in ethane receive the oxidation number -3. Formally, ethane is formed from two methyl radicals, which arise from two methane molecules by hydrogen abstraction. By oxidative linking with further methane molecules, longer-chain hydrocarbons are formed from the homologous series of alkanes, in which only the two terminal CH_3 groups still carry the oxidation number of -3 (Fig. 4.4). In the CH_2 groups, carbon receives the oxidation number -2. The C–C linking reactions are dehydrogenations, i.e. molecular hydrogen (H_2) is split off. In fact, in biochemical systems, hydrogen is never released directly, it would leave the Earth's atmosphere as a very light gas. As a result, the hydrogen concentration on Earth would decrease with unpredictable consequences for the chemistry of life. Therefore, it is logical that higher molecular hydrogen acceptors such as NAD^+ or FAD (Fig. 1.85) chemically bind hydrogen. In the most common case, and thus at the end of a long reaction sequence, the hydrogen acceptor is the ubiquitous oxygen, which is thus reduced to water. Even water, with its high boiling point, represents a stable "H_2 trap" from this point of view. The bound hydrogen retains its oxidation state of $+1$ during all transformations.

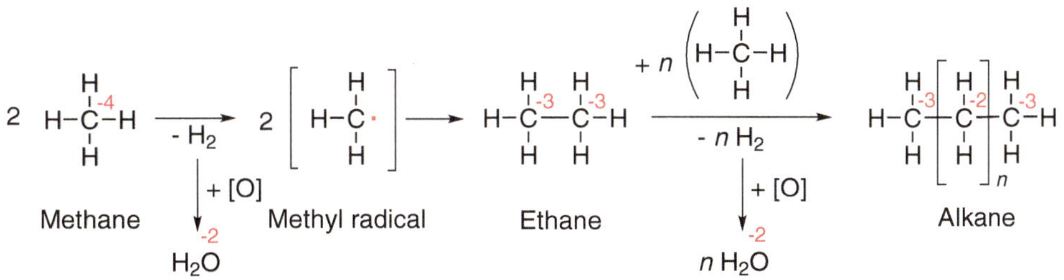

Fig. 4.4   C–C bond formation reactions as an oxidative process

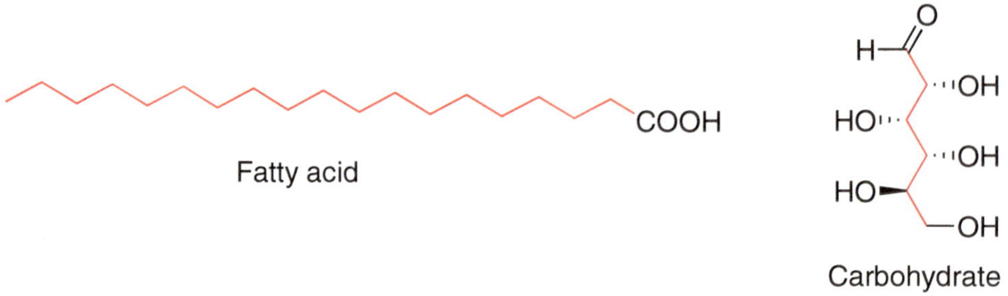

Fatty acid

Carbohydrate

Fig. 4.5 Examples of carbon chains in natural products

In this way, long carbon chains of slightly functionalized natural products are formally produced in the biochemical context, such as fatty acids, or those chains that form the backbone of highly functionalized compounds such as carbohydrates (Fig. 4.5).

C–C bond formation reactions not only pose a significant challenge in synthetic chemistry in the laboratory, as two identical elements have to be coupled and thus *a priori* no charge differences can come into effect (Fig. 4.6). In the biochemical context, coupling of hydrocarbon radicals hardly ever takes place. An exception to this is the rearrangement of methylmalonyl-CoA to succinyl-CoA mediated by vitamin B_{12} (cobalt complex!) (Fig. 2.16). However, in most cases, two carbon ends are coupled, which have different polarities or, in extreme cases, even opposite charges.

A positively charged or positively charged carbon atom (**carbocation**) in the vicinity of electronegative heteroatoms such as O or N is the normal case due to the intrinsically lower electronegativity of carbon. The generation of a partially negative or negatively charged carbon atom (**carbanion**) is more demanding, which also applies to laboratory chemistry. Such reactive species can only be generated by the involvement of functional groups in the vicinity. In some cases, of "Umpolung" (English: **"inversion of reactivity"**) is also spoken when the sign of charge is inverted at the carbon atom under consideration.

Fig. 4.6 Possibilities for C–C
bond formation

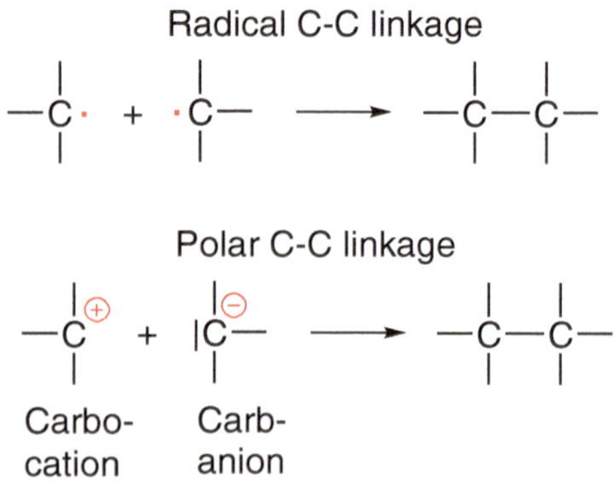

Radical C-C linkage

Polar C-C linkage

Carbo-
cation

Carb-
anion

An example of a polar C–C bond is the construction of citrate from acetyl-CoA and oxalacetate at the beginning of the citrate cycle (Fig. 4.7). A basic enzyme abstracts a proton from the methyl group of acetyl-CoA. As a result, the carbon atom concerned acquires a negative charge. The charge is delocalized over the adjacent carbonyl group and thus stabilized by mesomerism. The negatively charged carbon atom attacks that carbon atom in oxalacetate which, due to the adjacent oxygen atom and the two carboxylate groups, has a positive partial charge. As a result, a C–C bond is formed (shown in red in citryl-CoA). After cleavage of the acyl group carrier CoA-SH, the citrate named after the entire mechanism is formed.

Another example of polar C–C bond formation is the catalytic effect of vitamin B_1 (thiamine). Thiamine is a molecule consisting of two heterocycles which are linked to each other by a CH_2 bridge (Fig. 4.8). The biosynthesis of thiamine takes place via two completely different metabolic pathways, one in enterobacteria and one in yeast. The latter is also used by plants. Obviously, the biochemical effect of this structure is so unique and important that heterologous assembly mechanisms *had to* evolve independently of each other.

As will be discussed subsequently, only one of the two heterocycles is involved in the biochemical transformation. The different meaning of substructures for the actual chemical reaction is a characteristic of natural products. Often, molecular parts that are far away from the reactive center have no recognizable benefit *a priori*. There are various reasons for this: For example, binding interactions with protein-containing superstructures (enzymes) in which these reactive centers are embedded can justify their evolutionary necessity. Furthermore, polar sections can act as solubilizers. "Meaningless" molecular parts can also be relics of their own biosynthesis, where their presence was

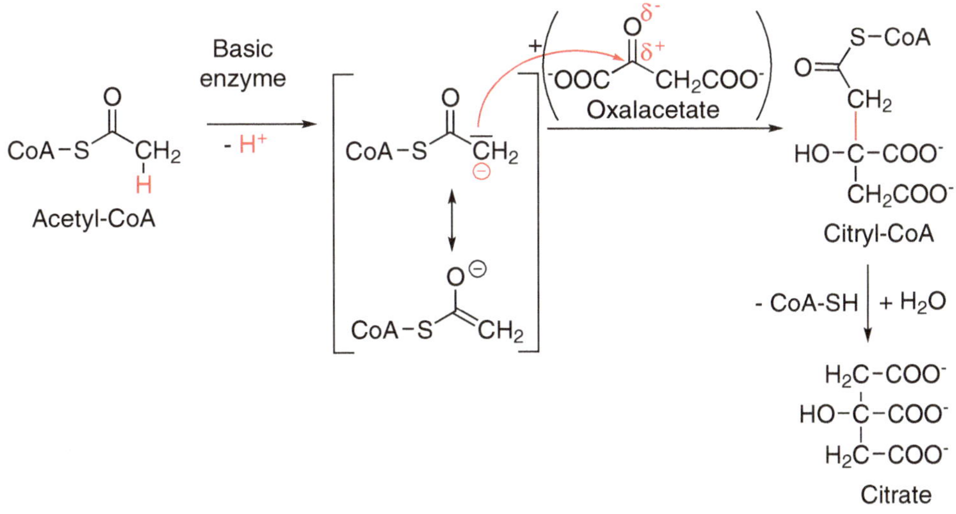

Fig. 4.7 The synthesis of citrate as an example of a polar C–C bond formation

essential. Another possibility is the origin of such "appendages" from central biochemical mechanisms, from which other biomolecules with other tasks emerge in parallel. In this case, the compound under consideration represents a structural and reactive compromise. A satisfactory answer can usually only be derived from a detailed and comprehensive analysis of chemical genesis and reaction requirements.

In the reaction context considered here, only the five-membered ring heterocycle with sulfur and nitrogen plays a role in vitamin B_1. As will be shown exemplarily in Sect. 4.1.7, part of the heterocycle can be traced back to the basic structure of formic acid, which is helpful for determining the oxidation number at the reaction-relevant carbon atom C^2.

Both in thiamine and in acetic acid, the carbon is positivated and thus unsuitable as a nucleophile for a polar C–C bond formation reaction (Fig. 4.9). The required negative charge on the C^2 atom is generated by the displacement of a proton by means of a base-acting enzyme. The negative charge is delocalized to the adjacent sulfur and nitrogen atoms in the five-membered ring, which explains their presence and their proximity in the molecule; with acetic acid itself, the reaction would not be possible in the biochemical context. It is a perfect "Umpolung" which is also clearly indicated by the negative charge on the C^2 in the resonance structures B and C. In chemistry, such structures are referred to as carbenes. The new C–C bond is formed by the attack of the polarized carbon atom in thiamine on the positivated carbon atom in the reaction partner pyruvate.

The product with two opposite charges forms the starting point for a series of synthesis and decomposition reactions (Fig. 4.10). By splitting off CO_2, a neutral compound A, is formed which, however, also has the potential to form a carbanion, as the charged mesomere structure B shows. This anion can, for example, react with a second pyruvate molecule, and another C–C bond is formed. After splitting off the catalytic thiamine, a highly functionalised carbon chain with four C atoms results. However, mainly the bond between the catalyst and the C_2 rest is cleaved. Formally, this results in acetaldehyde, which is oxidised to acetate. Both structures, which form the transition between glycolysis with pyruvate as a product and the citrate cycle with acetate as a substrate, do not occur in this form in the biochemical context, but are only shown for the sake of

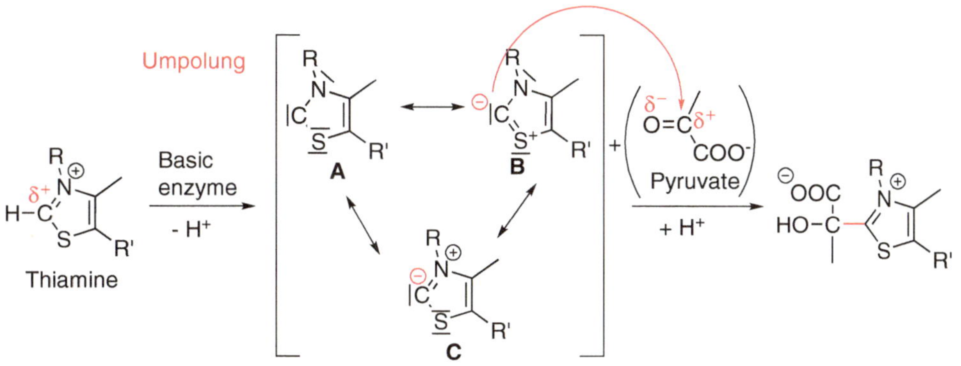

Fig. 4.9 C–C bond formation by a preceding "Umpolung" of reactivity

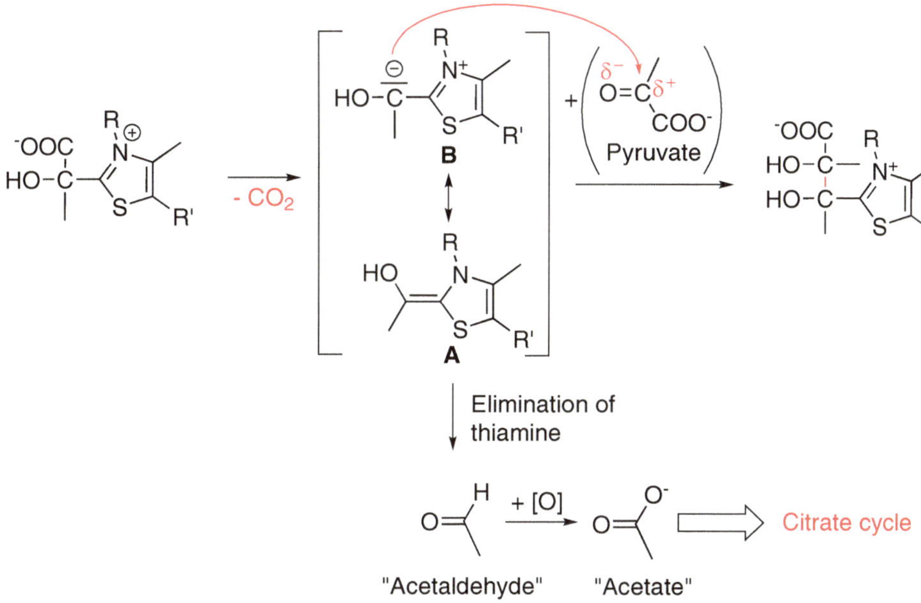

Fig. 4.10 A C–C coupling product as a substrate for construction and decomposition reactions

better understanding. In fact, the free acetaldehyde with a boiling point of 20 °C would immediately escape from the organism under physiological conditions, thus depriving the entire citrate cycle of its most important inflow and, as a result, its relevance. Alternatively, acetaldehyde could condense with amines, which would also explain its toxic effect during the breakdown of ethanol. Acetate does not occur in this context either; it is always linked to CoA. By thioester formation, the principle of molecular weight increase becomes effective and at the same time the acetate rest is activated for the subsequent coupling with oxalacetate.

4.1.2 Dehydrogenation of Alkanes

Alkenes

Alkenes are formed by the splitting off of molecular hydrogen from alkanes, as illustrated by the dehydrogenation of ethane to ethene.

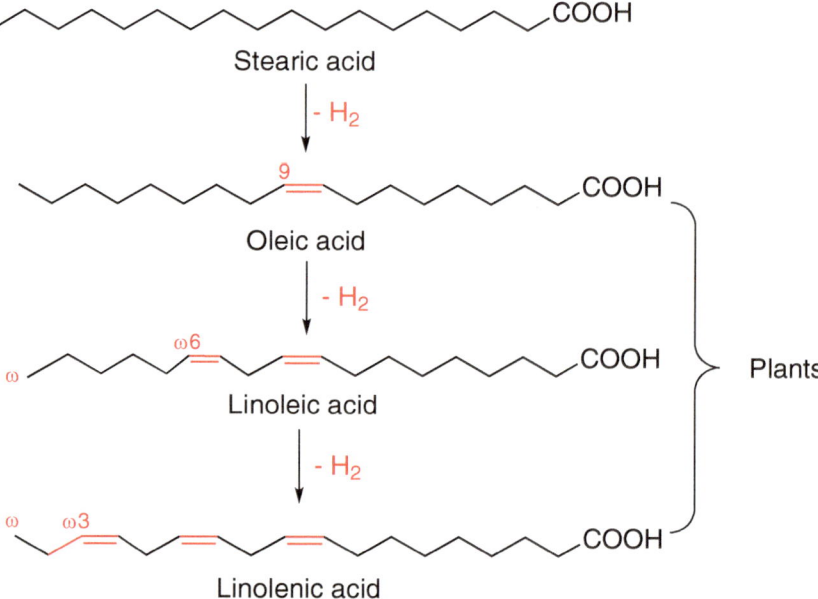

Typical compounds with C=C double bonds are numerous **unsaturated** fatty acids (Fig. 4.11). They are synthesized by dehydrogenation (desaturation) of saturated precursors. These in turn are formed by repeated linking of C_2 and C_3 units. Representatives are the plant acids oleic acid, linoleic acid or linolenic acid, all of which are derived from stearic acid and contain both unsaturated and also **saturated** sections. In the biochemistry of mammals, there is no mechanism for inserting double bonds beyond the ninth carbon atom, even though the compounds are necessary for further reactions and biological functions. For this reason, the so-called ω-6 or ω-3 fatty acids are essential fatty acids. They must be ingested with food. From a chemical point of view, this means

Fig. 4.11 Saturated and unsaturated fatty acids

that any food intake not only forms the basis, but always also an extension of the chemical resources and the biochemical mechanisms beyond the limits of one's own organism. The term "symbiosis" used in biology is partly based on this chemically based fact.

In biochemical systems, flavin adenine dinucleotide (FAD) acts as a hydrogen acceptor for the dehydrogenation of alkanes due to its "suitable" redox potential (Fig. 4.12). In the reduced form as $FADH_2$, the coenzyme is able to reduce alkenes. Such higher molecular weight hydrogen storage prevents the escape of H_2 from biochemical systems. This is a difference from synthetic reactions, where gas-phase H_2 can be easily worked with using pressure apparatus, although higher molecular weight hydrogen transfer reagents are also known.

Notably, direct transfer of H_2 does not take place in the biochemical context. H_2 is usually split into two protons and two electrons, which are transferred separately and often sequentially, with the partial reactions being mediated by specialized proton and electron carriers.

The dehydrogenation of saturated molecular moieties is often the starting point for a series of follow-up reactions. From an unreactive C–C single bond system, a reactive C=C double bond with numerous reaction alternatives arises.

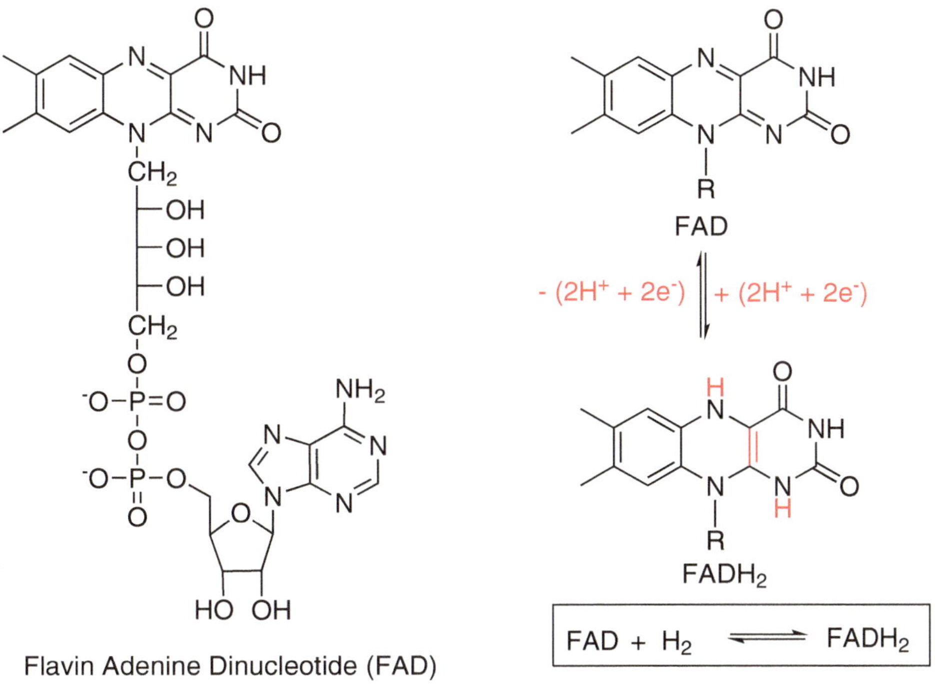

Flavin Adenine Dinucleotide (FAD)

Fig. 4.12   FAD and its function as a reversible hydrogen acceptor

A popular dehydrogenation reaction takes place towards the end of the citrate cycle, where succinate is converted to fumarate (Fig. 4.13). In a subsequent step, water is added to the double bond and L-malate is formed. When considering the reaction sequence, it should be noted that the first reaction is a dehydrogenation and, consequently, the oxidation state of the two central carbon atoms changes. However, in the transition from fumarate to L-malate, no electrons are transferred despite the change in oxidation numbers; one carbon atom has a lower and the other a higher oxidation state. In conclusion, it can be stated that the original alkane chain is functionalized by the dehydrogenation-hydration sequence with an HO group. In most cases, the oxidation of the alcohol to the carbonyl compound follows the addition of water, thus generating an even more reactive species. Oxaloacetate provides the conditions for linking with acetyl-S-CoA (Fig. 4.7), whereby citrate is formed and the citrate cycle can begin again.

The same principle applies at the beginning and then repeatedly throughout the breakdown of long-chain fatty acids (Fig. 4.14). First, an alkane chain is dehydrogenated in the vicinity of the functional group. The addition of water, followed by the oxidation of the newly formed hydroxy group, leads to a 1,3-dicarbonyl compound. Since the C=O group is formed in the β-position to the ester group, this process is called **β-oxidation**. The intervening C–C bonds of such structures are subject to easy cleavage reactions. In

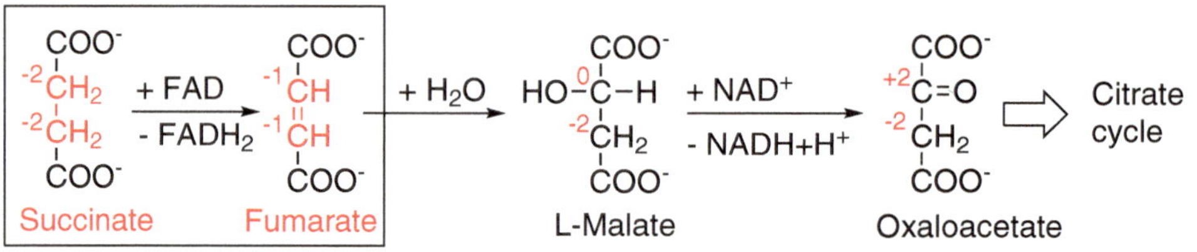

Fig. 4.13 Functionalization of alkanes by dehydrogenation

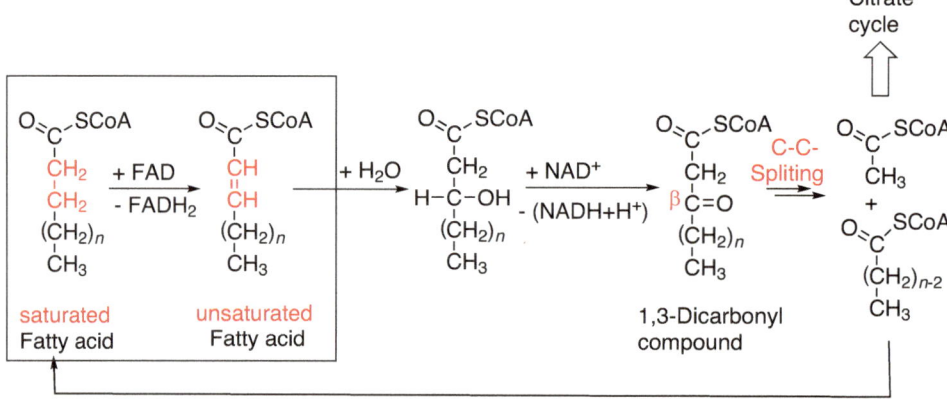

Fig. 4.14 Chain cleavage as a result of dehydrogenation

this way, C_2 units are successively cleaved from long-chain fatty acids, which are then used for energy generation, for example in the citrate cycle. The shortened chain is subject to the same mechanism each time.

Aromatic Systems

The dehydrogenation of cyclic alkanes can lead to aromatic systems, always when a conjugated double bond system is formed and the **Hückel rule** is fulfilled. The Hückel rule refers to the number of π-electrons and states: If $(4n + 2)$ π-electrons are contained in a ring with conjugated double bonds, it is an aromatic. They are more stable in comparison to other systems with multiple double bonds. In aromatics, the π-electrons are distributed over the entire system, as the example of benzene (with $n = 1$) illustrates, which can be described by two mesomeric structures. All C–C bonds are of equal length. Aromatics are planar.

Aromatics build weak attractive interactions with each other. The C–H bonds give the positively charged framework and the π-electrons in the interior form a negative electron cloud, which is often represented in the form of a circle. Two benzene rings can thus be arranged either parallel and slightly shifted or in the form of a "T". Such attractive forces play a role in the construction of proteins between aromatic rings of amino acids. For example, they give bird wings their elastic stability.

Attraction

In biotic systems, aromatization always proceeds via numerous intermediates, with the elimination of water (dehydration) from alcohols also playing a role. The best-known access to substituted benzene derivatives is described by the shikimate pathway (Fig. 4.15). At the beginning is the namesake shikimate, a compound with seven carbon atoms, which is synthesized from carbohydrates, thus polyalcohols. The carboxylate group at the top of the molecule activates the binding carbon atom of the ring and offers the potential to introduce functionalized residues.

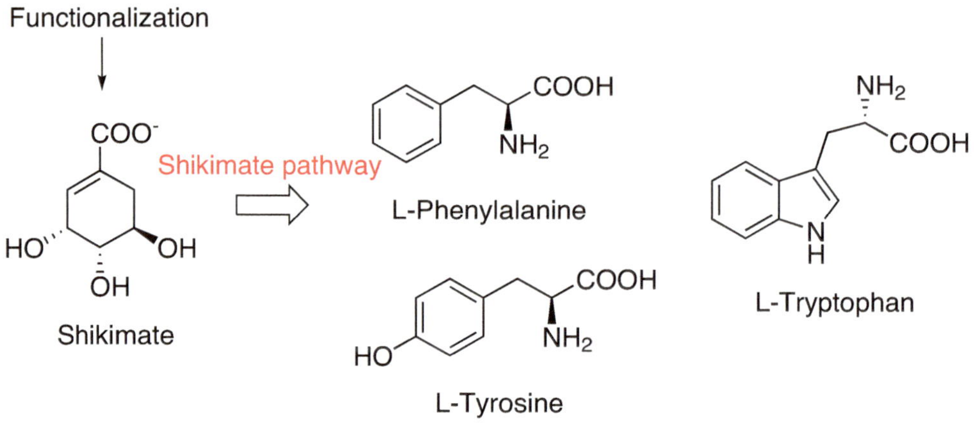

Functionalization

Fig. 4.15 Basic approach and examples of aromatic natural products

In plants, this is how the aromatic amino acids L-phenylalanine, L-tyrosine and L-tryptophan are synthesized. They belong to the essential amino acids, i.e. they must be taken in as part of human nutrition.

Aromatic systems can also contain heteroatoms. A basic structure frequently occurring in natural substances is pyrrole. Pyrrole is a secondary amine and should thus represent a particularly strong Lewis base. By involving the free electron pair in the ring, which has aromatic character due to six π electrons, the basicity is considerably weakened, which is the prerequisite that pyrrole derivatives play a role as free bases in biochemical context (neutral environment!) at all.

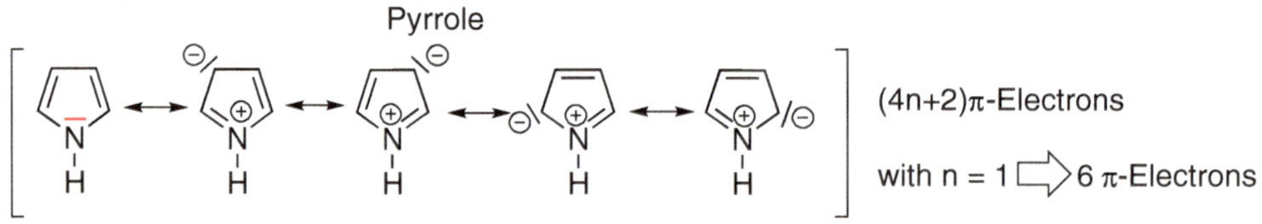

Hämoglobin, chlorophyll and vitamin B_{12} contain structures with four pyrrole rings, which are connected to each other for the most part by CH bridges (Fig. 4.16). In the porphyrin system of hemoglobin, this results in a conjugation of eleven double bonds over the entire molecule, which is supplemented by free N electron pairs. The structure is thus planar. The central metal iron changes between a position above the ring and exactly in the ring during oxygen transport in the blood, depending on whether it is bound to an oxygen molecule or not. Chlorin and corrin, on the other hand, are not completely flat due to saturated sections, which is advantageous for the electronic properties of the complexed metal ions magnesium or cobalt and their functions as oxygen carriers in plants or as vitamin B_{12}. The structure variants and their syntheses thus converged depending on the metal and the biochemical function, a typical evolutionary phenomenon.

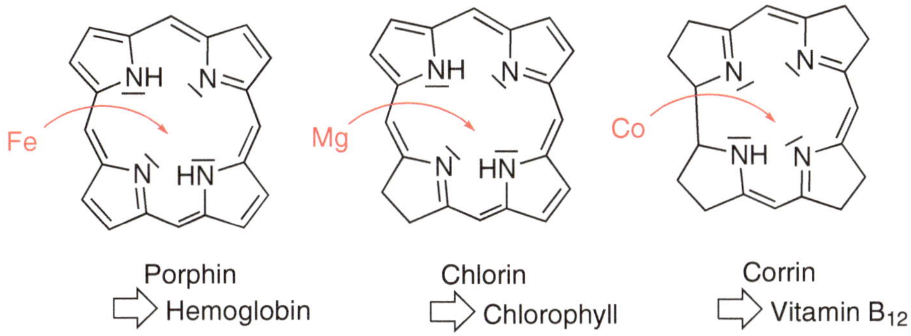

Porphin
⇨ Hemoglobin

Chlorin
⇨ Chlorophyll

Corrin
⇨ Vitamin B$_{12}$

Fig. 4.16 Natural substances based on pyrrole

Another biochemically important heterocycle is pyrimidine. The compound with ten π electrons also meets the Hückel rule. As with pyrrole, the basic character of the free electron pairs at the two nitrogen atoms is weakened, a prerequisite for its occurrence as a free base in biologically relevant compounds.

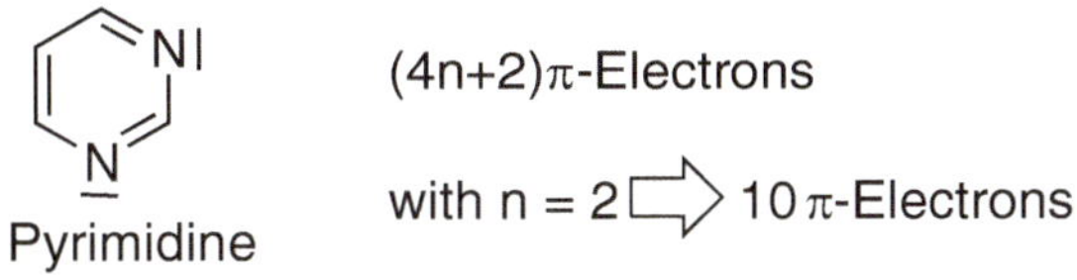

$(4n+2)\pi$-Electrons

with n = 2 ⇨ 10 π-Electrons

Pyrimidine

The basic structure of purines is formed by the combination (annulation) of a pyrrole with a pyrimidine ring (Fig. 4.17). The best-known natural product derived from this is adenine. In the biochemical context, adenine is usually bound to D-ribose. This compound is called adenosine. The esterification of the hydroxy group at $C^{5'}$ with phosphoric acid yields adenosine 5'-monophosphate (AMP). Other AMPs are formed by esterification at position $C^{2'}$ or $C^{3'}$ of the ribose.

If the aromatic system is disturbed, it easily returns to the aromatic starting state. An example was already given for the iodination of tyrosine (Fig. 1.13) This phenomenon is also found in redox-active catalyst systems, such as the NAD$^+$/NADH system, which is in action in catabolic H$_2$ transfers (Fig. 4.18). The system plays an especially important role in the oxidation of alcohols or in the reduction of carbonyl compounds. The pyridinium ring (pyridine with a positive charge on nitrogen) represents an aromatic with six π-electrons. By reaction with H$_2$, more precisely by uptake of a proton and two electrons (thus of hydride H$^-$), the C–H bond is linked to the ring and the ring is de-aromatized at the same time. By the reverse reaction, i.e. by oxidation of NADH, the aromatic system is restored.

At the end of the hydrogenation, one proton from the original H$_2$ molecule remains. If this reaction is repeated, the pH of the environment decreases. In mitochondria, where such oxidation reactions take place, these protons migrate from the interior to the

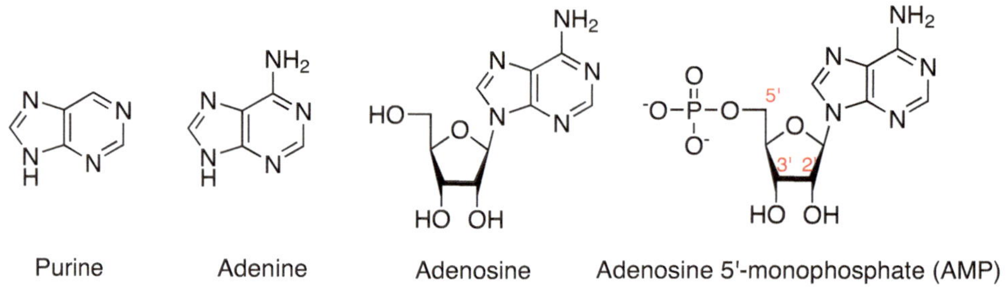

Fig. 4.17 Natural products derived from purines

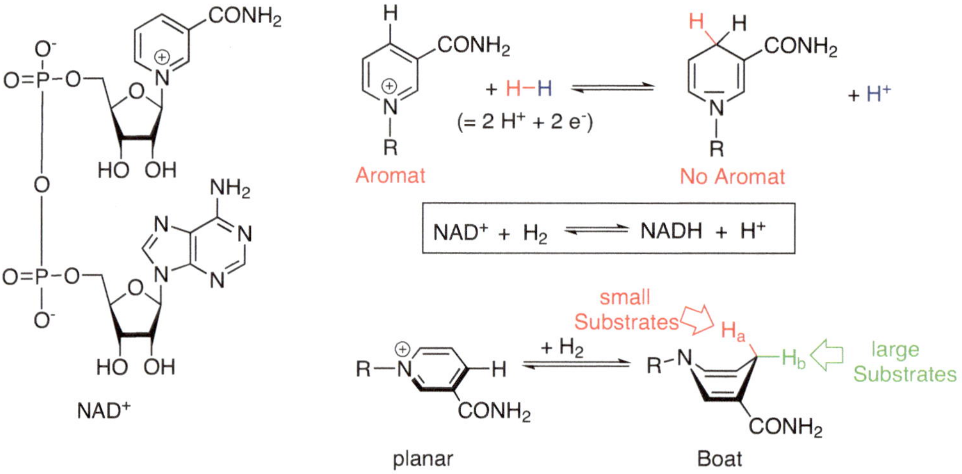

Fig. 4.18 The redox equilibrium of the hydrogen acceptor NAD $^+$

intermembrane space of the mitochondrial double membrane. A concentration gradient is built up. If the oxidation comes to a standstill due to the lack of oxidizable carbohydrates and fats, the migration of protons into the membrane space is also stopped and they migrate back. The concentration differences between the two sides of the membrane are subsequently equalized. In parallel, in a coupled reaction, the synthesis of ATP from the phosphoric ester AMP is driven, which is dependent on the H^+ ion concentration (Fig. 1.6). Therefore, it is self-explanatory that the generation of protons by oxidative degradation of energy-rich carbon compounds and the synthesis of ATP are also spatially coupled in mitochondria.

The dearomatization of NAD^+ by hydrogenation causes the originally planar aromatic pyridinium ring to pass into the three-dimensional boat geometry of the dihydropyridine at the same time. This results in the two H atoms (H_a and H_b) being in a different environment. When hydrogen (reduction) is transferred to differently sized substrates, this results in a selection option as to which of the two hydrides is transferred. This plays a role in particular when generating homochiral alcohols. From this overall view it becomes apparent that the pyridine substructure of the NAD system is optimally adapted to these complex relationships.

4.1.3   Organic Peroxides

Generation of oxygen radicals and their effect
It has been shown (Figs. 4.13 and 4.14), that the dehydrogenation of alkanes and the addition of water to the resulting C=C double bond represents a possibility for the synthesis of alcohols. Alcohols are alternatively formed by insertion of oxygen into a C–H bond. In the discussion at the beginning of Sect. 3.1 oxygen was introduced into the reaction equation pro forma as atomic oxygen [O]. But organisms take up molecular oxygen O_2, which structurally represents a diradical. Radicals are characterized by an unpaired electron. In this respect, hydrogen atoms also represent radicals. Radicals preferably interact with other radicals. Radical reactions are **chain reactions**. They are initiated by chain start, whereby a single bond is homolytically cleaved. The resulting radicals then propagate and generate new radicals. The chain reaction is terminated by recombination and thus pairing with other radicals.

The reaction sequence of methane with O_2 is shown below as an example, with different radical species being involved.

At the beginning of the reaction sequence is the cleavage of a C–H bond. The homolytic cleavage can be caused by external influences (increased temperatures or irradiation with high-energy light) or—much more frequently in the biochemical context—by reactive structures in the molecule, such as adjacent double bonds.

At the preliminary end of the radical chain reaction stands a peroxide. Peroxides are organic derivatives of hydrogen peroxide, i.e. in H_2O_2 both H atoms can be replaced by organic radicals. Deviating from the general rules for determining the oxidation numbers, oxygen in H_2O_2 is assigned the oxidation number -1, which is also consistent for organic derivatives.

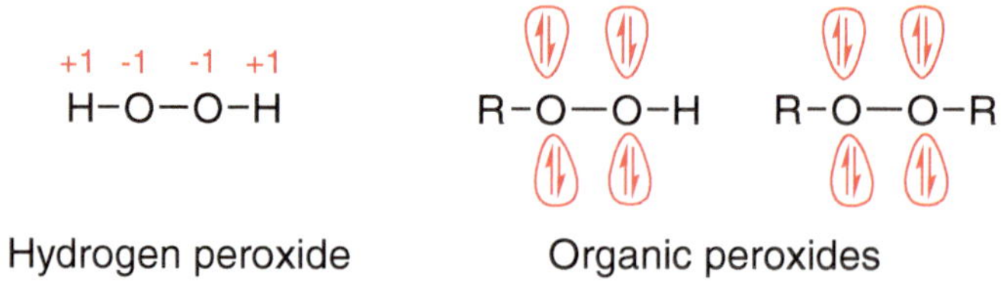

$$\underset{+1\ \ -1\ \ \ -1\ \ +1}{H-O-O-H}$$

R−O−O−H   R−O−O−R

Hydrogen peroxide   Organic peroxides

In the representation of organic peroxides, the four free electron pairs on the two oxygen atoms are highlighted. As already shown in the inorganic part of the book in Fig. 1.26, due to the small atomic radius of oxygen, which results in a short O–O bond distance, significant repulsion forces are effective in H_2O_2. Therefore, organic peroxides are also not stable and decompose quickly into radicals.

Peroxides play a key role in biochemical processes. On the one hand, they are the first products and thus a prerequisite for the functionalization of alkane chains. From this point of view, peroxides are at the beginning of the decelerated total oxidation of energy-rich carbon, which ultimately ends in carbon dioxide. The whole variety of vital functional groups is formally accessible through this opening reaction. As a result, not only structural and informational molecules are generated, but also any energy generation in the cell begins with variants of this reaction. These subsequent reactions are the content of the last part of the book.

On the other hand, vital structures are attacked by oxygen and their biological function is destroyed. In this respect, oxygen and its derivatives, which differ in oxidation states (Fig. 2.8), are cell poisons with the exception of water. For example, C–H bonds in the immediate vicinity of C=C double bonds in unsaturated fatty acid esters are attacked (Fig. 4.19). Also, C–H bonds adjacent to C=N bonds are subject to attack by oxygen. The result is, for example, the breakdown of nucleosides in DNA.

Often, the formation of peroxides in long-chain, multiple unsaturated fatty acids is accompanied by the shift (isomerization) of C=C double bonds (Fig. 4.20). Oxidation ultimately leads to the cleavage of the chain, which results in the loss of its function as a membrane former.

In principle, no clear differentiation between productive and counterproductive effects of oxygen radicals is possible. The assessment of which aspect predominates depends on

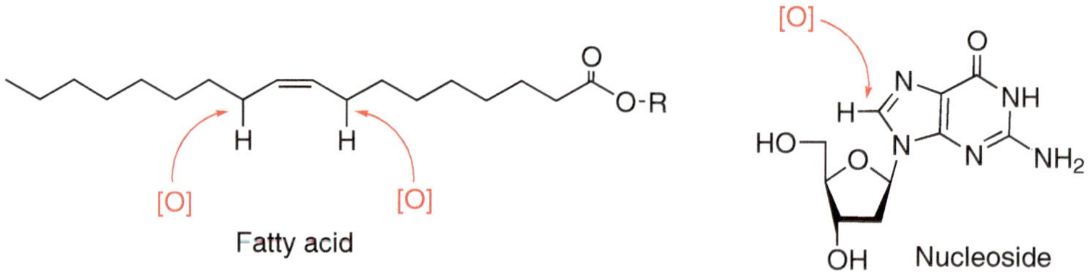

Fatty acid   Nucleoside

Fig. 4.19 The oxidative breakdown of unsaturated structures in natural products

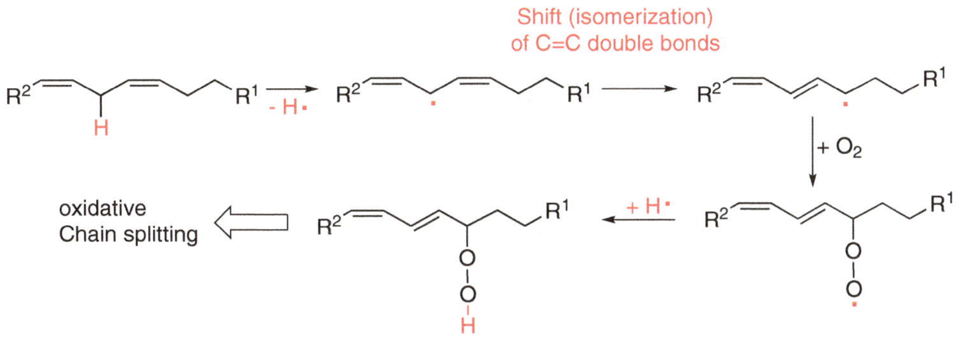

Fig. 4.20   The isomerization of a double bond in the course of radical oxidation

the cell location and the time. Constructive effects relate to the biosynthesis of natural products on the basis of oxidative processes. Pathogenic effects can be found, for example, in uncontrolled cell growth, i.e. in the development of cancer, which is initiated and propagated by oxygen. Every organism is a dynamic system in which construction and decomposition reactions, mediated by oxygen, balance each other for a certain lifetime. If the negative effects predominate, individual organelles, cells or organs fail and, at the end, the death of the organism occurs.

In biological systems, there are two mechanisms that limit the negative oxidation effect of oxygen and thus cause a deceleration effect:

1. the diversion of the oxygen attack to internal "sacrificial compounds" or to external structures,
2. the decomposition of hydrogen peroxide and inorganic oxygen radicals and organic peroxides before they initiate destructive chain reactions.

Diversion to internal "sacrificial compounds" or to external structures
The destructive radical effect of oxygen species can be diverted to other reactive compounds, so-called "sacrificial compounds". This prevents the attack on vital biomolecules. Often these compounds contain very many **conjugated double bonds**, these are systems in which single and double bonds alternate. Examples are β-carotene (carrots), astaxanthin (flamingos, salmon) or canthaxanthin (chickens) (Fig. 4.21). They have a large number of attack points for oxygen radicals. Due to the conjugated double bonds, these compounds are colored and thus serve as biological markers for animals. The color fades during oxidation. They thus have, in addition to the causal chemical antioxidant effect, an evolutionary biological fitness effect for the species or for the individual. As shown in Sect. 4.1.4.2, all aromatic alcohols are also radical scavengers.

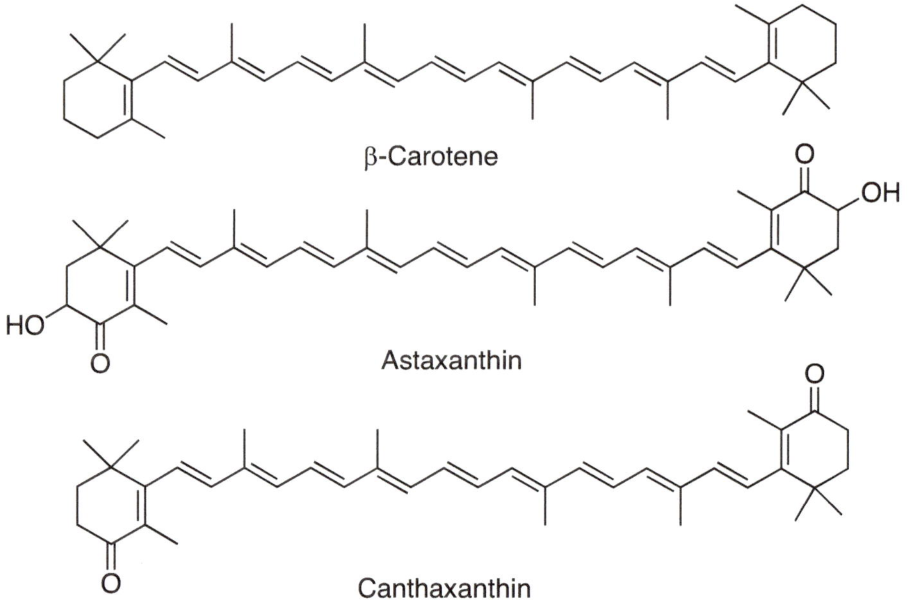

β-Carotene

Astaxanthin

Canthaxanthin

Fig. 4.21 Conjugated double bond systems as "sacrificial compounds" against oxygen radicals

If the destructive effect is diverted to vital structures of other organisms, a biochemical protection results. Organic peroxides then help to defend against parasites, fungi or bacteria and thus preserve the chemical integrity of the producing organism.

A well-known example is artemisinin, which is found in leaves and flowers of annual mugwort *(Artemisia annua)* (Fig. 4.22). Structurally, it is a dialkyl peroxide, which is converted into a radical by the action of iron (II) ions and in this form destroys the membranes of bacteria. Since iron ions occur in particularly high concentrations in erythrocytes (red blood cells), this mechanism could be responsible for the defense against *Plasmodium falciparum,* the causative agent of malaria, for which artemisinin is now used worldwide as a medication.

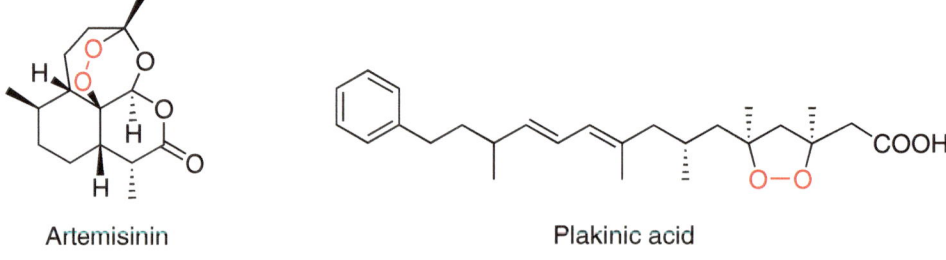

Artemisinin   Plakinic acid

Fig. 4.22 Peroxides as protective compounds against biological attacks

Sponges of the genus *Plakinastrella* produce a number of cyclic peroxides of plakinic acid partly in symbiosis with other sponges, which have a deterrent effect on fungi.

The rapid decomposition of radical oxygen compounds and their deceleration
In addition to the diversion of free radicals to sacrificial compounds, their rapid decomposition is the most common form of biochemical defense. Hydrogen peroxide, superoxide or hydroxyl radicals and organic peroxides are destroyed by numerous redox-active metal ions. In the test tube, the reaction usually only takes place in basic media. The best-known inorganic reaction is the decomposition of hydrogen peroxide in the presence of catalytic Fe^{3+} ions, which leads to water and oxygen. In the simplest case, the catalyst is $FeCl_3$.

$$2\ H_2O_2 \xrightarrow{\ FeCl_3\ } 2\ H_2O + O_2$$

Manganese ions have a similar effect. It is therefore evolutionarily consistent that enzymes that break down peroxides primarily contain these two metals, especially since they are also very common in the Earth's crust. The best known are catalases. They are among the most productive enzymes that have been known so far. They can convert up to ten million substrate molecules per second. The non-catalyzed reaction is however a billion times slower. Catalysts are evolutionarily very old enzymes, which is logical because "oxygen management" has always played a major role since the beginning of life on Earth and the emergence of free oxygen. Catalases develop their effect under neutral conditions, unlike simple iron salts.

Due to the embedding of the redox-active metal ion in metal complexes with varying organic ligands, catalysts with a wide range of, but also graded, function spectra evolve. This benefits the speed and selectivity of biochemical reactions. Only by modification with organic ligands is the "destructive" effect of the original metal salt available as an enzyme for the "constructive" biochemical application (Fig. 4.23).

This principle results in a slowdown in the decomposition of aggressive oxygen species. Catalysts not only break down oxygen radicals, but they are also able to transport oxygen. This means they stabilize reactive oxygen species for a certain period of time and activate them for selective further reactions at the same time. One of the best known examples is hemoglobin, an iron-containing enzyme that takes over oxygen transport in mammals. Manganese complexes play a similar role in plant photosynthesis, namely in the oxidation of water to oxygen in enzymes of the photosystem II.

4.1.4 Alcohols

4.1.4.1 Possibilities of Formation
After oxygen is introduced into a C–H bond, the resulting radicals recombine to form an alcohol. In the special case discussed below, methane is converted to methanol via the corresponding peroxide (Fig. 4.24).

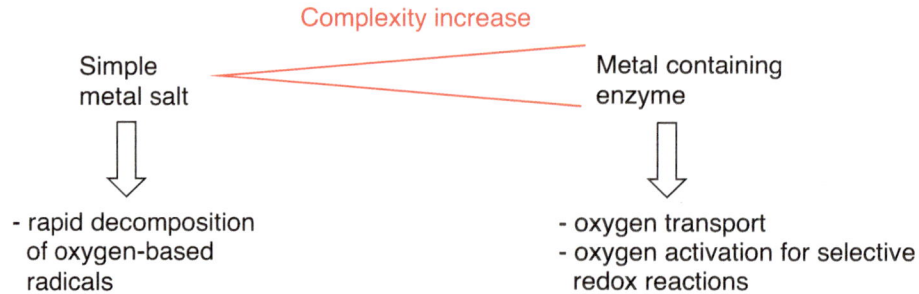

Complexity increase

Simple   Metal containing
metal salt   enzyme

- rapid decomposition   - oxygen transport
  of oxygen-based   - oxygen activation for selective
  radicals   redox reactions

Fig. 4.23 Only by increasing complexity are selective transformations possible

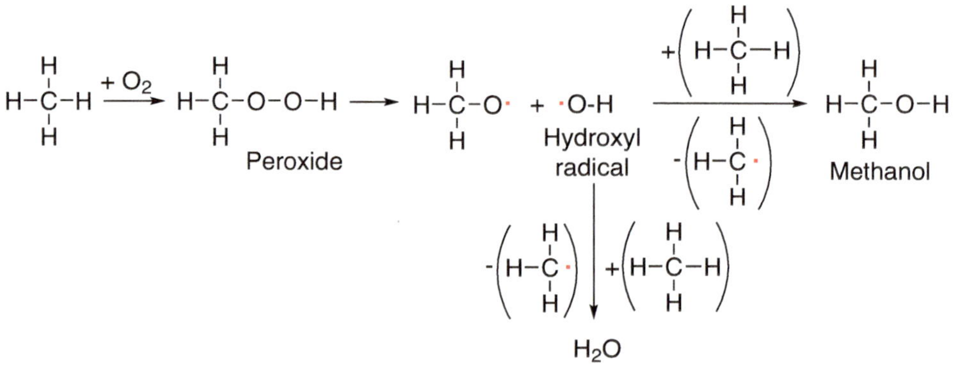

Fig. 4.24 The formation of an alcohol via a peroxide as an unstable intermediate

In this respect, alcohols represent the stable "entry point" into the chemistry of life, since they are much more long-lived than their predecessors, the peroxides (Fig. 4.25). Compared to the original nonpolar C–H bond, a C–OH group is polar, which increases the water solubility of the corresponding compound. The more HO groups there are, the faster the compound is flushed out of the organism with water. With the liver, a special oxidizing organ has evolved in many animals, which is primarily associated with the detoxification of toxins. The liver can even adapt to an increased oxygen demand by changing the oxygen uptake from the blood. Despite these beneficial effects, the construction of biochemical and biological structures is always a "race" against the increasing water solubility of biochemically important compounds as a result of oxidation reactions.

Alternatively, alcohols are formed by the addition of water to C=C double bonds. The hydrogenation of carbonyl compounds (aldehydes, ketones) also leads to alcohols.

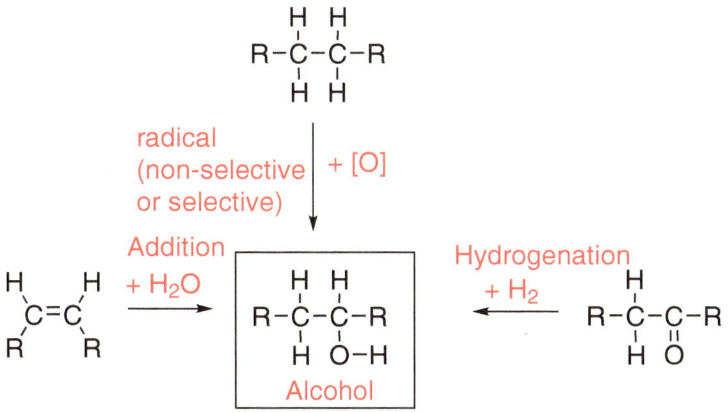

Fig. 4.25 Possibilities of the formation of alcohols in a biochemical context

Radical oxidation of C–H bonds

Radical oxidations of C–H bonds in the absence of enzymes have a destructive effect, because they proceed unselectively and destroy vital biological structures. Such "side reactions" are associated with the occurrence of "free radicals". The lack of direction leads to a variety of and diversity of oxidized products, for which no established biochemical utilization mechanisms exist. Nevertheless, an evolutionary advantage can be derived from this phenomenon, which consists in that less fit organelles, cells or even whole organisms are destroyed and thus renewal processes are initiated.

In biology, the term **homeostasis** is sometimes used to express the long-term stability of biological organisms. However, since homoeostasis implies stagnation and no change, as is typical for life, it is now being supplemented by the term **homeodynamic**. Homeodynamics may be the result of changed environmental conditions or the result of chaotic processes within the organism itself. The latter include primarily the unselective oxidation by free radicals.

Selectivity in the biochemical context, however, is generated by enzymes. Therefore, it is remarkable when enzymes, such as the manganese-based laccase, are able to attack all types of C–H bonds indiscriminately. The enzyme is found in white rot fungi, which break down lignin (Fig. 4.53), a particularly heterogeneous polymeric aromatic system, to coal.

However, the enzyme-mediated selective insertion of oxygen into selected aromatic or aliphatic C–H bonds is the "normal case" and belongs to the biochemical basic repertoire in the context of the slowed-down total oxidation of energy-rich carbon. The reaction is also referred to as hydroxylation. The enzymes are often characterized by iron complexes that assume various oxidation states in the catalytic cycle. For example, hydroxylation at C^4 results in the proteinogenic amino acid L-phenylalanine being converted into another, namely L-tyrosine. In a protein chain, an amino acid therefore already coexists with its own oxidative degradation product. Similar relationships also exist for nucleic acids (Fig. 4.126). L-Tyrosine is selectively oxidized at C^3 to L-DOPA, which is no longer incorporated into proteins.

L-Phenylalanine   L-Tyrosine   L-DOPA

Enzymatically generated oxidation products fulfill various biochemical functions that can even vary between organisms. The monophenol L-tyrosine belongs to the proteinogenic amino acids and is thus ubiquitous in the animate nature. The diphenol L-DOPA is a precursor of the neurotransmitters adrenaline and noradrenaline, which occur in all vertebrates, but only in some invertebrates (Fig. 4.26). They are important in the conduction of stimuli. Mussels also form L-DOPA. The function of L-DOPA in this context is to increase the hydrophilicity of the protein chain, which is the actual adhesive, which helps to adhere to solid surfaces. The two HO groups contribute to this. Noradrenaline is formed by the hydroxylation of an aliphatic C–H bond in dopamine. By N-methylation, adrenaline is formed. With L-DOPA it is an aromatic diol, which could be immediately further oxidized to benzochinone (Sect. 4.1.4.2). This reaction pathway is prevented by monomethylation of an HO group and represents an actually *in vivo* observed alternative to the formation of noradrenaline and adrenaline.

The aromatic amino acid L-tryptophan is also the starting point for numerous selective oxidation reactions (Fig. 4.27). Depending on whether oxygen is inserted into the aromatic ring at positions C^4, C^5 or C^6, natural substances with different physiological effects are produced: Oxidation at the C^5 atom of tryptophan followed by decarboxylation leads via 5-hydroxy-L-tryptophan (5-HTP) to serotonin. The compound is

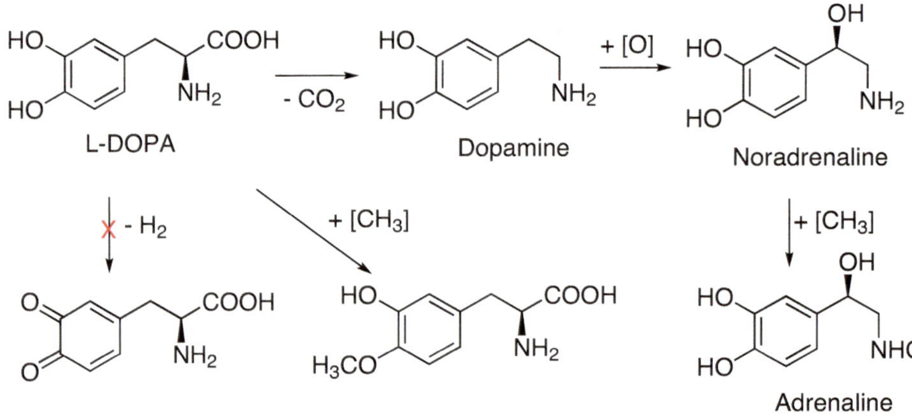

L-DOPA   Dopamine   Noradrenaline

Adrenaline

Fig. 4.26 The synthesis of adrenaline by a selective oxidation reaction

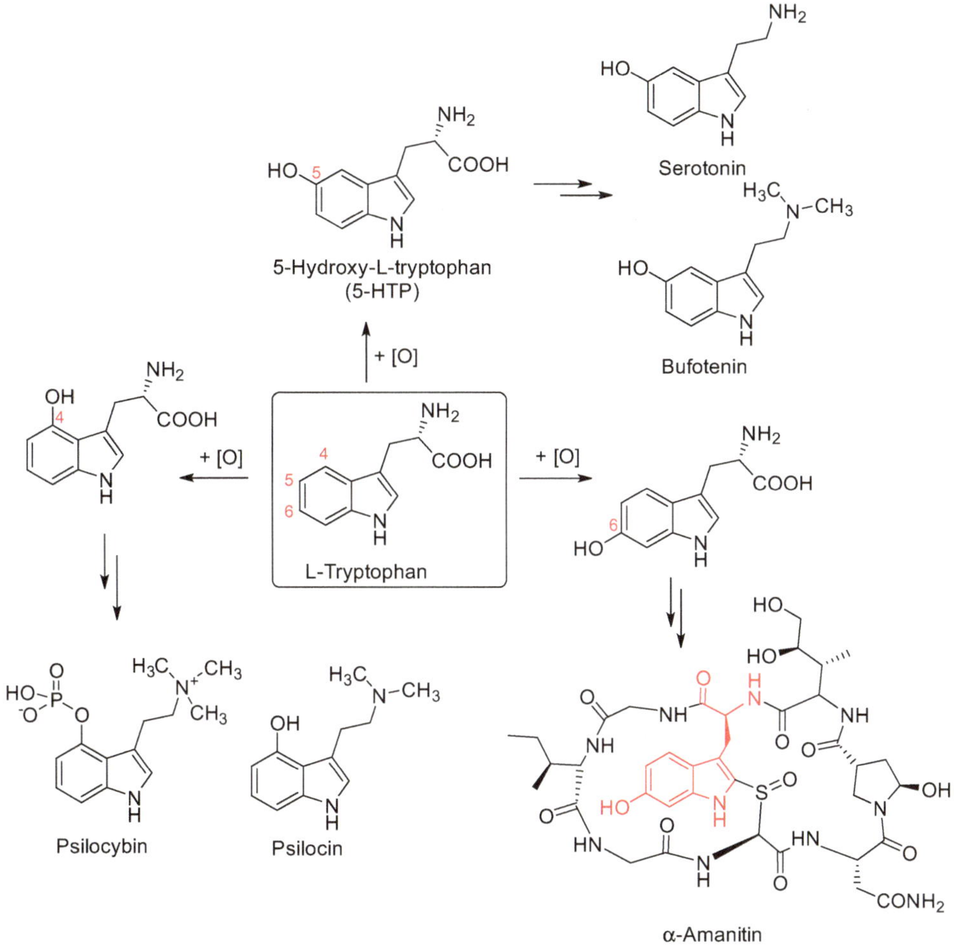

Fig. 4.27 Natural substances by selective oxidation of L-tryptophan

widespread in the plant and animal kingdom. Some of the best-known effects on the human central nervous system include influencing mood. Bufotenin is found in the skin secretion of various toads *(Bufo marinus)*. In addition, the substance has been detected in small amounts in human urine, suggesting that it is also a normal breakdown product of human metabolism.

Psilocybin and psilocin are formed by oxidation at C^4 of the aromatic ring. They are found in fungi of the genus *Hymenogastraceae*. The insertion of oxygen into the C–H bond of C^6 leads to a phenol derivative, which is incorporated into the polypeptide α-amanitin via numerous synthesis steps. α-Amanitin is one of the main toxins of the fly agaric mushroom *(Amanita phalloides)*.

The selectivity in the oxidation of C–H bonds in natural products not only generates primary natural products such as proteinogenic amino acids, but also contributes to the great diversity of biological species. Many of these natural products are characteristic of the respective organisms, but not essential for life. They are therefore referred to as secondary natural products. Secondary natural products based on aromatic phenols are the active ingredients of green tea, such as (−)-catechin and gallotannin (Fig. 4.28). The latter belongs to the class of tannins and gives the respective organism a protective effect against predators due to the bitter taste.

Many flower pigments also have a phenol structure. Examples are apigenin, which occurs in many vegetables such as celery or parsley, and pelargonidin, which is widespread in higher plants and causes the reddish color of many flower petals and fruits (Fig. 4.29). Cyanidin is also a polyphenol with a distribution in red roses, hibiscus and numerous berry species.

All of these substituted phenols are produced by oxidation of C–H bonds in the corresponding precursors. It is striking that these compounds not only have an evolutionary

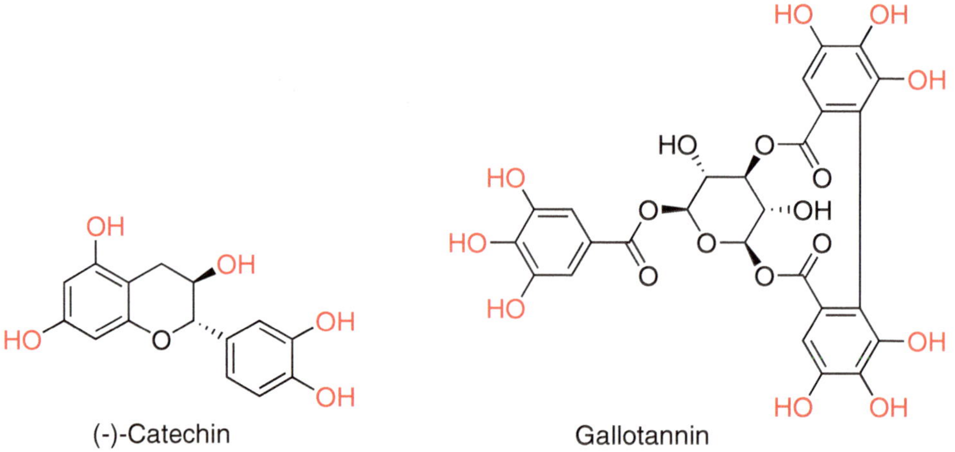

Fig. 4.28 Aromatic polyols as products of oxidation of C–H bonds

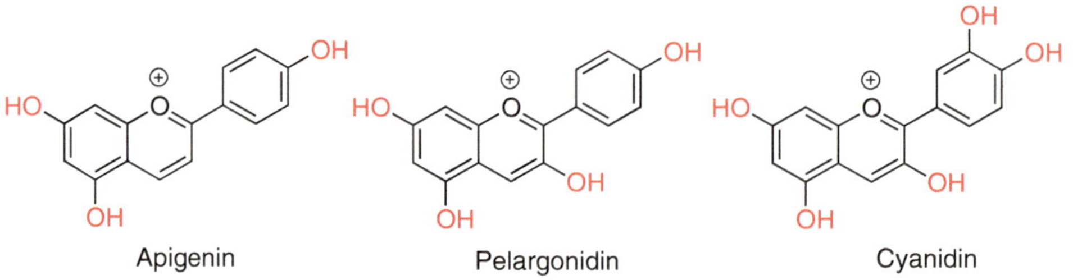

Fig. 4.29 Flower pigments with phenol structure

function such as color or deterrent effect on other organisms, but that they *a priori* have a chemical function: As will be shown below, many of them act as radical scavengers (Sect. 4.1.3) and slow down the total oxidation of the energy-rich carbon.

Hydration of olefins

Hydration, i.e. the addition of water to C=C double bonds, is a reaction that leads to aliphatic alcohols. An example is the synthesis of L-malate from succinate already discussed above (Fig. 4.30).

Another hydration is the conversion of unsaturated fatty acids into the corresponding β-hydroxycarboxylic acids during fatty acid degradation (Fig. 4.31).

The aliphatic alcohols are educts for oxidation to aldehydes and ketones. The latter are chemical "switching centers" for construction and degradation processes.

4.1.4.2 The Oxidation of Alcohols and the Formation of Carbonyl Compounds

Primary, secondary and tertiary alcohols

The carbonyl group in **aldehydes** and **ketones** is formed by dehydrogenation of aliphatic alcohols. By the loss of molecular hydrogen, the oxidation number is increased by two units in the transition from alcohol to carbonyl compound. Conversely, a carbonyl group can be hydrogenated to alcohol.

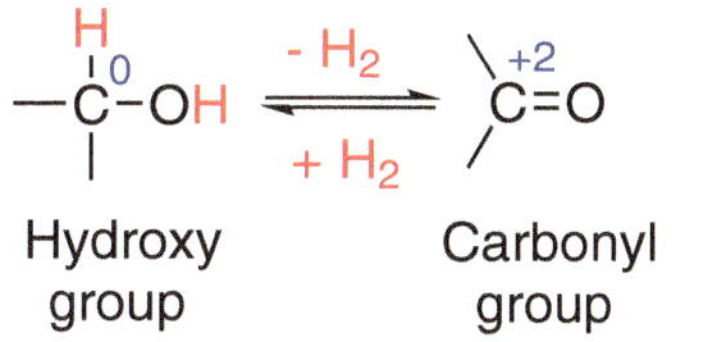

Hydroxy group   Carbonyl group

The dehydrogenation can be described formally as oxidation and as an intermediate stage a hydrate (geminal diol) is formulated, which according to the Erlenmeyer rule splits off water (Fig. 4.32).

Among the alcohols, one distinguishes between primary, secondary and tertiary alcohols. Aldehydes are formed by oxidation of primary alcohols. Ketones are formed from secondary alcohols.

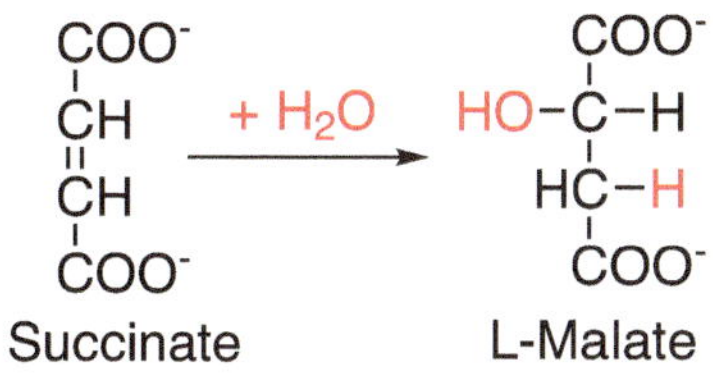

Fig. 4.30 The synthesis of L-malate by hydration of succinate

Succinate   L-Malate

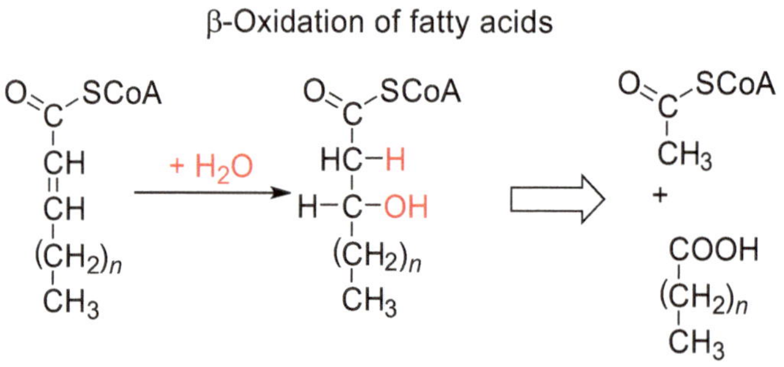

β-Oxidation of fatty acids

Fig. 4.31 Hydration of olefins as a precursor to the cleavage of C–C bonds

Fig. 4.32 The dehydrogenation of an HO- to a C=O group by formulating a hydrate as an intermediate stage

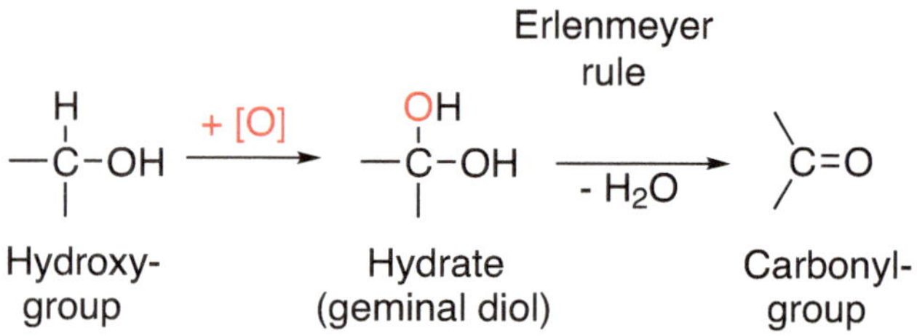

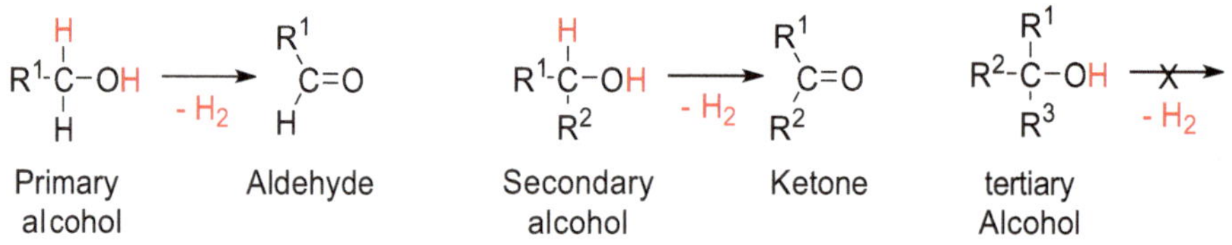

Since there is no hydrogen atom bonded to the adjacent carbon atom in tertiary alcohols, they are usually not oxidized. They lack the second hydrogen atom that is required to form H_2. Therefore, the oxidation of tertiary alcohols must be preceded by a shift of the HO group in biochemical degradation processes. As an example, the isomerization of citrate to isocitrate at the beginning of the citrate cycle (Fig. 4.33) is shown. Isocitrate is formed from *cis*-aconitate as an intermediate compound by a dehydrogenation-hydration sequence. In contrast to citrate, isocitrate is a secondary alcohol and is subject to oxidation. In the case shown here, oxalosuccinate is formed, and the citrate cycle can only be continued after this modification.

Aromatic alcohols and enols

Aromatic alcohols, usually referred to as phenols, occupy a special position in the oxidation of alcohols. By definition, they should not be oxidizable as tertiary alcohols. However, the proximity to an aromatic system and the presence of other alcoholic groups

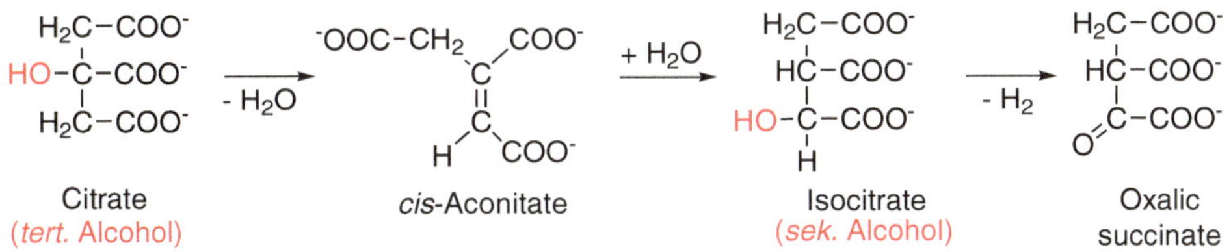

Fig. 4.33   The isomerization of an HO group as a prerequisite for oxidizability

on the aromatic compound lead to an exceptional situation. In principle, mono-, di- and polyphenols are distinguished in the class of aromatic alcohols.

In the oxidation of certain diphenols (dihydroxyphenols), such as catechol, the second hydrogen atom for the formation of H_2 comes from a second HO group. The two hydroxy groups are adjacent, which is referred to as **ortho**. The aromatic ring of brenzcatechol is subsequently de-aromatized. However, since a conjugated double bond system together with the two carbonyl groups is formed in 1,2-benzoquinone, both compounds are energetically almost equal. A similar situation is found in the oxidation of a **para**-substituted diphenol such as hydroquinone, which is dehydrogenated to quinone. In contrast, no molecular hydrogen is split off from resorcinol (**meta**) since the product cannot be stabilized by conjugation.

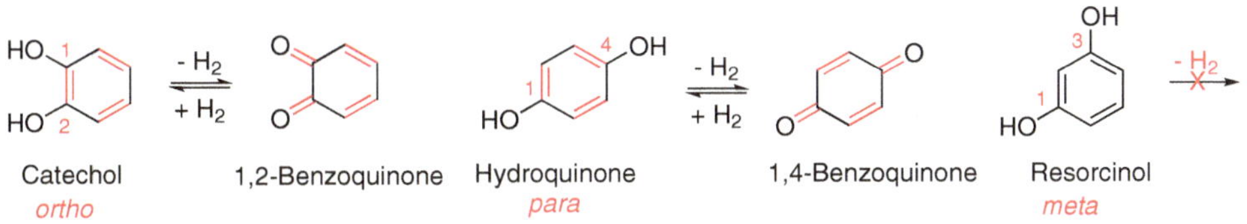

The redox equilibrium between quinones and hydroquinones plays a role in reversible electron transfer processes in living nature and thus in redox-active enzymes. A typical example is the interplay between ubiquinone and ubihydroquinone (Fig. 4.34). The trivial name is derived from ubiquitous (occurring everywhere), a semantic hint that ubiquinones occur in all living beings. The two CH_3O groups are striking in both structures. They are formed by methylation of two hydroxy groups and thus withdraw them from the redox reaction (Fig. 4.40).

In exceptional cases, the second hydrogen atom may come from a neighboring methyl group that is adjacent to the phenolic hydroxy group. Then pseudo-quinoid structures are formed, as will be shown below using the example of vitamin E (Fig. 4.39).

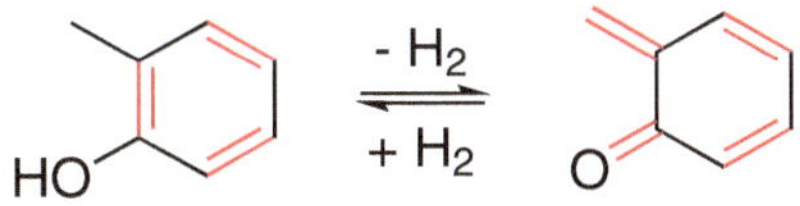

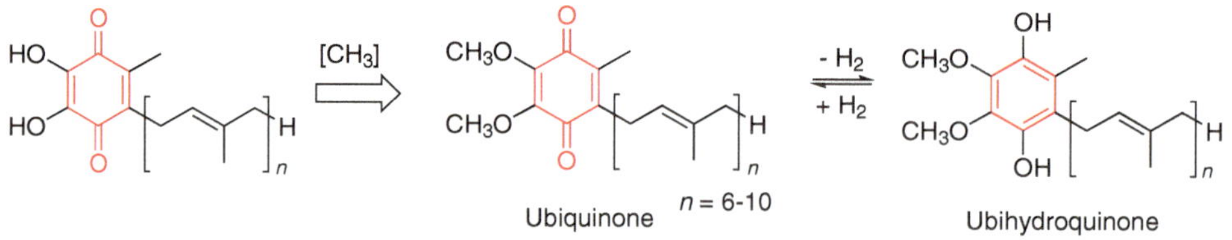

Fig. 4.34 Methylation protects against oxidation

The oxidation of hydroquinone systems does not necessarily have to stop at the quinone stage if other oxidation-sensitive structures are present in the compound. An example of a reversible further reaction is the enzymatic system with the blood clotting factor vitamin K in the center (Fig. 4.35). The isolated C=C double bond in vitamin K, which is formed by dehydrogenation of vitamin KH_2, is further oxidized by oxygen radicals to the epoxide (vitamin KO). An epoxide is a strained three-membered ring system and thus represents an energetically demanding system that is in equilibrium with its precursor, vitamin K. This is also the prerequisite for its function as a coenzyme.

Dehydrogation and hydrogenation are reversible under biotic conditions due to the low energy barrier for both *ortho*- and *para*-diphenols. The reaction of catechol to 1,2-benzoquinone can also be formulated as a multistep reaction, in which one hydrogen atom is successively cleaved off (Fig. 4.36). This results in radicals as intermediates. The loss of a hydrogen atom can be regarded as more differentiated than the loss of a proton (H^+) and an electron (e^-). In fact, in biochemical systems (e.g. in ubiquinones), both processes, proton transfer and electron transfer, often take place separately, which requires separate carrier systems. At the same time, this regulates the overall reaction in the cell via the pH value.

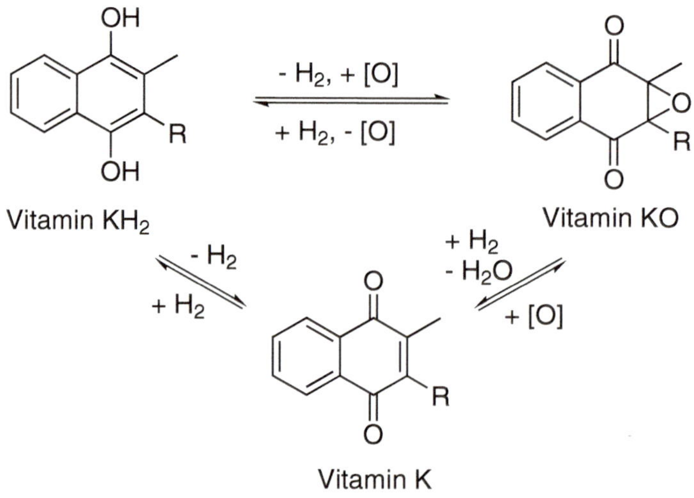

Fig. 4.35 The redox-reversible system of vitamin K

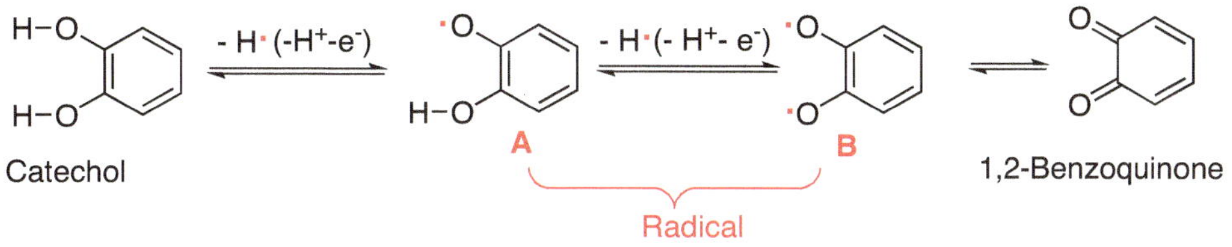

Fig. 4.36 Stepwise oxidation of catechol

In phenol, there is only one HO group, and thus the intramolecular cleavage of H_2 is not possible. Nevertheless, one H atom can be cleaved off and a radical is formed. Radicals from monophenols, in contrast to those from aliphatic alcohols, are stabilized by the aromatic ring. In the extreme case, radicals only temporarily become stable through a single adjacent C=C double bond, as in the case of an enol.

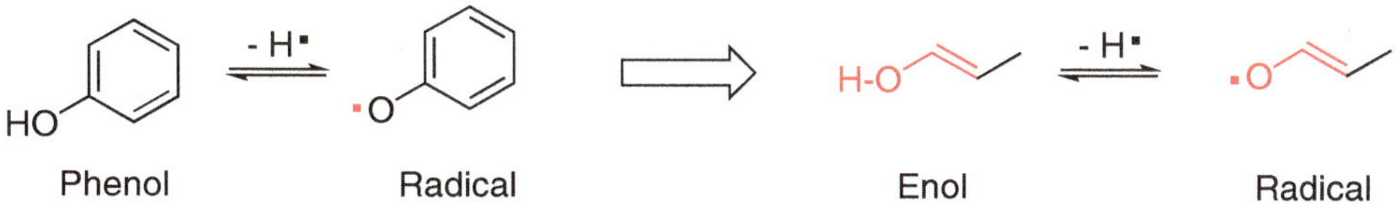

Many structurally important molecules that are formed by oxidation of aromatic C–H bonds have the potential to form radicals and therefore act as radical scavengers by terminating chain reactions. These include the female sex hormone estradiol and other estrogens, such as 2-hydroxyestrone and 2-methoxyestrone (Fig. 4.37). The phenols have an antioxidant effect. Estrogens occur in all vertebrates and some insects, such as female flesh flies (*Sarcophaga bullata*), indicating that they are derived from a common biochemical ancestor.

A particularly well-known example of an antioxidant is vitamin C, which structurally represents an endiol (Fig. 4.38). It can gradually give off hydrogen, with the intermediate formed monoradical being particularly effectively stabilized by two hydrogen bridges. The hydrogen, for example, combines with an oxygen molecule to form hydrogen peroxide. When hydrogen peroxide decomposes, further radicals are formed, such as the

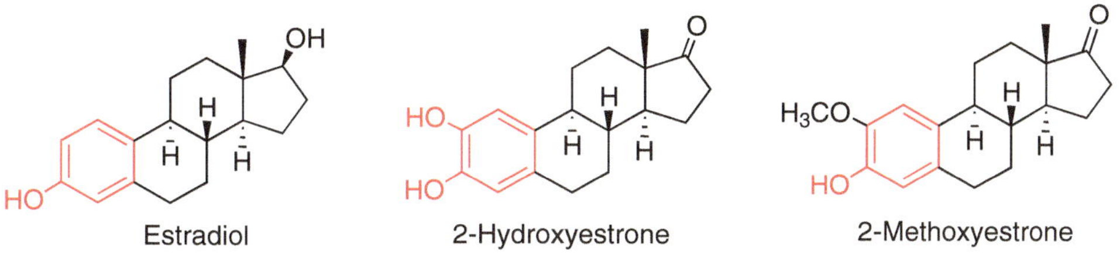

Fig. 4.37 Sexual hormones as antioxidants

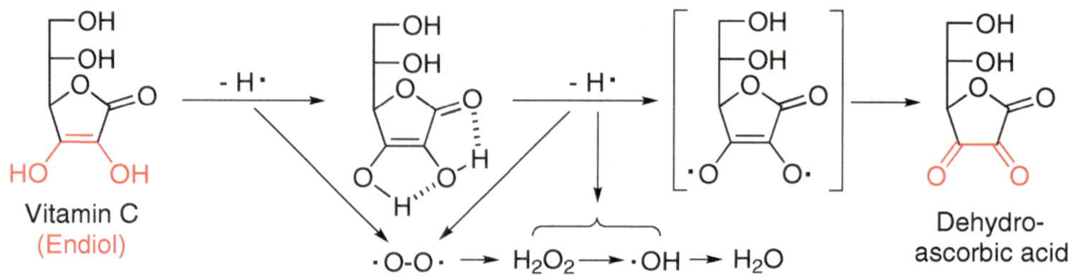

Fig. 4.38 The effect of vitamin C as an antioxidant

hydroxyl radical (Sect. 2.2.2), whose destructive effect is also slowed down by vitamin C. In the end, water always results.

Vitamin C (L-ascorbic acid) is converted to dehydroascorbic acid, which, if it is not excreted by the organism as a water-soluble compound, is rehydrogenated to vitamin C.

The antioxidative effect of vitamin E is quite similar (Fig. 4.39). The compound is characterized by its long unpolar alkyl chain, which gives it a high lipophilicity. This property causes the spatial proximity to membranes, which are made up of long hydrophobic alkyl chains. The attractive interactions are caused by van der Waals forces. Therefore, vitamin E is the first station in a cascade with other radical scavengers (vitamin C, glutathione). For the dehydrogenated form of vitamin E, there are two possibilities for stabilization: Either one of the two adjacent methyl groups participates in the conjugation of the double bonds, resulting in two structures A, or alternatively, quinone C is formed. However, for the latter, the cyclic six-membered ring, an ether, must be opened with water, releasing the second hydroxy group. The way from B to C increases the water solubility (hydrophilicity) of vitamin E and thus improves the interaction with the next radical scavenger, the (water-soluble) vitamin C.

The discussed cleavage of the ether ring in vitamin E leads to another phenomenon in the radical course of the cell: If phenolic hydroxy groups are etherified, dehydrogenation

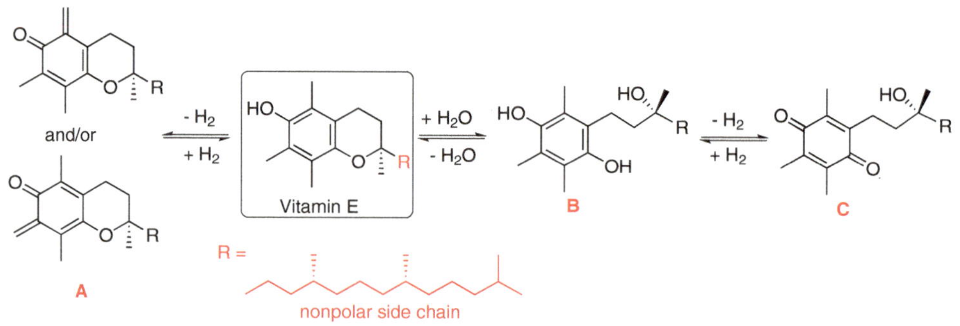

Fig. 4.39 The effect of vitamin E as an antioxidant

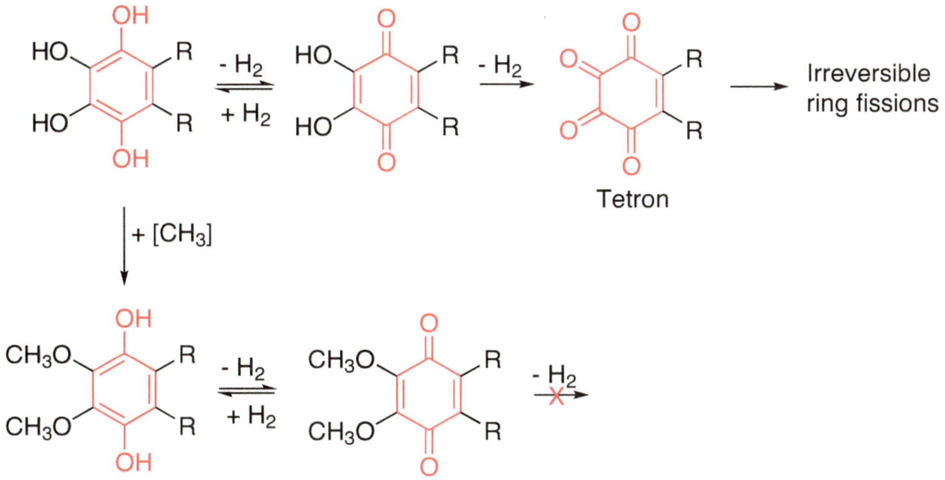

Fig. 4.40 Dimethylation prevents dehydration

is stopped. Ethers thus act as protection for hydroxy groups against oxidation. This protective effect was already recognizable with the two methyl ether groups of ubiquinone (Fig. 4.40). Without these, oxidation would take place via the quinone stage to a tetrone, which is structurally very unstable and immediately undergoes an irreversible ring cleavage. In other words, without the two methyl ether groups, a corresponding ubiquinone derivative would not be reversibly hydrogenated and would no longer act as a coenzyme.

Comparably effective examples of protection through etherification can be found in the anthocyanins, a large class of flower pigments (Fig. 4.41). Delphinidin has six hydroxy groups, all of which are subject to oxidation. In petunidin, one of them is protected as a methyl ether and in malvidin, two. They are no longer oxidized.

Since the relevant hydroxy groups are no longer involved in the coloration through the conjugation of the numerous double bonds at the same time, the methylation also changes the absorption maximum. The change in oxidation properties results in a biological fitness effect that manifests itself in the individuality of the associated flower plants.

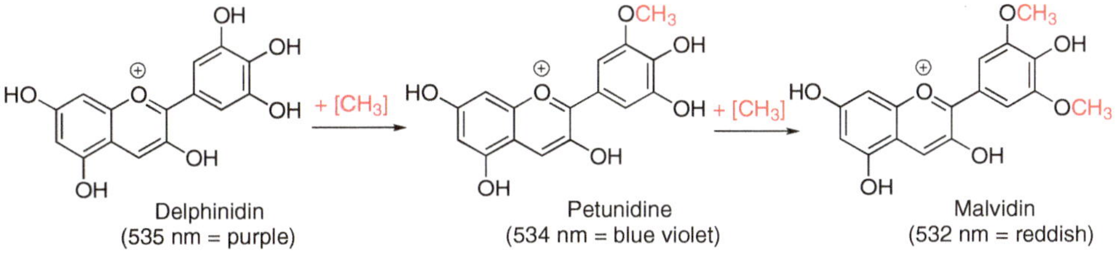

Fig. 4.41 Methoxy- and hydroxy groups are responsible for different flower pigments

This is once again evidence for the fundamental thesis of this book that a large part of the chemistry of life is based on the slowed-down oxidation of energy-rich carbon, on which biological phenomena are based.

4.1.5   Carbonyl Compounds—Aldehydes and Ketones

4.1.5.1 Formation and Properties of the Carbonyl Group

Aldehydes are formed by oxidation of primary alcohols. Ketones are formed by oxidation of secondary alcohols. The carbonyl group, which is common to both classes of compounds, occupies a prominent position among all other functional groups due to its special properties. In the carbonyl group, the same electronegativity difference between C and O is found as in an alcohol (Fig. 4.42a). At the same time, the oxygen atom is linked to the carbon atom via a π bond in addition to the σ bond. There are four electrons in this area, and thus there is an accumulation of negative charge. This situation is reminiscent of a C=C double bond. The reactivity is potentiated by the combination of both bonding phenomena in a functional group. In contrast to an aliphatic alcohol, the carbon atom is only surrounded by three substituents. The functional group is therefore planar, which facilitates the attack of nucleophiles (Fig. 4.42b). The bond of the carbon atom to the electronegative oxygen is highly polarized, which is reflected in the charged mesomere boundary structure (Fig. 4.42c). As a result, the entire carbonyl group has a strong electron-withdrawing effect on the surroundings at the same time. By protonation, which is realized by acid-acting enzymes, the separation of charges between C and O is further increased (Fig. 4.42d).

4.1.5.2 Reactions of Carbonyl Compounds

Hydrates and hemiacetals
The reactions of a carbonyl group can be generally described by the following scheme.

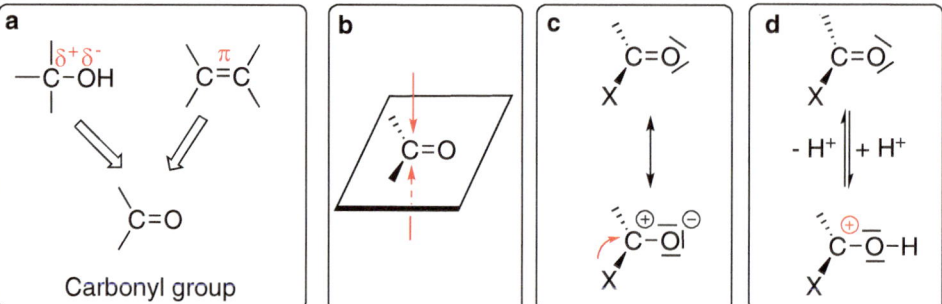

Fig. 4.42 Properties of a carbonyl group

$$\underset{R}{\overset{R}{\diagdown}}\overset{\delta+}{C}=\overset{\delta-}{O} \quad + \quad \text{H-Nu} \quad \rightleftharpoons \quad \underset{R}{\overset{R}{\diagdown}}C\overset{O-H}{\underset{Nu}{\diagup}}$$

A nucleophile (Nu), which is usually linked to an acidic H-atom, attacks the carbonyl carbon atom, which is positively charged, and a C–Nu bond is formed. The proton is linked to the oxygen atom to form a hydroxy group. A four-valent carbon atom with approximately tetrahedral geometry is formed. With the transformation of the very reactive C=O double bond, the original trigonal geometry is also lost. A two-dimensional structure is transformed into a three-dimensional one, which affects the environment around the central carbon atom. The reaction is a chemical equilibrium, the position of which depends on the electronic and steric properties of the carbonyl compound, the nucleophile and the product.

If water is the attacking reagent, an organic hydrate is formed, sometimes also referred to as a geminal diol. The equilibrium is located far to the left in most cases. This also explains the Erlenmeyer rule, which has been quoted several times before, according to which most hydrates are not stable and therefore have no relevance in biochemistry. If it were otherwise, all subsequent reactions discussed below would be blocked in the biological solvent water and, as a result, in the extremely excess of water. Aldehydes and ketones would act as "water sponges". Even more, a large part of biochemical follow-up reactions would be eliminated, which would have fundamental consequences for the material basis of biology. It should be remembered that the Erlenmeyer rule is a consequence of the relatively small atomic radius of the carbon atom.

$$\underset{R}{\overset{R}{\diagdown}}C=O \quad + \quad \text{H-OH} \quad \rightleftharpoons \quad \underset{R}{\overset{R}{\diagdown}}C\overset{O-H}{\underset{OH}{\diagup}}$$

Hydrate
(geminal diol)

There are only very few stable hydrates, such as ninhydrin and chloral hydrate. Both can only be produced synthetically. The structure of the two compounds makes it clear that the organic backbone is responsible for the stability of the hydrates. In both cases, it is the electron-withdrawing effects of the substituents C=O and Cl that make it possible for two hydroxy groups to be bonded to the (small) carbon atom. The lack of such structures in biochemical systems suggests that organic hydrates would be "dead-ends" and that they were therefore "sorted out" by biochemical evolution.

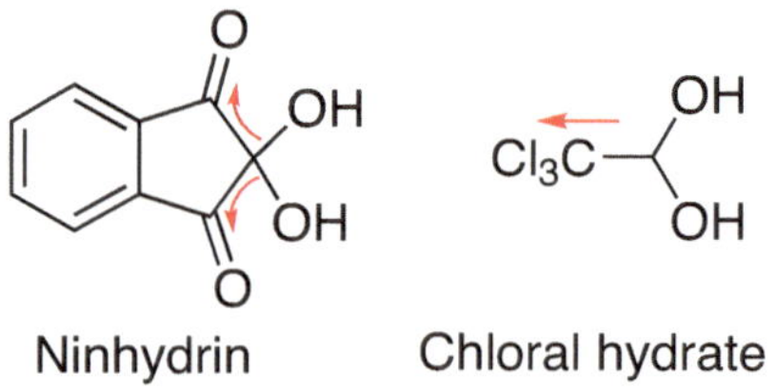

Ninhydrin   Chloral hydrate

The existence of both hydrates is a first indication that certain organic residues can over-ride the Erlenmeyer rule, that is, the rule that plays such a dominant role in the formation of carbon dioxide from carbonic acid. The two examples show how organic chemistry can extend the properties of inorganic compounds. This quality jump is particularly evident in the addition of alcohols to carbonyl compounds discussed below. This reaction also proceeds according to the same mechanism. This results in hemiacetals (in case of ketones: hemiketals), for which the Erlenmeyer rule applies in principle, i.e. the equilibrium is generally on the left side.

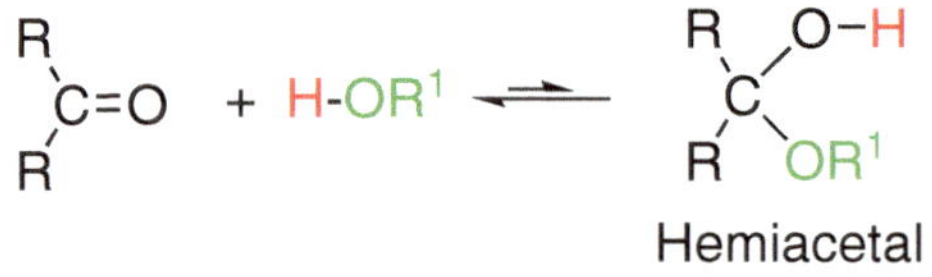

Hemiacetal

Biochemically extremely important are the stable exceptions that are only made possible by the organic residue. The most important exception concerns the ring formation of D-glucose, in which a ring is formed by nucleophilic attack of the HO group on C^5 and predominantly the β-D-glucopyranose is formed (Fig. 4.43).

The equilibrium of this special hemiacetal formation is unusually far to the right. The suspension of the Erlenmeyer rule has three causes (Fig. 4.44):

1. The β-D-glucopyranose is based on a saturated six-membered ring. Such six-membered rings have, among all other ring systems, the greatest stability due to the lowest tension and minimal repulsive forces between the substituents (Fig. 4.44a).
2. The six-membered ring assumes the energetically particularly favorable chair conformation (Fig. 4.44b).

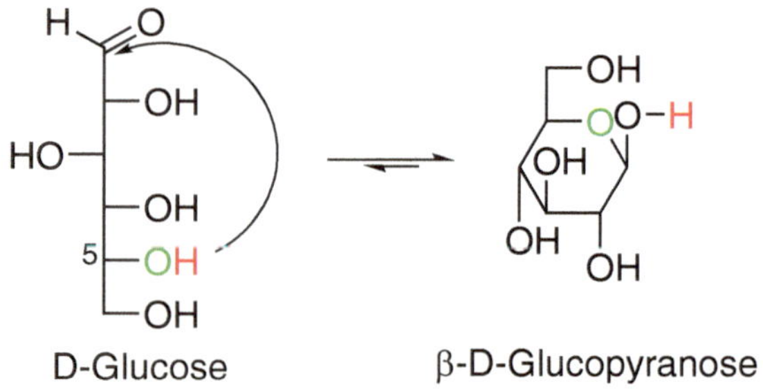

Fig. 4.43 Formation of hemiacetals in the D-Glucose

D-Glucose   β-D-Glucopyranose

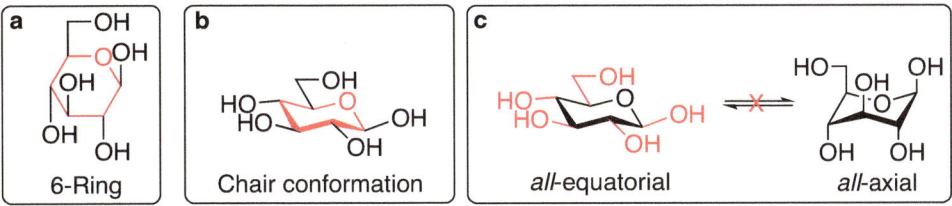

Fig. 4.44   Reasons why D-glucose forms such stable hemiacetals

3. All substituents are aligned in the energetically stable *all*-equatorial position, which results from the *gluco*-configuration of the monosaccharide. An equilibrium with that conformer, in which all large substituents are axial, does not exist (Fig. 4.44c).

If only one of these three criteria is set aside, the hemiacetal loses stability. This occurs, for example, when one or more substituents occupy the axial position in the six-membered ring. Axial HO groups in 2- or 3-position are particularly destabilizing. D-Mannose is an example of this. The formation of two basic structures that are in equilibrium with each other is an indication of this. The equilibrium has an influence on the stability of polymers and their chain length (Fig. 4.63). The biochemical evolutionary potential of D-mannose is therefore smaller than that of D-glucose.

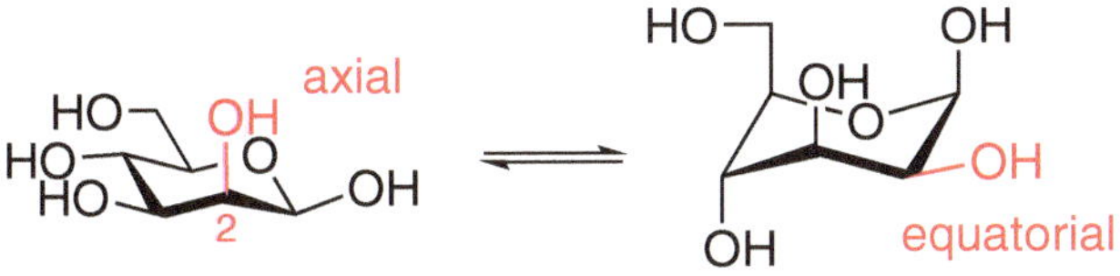

D-Mannose

A change in ring size also has an effect. Such a less stable system is, for example, β-D-glucopyranose, which is formed by nucleophilic attack of the HO group at C^4 and hemiacetal formation (Fig. 4.45).

Fig. 4.45   β-D-Glucopyranose, the hemiacetal of a five-membered ring

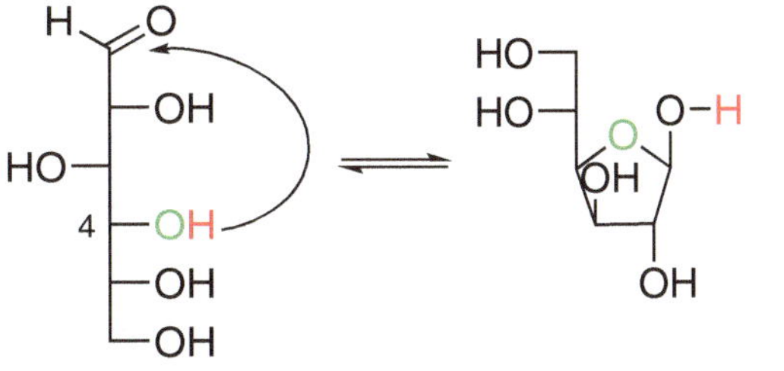

D-Glucose   β-D-Glucofuranose

The product is a five-membered ring (Fig. 4.46). In the envelope conformation preferred for five-membered ring, all substituents do not take up an ideal equatorial orientation like in the six-membered ring of D-glucose, but are pseudo-equatorially oriented.

This less stable five-membered ringg is also formed as the hemiacetal of D-fructose (in addition to the six-membered ring) and is here referred to as β-D- or α-fructofuranose (Fig. 4.47). All forms are in equilibrium with the open-chain aldehyde form.

As shown below (Fig. 4.62), the formation of a hemiacetal from sugars leads to a first deceleration effect on their breakdown. Any destabilization of the hemiacetal leads to faster decomposition. In other words, the overwhelming dominance of D-glucose in the living world over all other sugars is due, inter alia, to the three aforementioned causes: six-membered ring formation, chair conformation and *all*-equatorial orientation of all large substituents during the formation of the hemiacetal. In principle, the entire carbohydrate chemistry is evidence of how dramatically organic residues modify the original properties of functional groups (Fig. 4.48): Glucose does not form a hydrate with water, which is in accordance with the Erlenmeyer rule, but an intramolecular hemiacetal is formed contrary to the Erlenmeyer rule. This opens up new reaction channels, for example, leading to the formation of acetals.

Acetals

Hemiacetals react with an excess of the same or another alcohol to form acetals (Fig. 4.49). The establishment of chemical equilibrium is accelerated by acid catalysis,

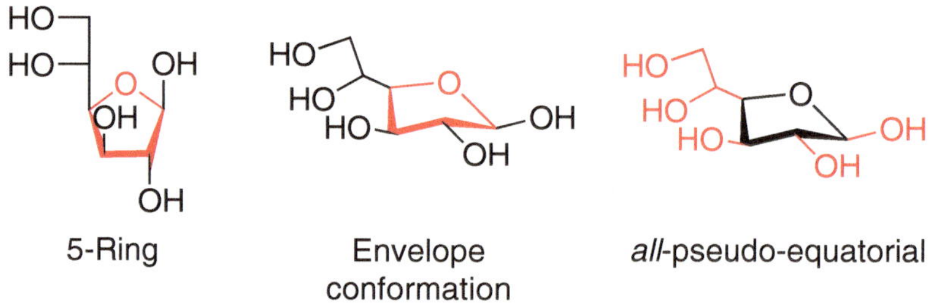

5-Ring   Envelope   *all*-pseudo-equatorial
                          conformation

Fig. 4.46 Geometrical properties of a five-membered ring

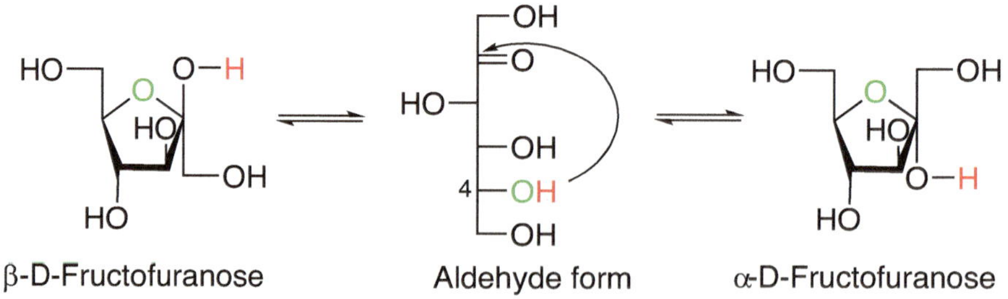

β-D-Fructofuranose   Aldehyde form   α-D-Fructofuranose

Fig. 4.47 Hemiacetal forms of D-fructofuranose

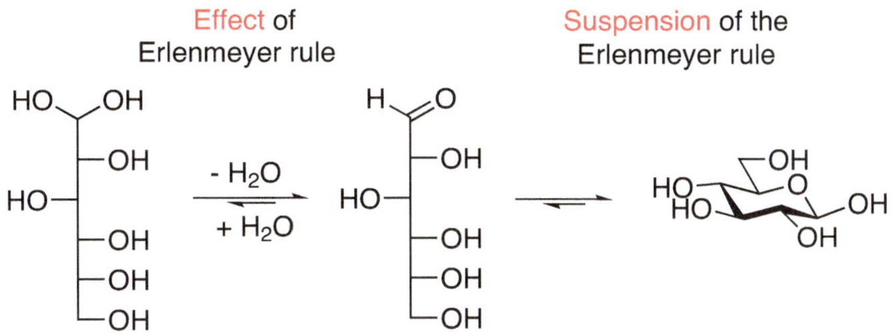

Fig. 4.48 Hydration *versus* formation of a hemiacetal in the case of D-glucose

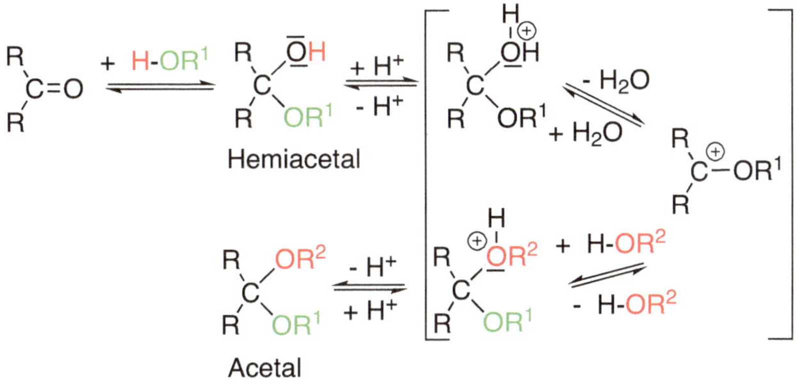

Fig. 4.49 Mechanism of the formation of acetals from hemiacetals

in the simplest case protons. The reverse reaction also takes place under these conditions and requires the presence of water, which was split off during the forward reaction.

The formation of an acetal prevents the reverse reaction to the aldehyde or ketone, as discussed for the hemiacetal, and thus prevents it from being oxidized over the corresponding carbonyl compounds to carboxylic acids (Fig. 4.50).

This fact is used in the Fehling's test (with Cu(II) as oxidizing agent) to check whether sugars are in the form of a hemiacetal or an acetal. Only hemiacetals are in equilibrium with the oxidizable aldehyde form and are classified as reducing sugars.

Glycosides, disaccharides and oligosaccharides

In principle, any alcohol can react with a hemiacetal in the course of a biological cell to form an acetal. Derivatives of sugars, whose acetals are referred to as *O*-**glycosides**, are of great importance. They consist of a sugar fragment and a sugar-free part, the aglycon.

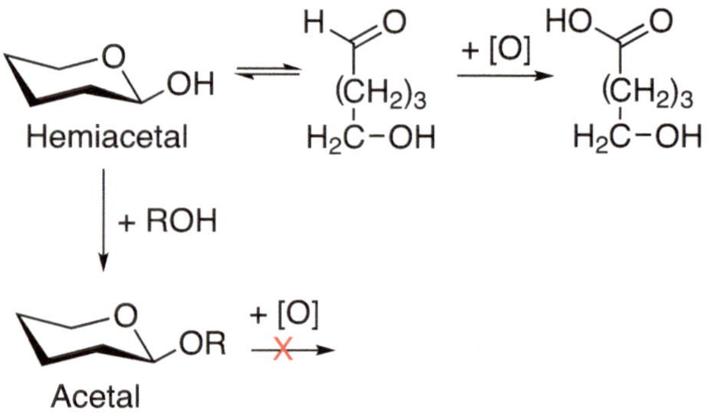

Fig. 4.50 Acetal formation prevents oxidation of an aldehyde group

The precursor of vanillin, vanilloside (Fig. 4.51), is exemplary. The building blocks of lignin, the monolignols cumaryl alcohol, coniferyl alcohol and sinapyl alcohol, which differ in their degree of oxidation or methylation, also represent aglycones.

Since the aglycon is often not water-soluble, a water-soluble compound is formed by linking it to the polar sugar. The monolignols are thus only transported in plants from the place of origin to the tips of branches and branches via the water-conducting cells of the phloem as a result of this modification (Fig. 4.52). There, the acetal is split into water under the catalytic influence of an acid-acting enzyme. The glucose part is incorporated into a growing starch or cellulose chain via an acetalization reaction and the sugar-free part is incorporated into a growing lignin chain via a radical mechanism. Both products are insoluble in water. Due to the common transport phenomenon, the ratio between the polymers in the wood is approximately equal. At the same time, the three-dimensional

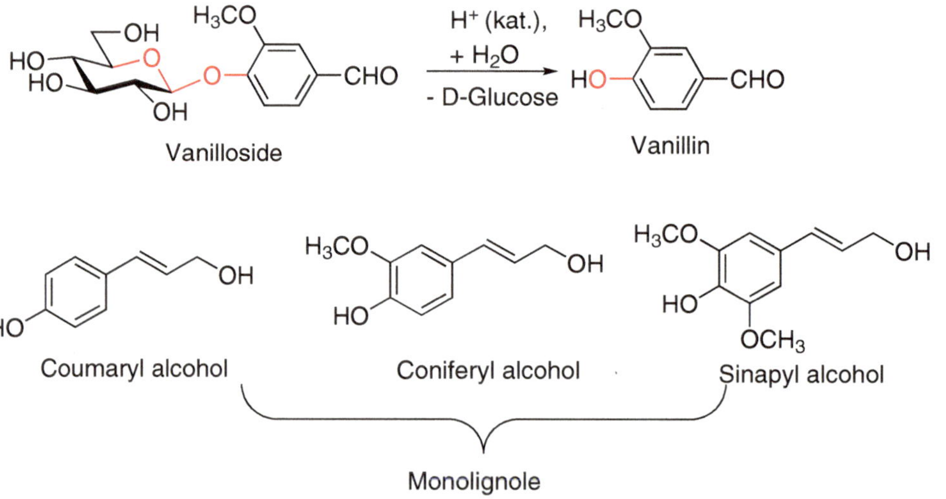

Fig. 4.51 Examples of biochemically important *O*-glycosides

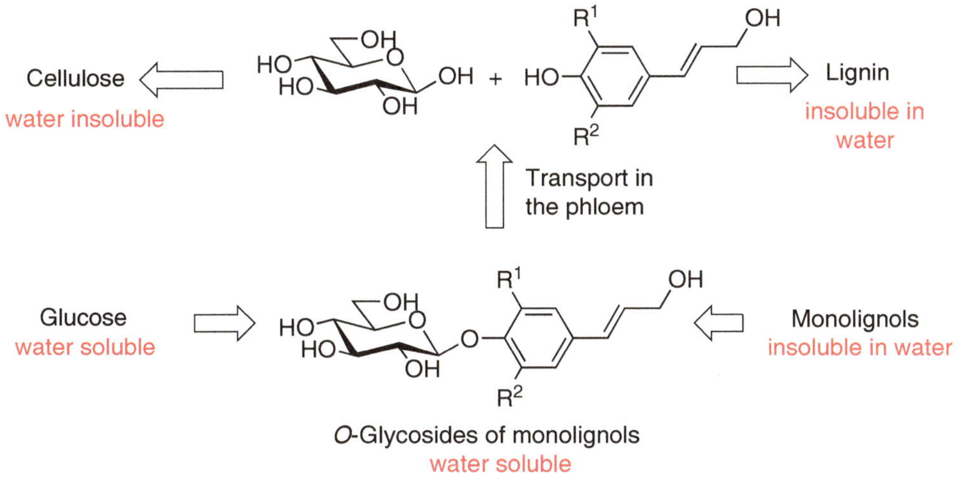

Fig. 4.52 *O*-Glycosylation is a prerequisite for transport with water

structures of cellulose and lignin are locally closely adjacent, partly even covalently linked, which biologically has the additional stabilizing effect of preventing degradation.

Lignins of different chain lengths and compositions are responsible for the strength of plant tissue, especially for its compression strength (Fig. 4.53). The embedded cellulose fibers ensure the tensile strength. The composition of lignins depends mainly on the ratio of monolignols to each other. In dicotyledonous plants, lignin is mainly composed of sinapyl and coniferyl alcohols. In monocotyledonous plants, coniferyl alcohol predominates. All three monolignols are found in grasses. It is estimated that 2×10^{10} t of lignin are built up in plants on Earth every year.

In addition to sugar-free alcohols, other sugars can also be used as alcohol components for acetals. Typical coupling products of D-glucose are cellobiose, maltose and gentiobiose (Fig. 4.54). These disaccharides only differ in the type of linkage, i.e. 1,4- or 1,6- or the orientation of the oxygen atom at C^1-atom.

In cellobiose and gentiobiose, the exocyclic oxygen atom of the acetal group is arranged in the sterically preferred equatorial position. In contrast, the same oxygen atom in maltose points downwards, i.e. it is in the axial position on the six-membered ring. These different orientations are indicated by β and α (Fig. 4.47). Their formation can already be observed when D-glucose is dissolved in water, whereby, in addition to the aldehyde form, both α- and β-D-glucopyranose are formed (Fig. 4.55). The cause of the unusually large concentration of α-D-glucopyranose (36%) in equilibrium with the more stable β-D-glucopyranose (64%) and the hardly observed aldehyde form (0.02%) is the ring oxygen. The free electron pairs of the latter lead to repulsive interactions with the electron pairs of the neighbouring hydroxy group in the β-form and favour the axial orientation in the α-form.

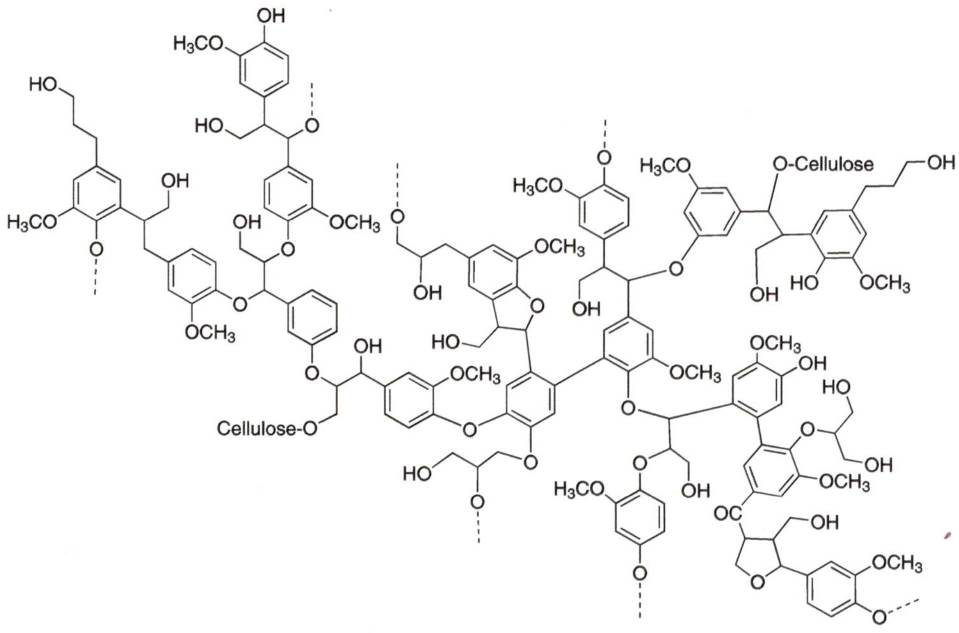

Fig. 4.53 The structure of lignin (detail)

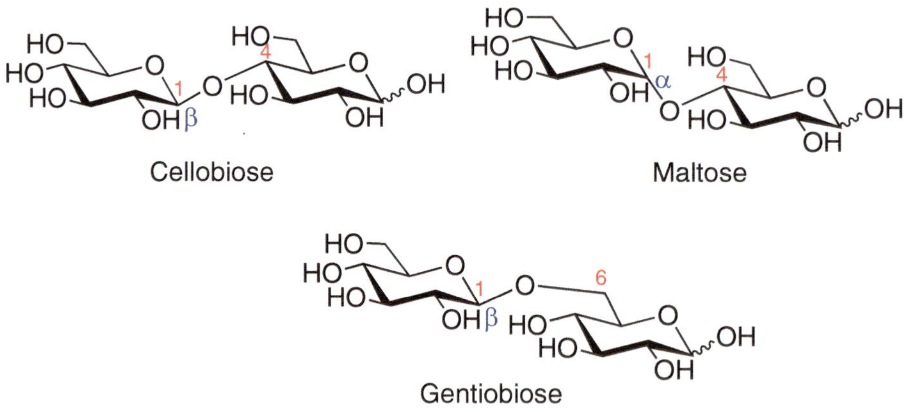

Cellobiose

Maltose

Gentiobiose

Fig. 4.54 Disaccharides based on D-glucose

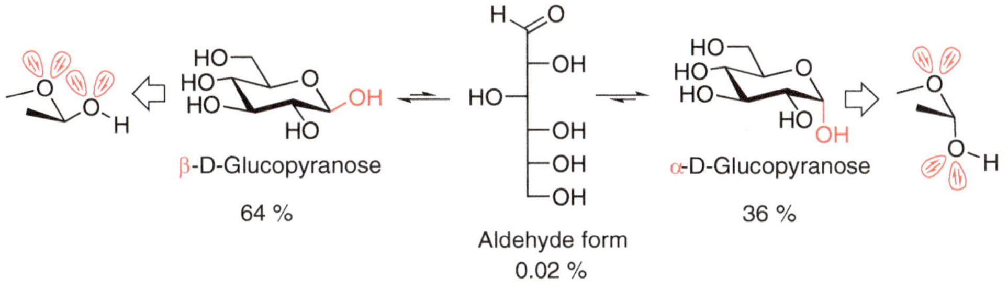

β-D-Glucopyranose

64 %

Aldehyde form
0.02 %

α-D-Glucopyranose

36 %

Fig. 4.55 The equilibrium between β- and α-D-glucopyranose

The presence of both diastereomeric forms is indicated by a helical line.

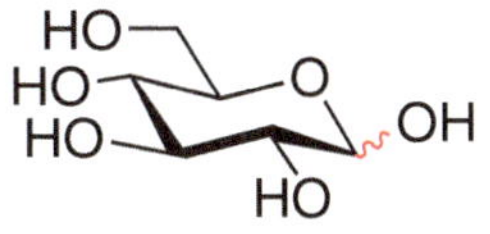

αβ-D-Glucopyranose

Since this effect is found primarily in hemiacetals and acetals of carbohydrates at the so-called anomeric center, it is also referred to as **anomeric effect**. The anomeric effect can act as a stabilizing or destabilizing force, depending on the geometric orientation of the participating heteroatoms. In principle, it occurs in all structures in which electronegative substituents X and Y with free electron pairs, which are connected to each other by a common bridging atom Z, face each other (Fig. 4.56).

An example of destabilization is cAMP (Fig. 1.79), in which the cyclic phosphoric diester is easily cleaved and thus the equilibrium with AMP is established. In α-D-glucopyranose, the anomeric effect is stabilizing.

Another coupling product of two monosaccharides results from the linking of α-D-glucopyranose and β-D-fructofuranose to sucrose (table sugar).

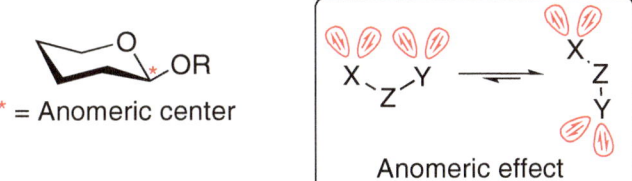

* = Anomeric center

Anomeric effect

Fig. 4.56 The anomeric effect

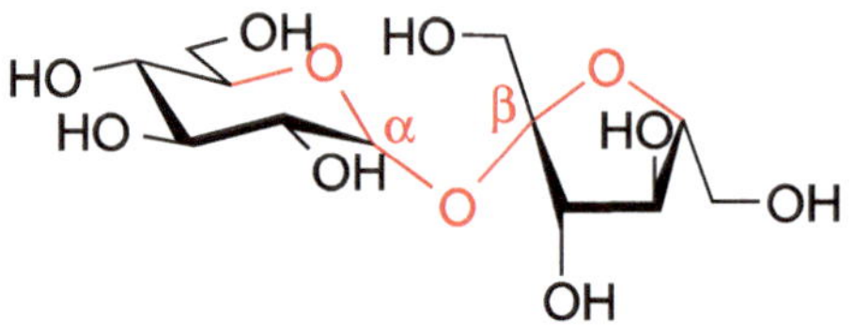

In sucrose, two acetals can be assigned. Sucrose therefore belongs to the non-reducing carbohydrates, which gives the overall structure increased chemical robustness. However, the two sugar substructures differ in terms of their individual stability. The glucose unit is based on a stable six-membered ring in the energetically favorable chair conformation, while the fructose part is based on the less stable five-membered ring. In fact, D-fructose is a degradation product of D-glucose. It is formed via a double keto-enol tautomerism (Fig. 4.57).

The breakdown of glucose to fructose is one of the first steps of glycolysis, a catabolic mechanism already mentioned, which describes the cleavage of the chain of six carbon atoms into two equal C_3 fragments (Fig. 4.58). By converting a six-membered ring into a five-membered ring, the structure is destabilized and the chain cleavage between C^3 and C^4 is prepared. At the same time, fructose is more symmetrical than glucose, which is particularly evident in the formula of α-D-fructofuranose, which is in equilibrium with other hemiacetals. The symmetrization has consequences for the bioeconomy of the subsequent reactions: In the course of glycolysis, two only marginally

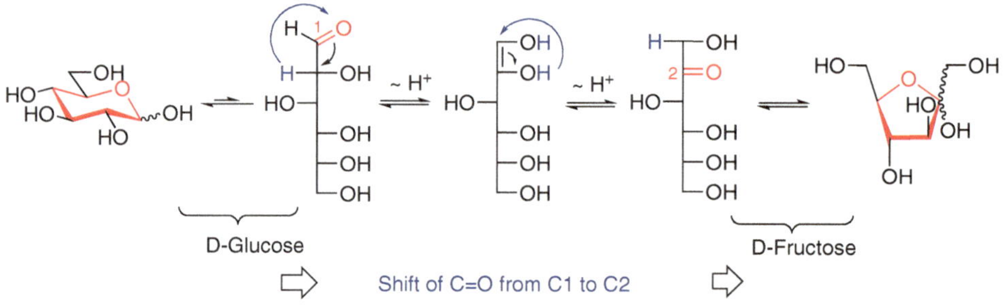

Fig. 4.57 The equilibrium between D-glucose and D-fructose

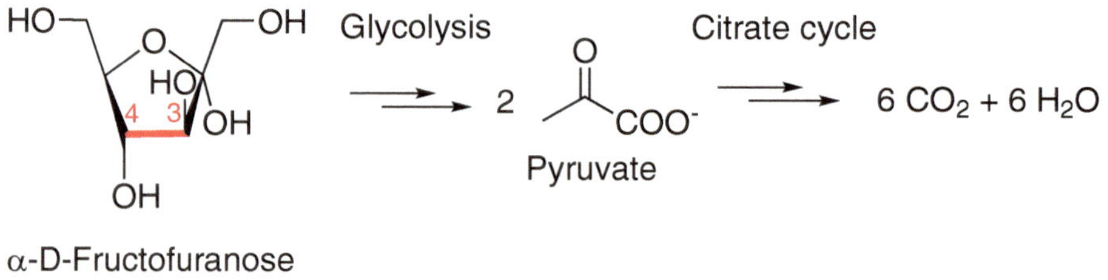

α-D-Fructofuranose

Fig. 4.58 α-D-Fructofuranose as the structurally most favorable starting structure for C–C cleavage mechanisms

different molecules are formed by cleavage, the unification of which to pyruvate requires only fewer transformations. The end product pyruvate can therefore be oxidized to carbon dioxide and water in a subsequent uniform mechanism, the citrate cycle. Separate degradation pathways with separate enzyme equipment are not required. This is an illustrative example of how several degradation mechanisms have been brought together for the benefit of "economically favorable" solutions and how this results in an evolutionary advantage.

The conversion of glucose to fructose has another chemical driving force: The structural prerequisite for almost all C–C cleavages is a neighboring carbonyl group. In glucose, this is at the top of the molecule and thus relatively far away from the future cleavage site. By the double tautomerism between glucose and fructose, this carbonyl group is shifted to C^2 and thus to the middle of the molecule. In this way, the formation of the two C_3 building blocks is prepared. The cleavage of the C^3–C^4 bond is the best-known degradation reaction of glucose. In fact, shorter-chain carbohydrates, such as D-ribose, are generated by similar mechanisms, which are integrated into RNA or DNA in the form of their nucleosides (Fig. 4.109).

If several sugar building blocks are linked together, this is called oligosaccharides. Due to the large and diverse range of cells, different sugars are also coupled, as illustrated by the example of raffinose, a trisaccharide of D-galactose, D-glucose and D-fructose (Fig. 4.59). Raffinose is thus a saccharose extended by D-galactose. Like the latter, it belongs to the non-reducing sugars. The two acetal groups that link the three monosaccharides together do not allow the formation of hemiacetals or aldehydes and thus prevent the oxidation of the particularly vulnerable aldehyde carbon atom.

In some plants, raffinose replaces starch as a storage carbohydrate. It is found mainly in legumes, sugar cane and sugar beets. Since the reactive anomeric centers are blocked as acetals and the trisaccharide is structurally very heterogeneous, consisting of three different sugars, the formation of polysaccharides does not take place. The biochemical evolution to higher and thus more stable aggregates has come to an end at this stage.

Polysaccharides

In addition to the reaction of two or three carbohydrates to di- or trisaccharides by means of acetal formation, the reaction to polysaccharides plays a central role in the generation

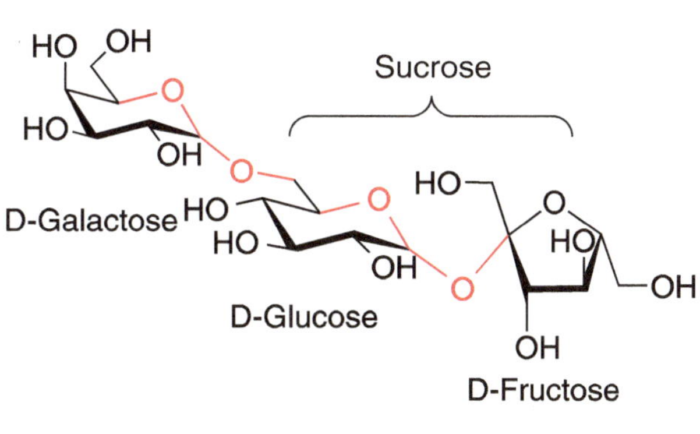

Fig. 4.59 The structure of raffinose

of biologically important scaffold and storage compounds. In principle, a uniform structure of the monomeric building blocks is required for the construction of carbohydrate polymers. As already discussed, saccharose and raffinose, which are composed of different monosaccharides, do not form polysaccharides. However, they still occur in large quantities in specific plants. On the other hand, polysaccharides based exclusively on D-glucose are almost exclusively found in all biological organisms due to the uniform and stable form of the monomer.

Depending on the biological species, different polysaccharides are formed on the basis of D-glucose. Cellulose, consisting of up to 10,000 building blocks, is formed by 1,4-linking of β-D-glucose monomers (Fig. 4.60). 1,4-Linking is also found in amylose, in which the glucose building blocks are α-linked. Chain extension and 1,6-branching lead to amylopectin or glycogen, which contain up to one million glucose monomers. Amylose and amylopectin together form the starch of plants. Glycogen is the most important storage carbohydrate in animals.

The linkages are not limited to 1,4 or 1,6. Dextrans are, for example, such highly branched polysaccharides. The glycosidic bond to the adjacent glucose molecules is realized via 1,6-, 1,4- or 1,3-, rarely also via 1,2-linkage. Dextrans serve as reserve substances in yeast and bacteria.

In lentinan, there are two β-1,6-glycosidic branchings on each of the five straight-chain β-1,3-glycosidically linked monomers (Fig. 4.61). Lentinan is a polysaccharide isolated from the shiitake mushroom (*Lentinula edodes*).

The formation of hemiacetals and even more acetals protects the extremely reactive carbonyl group of glucose in the aldehyde form and thus inhibits its decomposition via glycolysis and the citrate cycle ultimately into carbon dioxide and water (Fig. 4.62). Since carbon dioxide, as has been emphasized several times, is a gas, the chemical

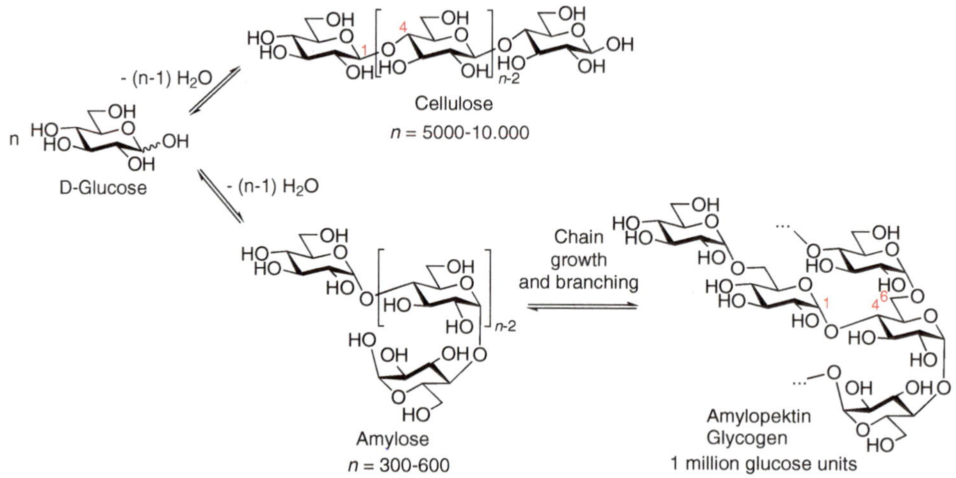

Fig. 4.60 Polysaccharides based on D-glucopyranose

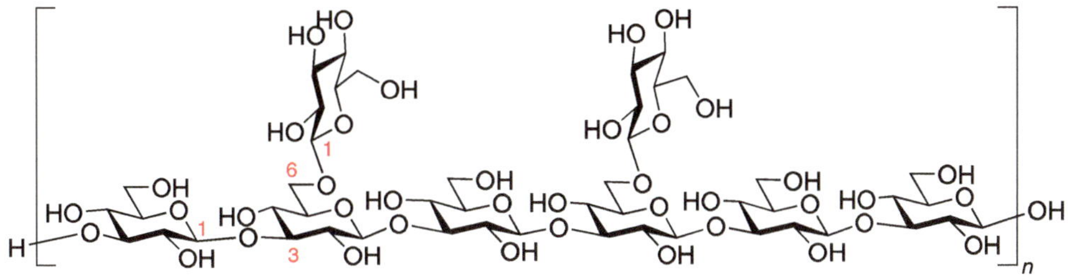

Fig. 4.61   The structure of lentinan

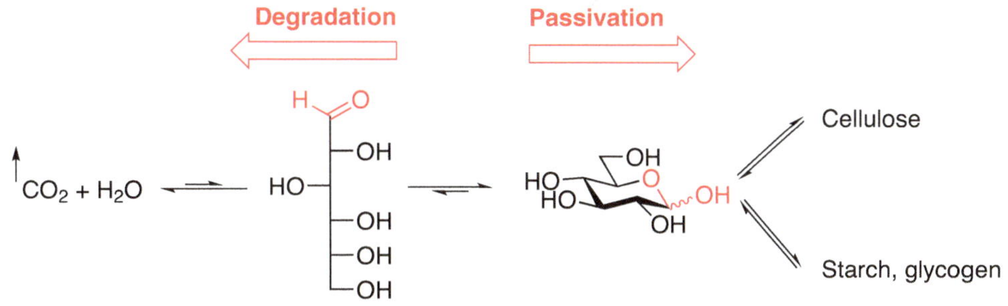

Fig. 4.62   Hemiacetal and acetal formation as a barrier to total oxidation

equilibrium is always shifted in its direction. The longer and more stable the chains are, the more energy is required for their cleavage and the more this process is delayed in time. Polysaccharides thus represent extreme "stability islands" in the context of the slowed total oxidation of energy-rich carbon.

The driving force for the formation of macromolecules from monomeric glucose or their cleavage results from the Gibbs-Helmholtz equation:

$$\Delta G = \Delta H - T\Delta S$$

This physical-chemical equation describes the relationship between the change in Gibbs energy (ΔG), the change in enthalpy (ΔH) and the change in entropy (ΔS) at a certain temperature (T). The enthalpy can be simplified as the heat of formation. The linking of bonds and thus the formation of polysaccharides is an energy-providing process and thus exothermic. The formation of chains is a process that leads to an increase in order and thus to a decrease in entropy in view of the countless individual sugar monomers. For the construction of macromolecules, regardless of their individual components, the following relationship results for the Gibbs energy:

$\Delta G = -$ (exergon): chain growth
$\Delta G = 0$ (equilibrium): chain growth and chain degradation in equilibrium
$\Delta G = +$ (endergon): chain degradation

Chain growth results in a negative sign for ΔG. Chain construction and degradation are in equilibrium, which occurs at a certain temperature, the so-called **Ceiling Temperature**. For reactions in living nature that take place at approximately the same temperatures, it can be neglected. If the entropy difference and thus this parameter in the Gibbs-Helmholtz equation become too large, chain degradation prevails over chain formation.

The position of the enzyme-catalyzed equilibria depends on the stability and concentration of all components. As has been emphasized several times, the formation of carbon dioxide (gas!) always exerts a decomposition pressure not only on glucose, but also on its polymers. However, due to its particularly stable structure, glucose is best suited to resist this decomposition pressure.

A change in the properties of monomers always has an effect on the chain length of polymers. This is shown by those polysaccharides that are not made from D-glucose (Fig. 4.63). D-Mannose, a hexose, and L-arabinose, a pentose, belong to this group. In the D-mannose, in the chair form, there is an HO group axially (Sect. 4.1.5.2), and the L-arabinose forms a five-membered ring. Thus, the corresponding cyclic monomers are less stable than those of D-glucose, from which both arise by chemical degradation reactions. Consequently, the corresponding polymers D-mannan or L-arabinan are subject to greater degradation pressure; they consist of fewer building blocks. They are referred to as hemicelluloses (hemi = Greek "half") because of their shorter chain lengths and are found in

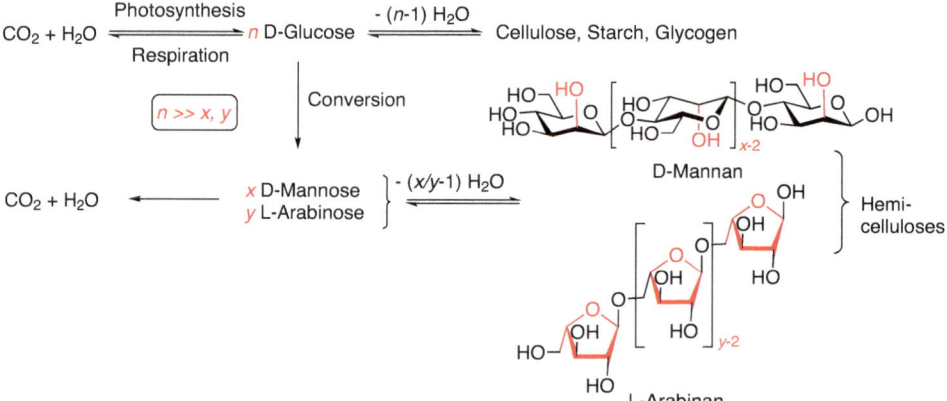

Fig. 4.63 Energetically less favorable monomers produce shorter polymer chains

the cell walls of higher plants. This analysis also explains why the trisaccharide raffinose with three different monosaccharides is not the basis for polysaccharides.

Not only the number of monomers in a polymer determines its stability, but also the interactions within a chain or with other chains, thus adding another chemical evolution criterion. In the case of polysaccharides, the HO groups are responsible for this. In cellulose, there are numerous inter- and intramolecular hydrogen bonds between the highly ordered long chains (Fig. 4.64). Almost no HO group is excluded from this, which at the same time prevents water molecules from being included. Due to this highly ordered structure, cellulose can even crystallize.

In addition, cellulose is interwoven with the non-carbohydrate lignin, another polymer, to form a composite material of great strength. Plants produce about 180 billion tons of cellulose per year. This means that approximately half of the organic carbon on Earth is in the form of high-molecular cellulose.

The chains of starch and glycogen are characterized mainly by intramolecular hydrogen bonds (Fig. 4.65). Most of the hydroxy groups are engaged. This creates helical structures. In contrast to cellulose, the numerous types of starch contain different amounts of water. The additional branching (in addition to 1,4- also 1,6-linkages) via covalent bonds not only increases the molar mass, but also the frequency of occurrence, as can be seen in the

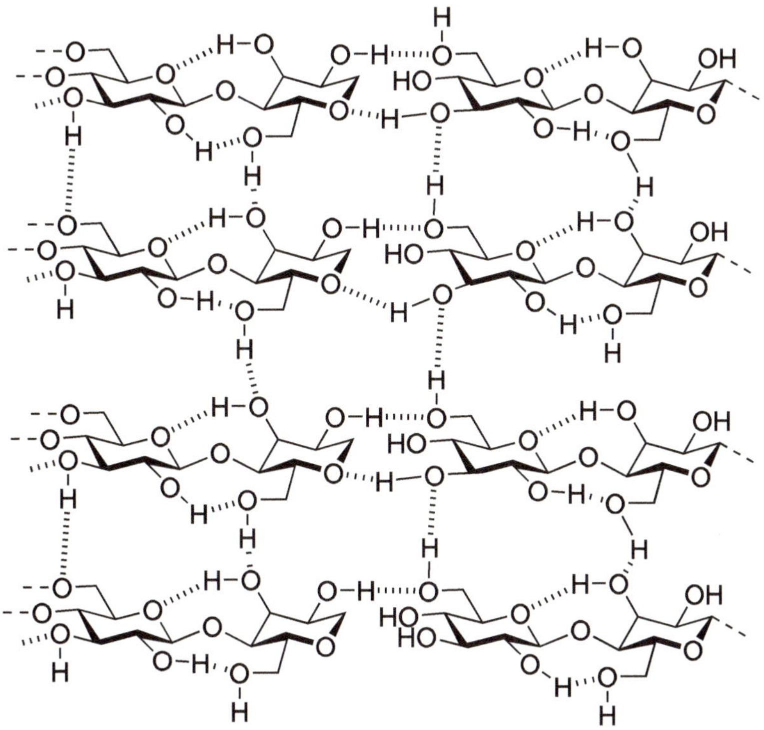

Fig. 4.64 The structure of cellulose with hydrogen bonds

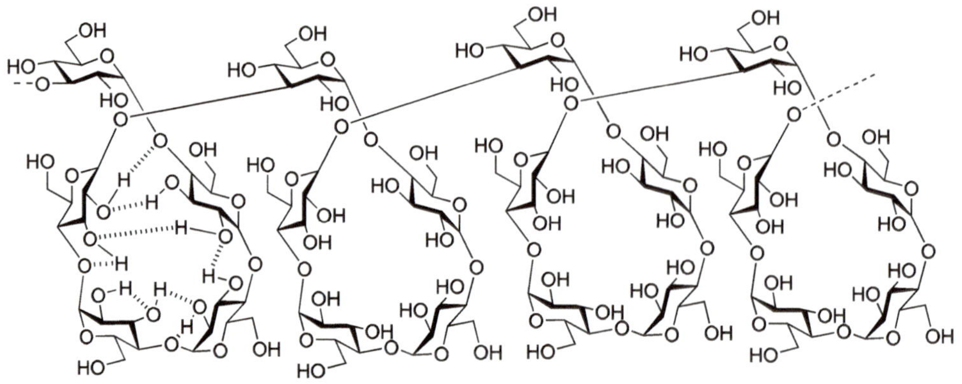

Fig. 4.65 The structure of amylopectin (with hydrogen bonds only depicted in the left part)

comparison between amylose and amylopectin: The shorter-chain amylose is only present to 10-30% in the starch, while the proportion of highly molecular amylopectin is 70–90%.

In contrast to the monomeric and water-soluble glucose, the numerous hydrogen bonds convert the polymers into an osmotically inactive form. This also prevents the storage of those water molecules that are formed during the polycondensation of the glucose molecules. But even external water is kept away. Water is necessary to release individual glucose monomers from the chain. Since the cleavage of acetals is catalyzed by acids (Fig. 4.49), it is logical that life can only develop in an approximately neutral environment. Acids in combination with water would immediately destroy a large number of these vital biomolecules, including scaffold components such as cellulose. The absence of water in cellulose has another fitness advantage: Especially in shrubs and trees, which are often exposed to temperatures below 0 °C, water would crystallize and the water crystals would cause severe mechanical damage. The (water-free) cellulose is thus a prerequisite for land plants to colonize colder areas of this earth.

Starting from the structure of D-glucose with six carbon atoms, the *gluco*-arrangement of the hydroxy groups, the possibility of forming a stable six-membered ring in which all substituents (β-D-glucopyranose) or with the exception of one substituent (α-D-glucopyranose) are equatorially aligned, results in an increase in stability at each level (Fig. 4.66). In addition, there is stabilization in the form of polymers such as cellulose and starch or glycogen. This makes the singular role of D-glucose in living nature clear.

These properties are also the reason why only D-glucose is the sugar that dominates as a product of photosynthesis. Other, less stable sugars that could also be produced by similar construction reactions are either immediately degraded or transformed into D-glucose. And finally, the polysaccharides based on D-glucose already create an evolutionary pressure on the products of photosynthesis. Since the uniformity of the corresponding polysaccharides such as cellulose or starch is only guaranteed by the incorporation of

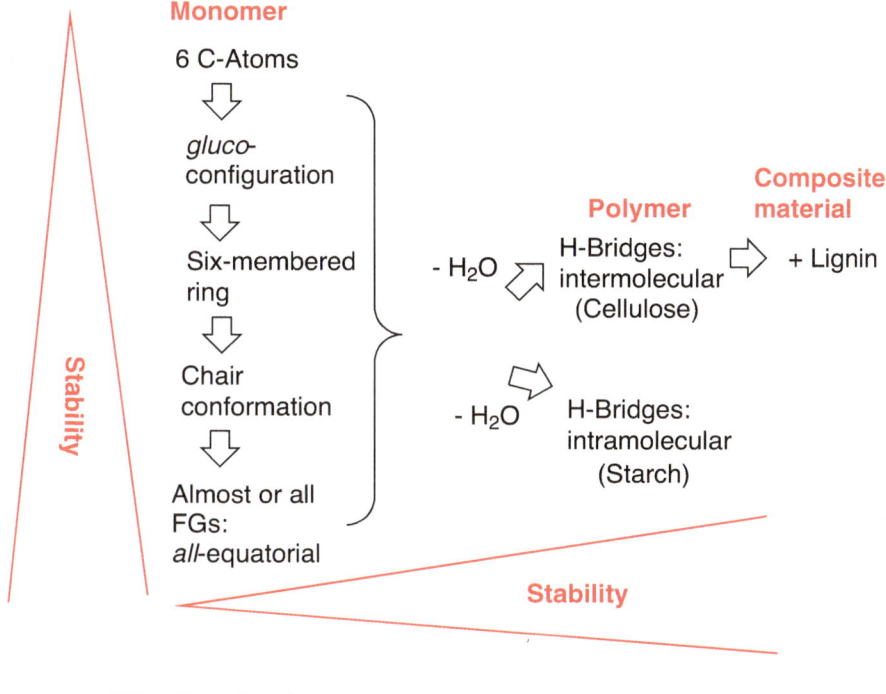

FG = Functional group

Fig. 4.66   Complexity increase and stability using the example of D-glucose

D-glucose, it becomes clear why only the D-form and not the L-form plays a role. Only the selection at a higher molecular level explains the homochirality in living nature.

The formation of polysaccharides from monosaccharides has a vital effect on plant cells. With increasing glucose concentration, the osmotic pressure rises due to the photosynthetic activity of chlorophyll in the chloroplasts. They would burst, which is prevented by the formation of the water-repellent polysaccharides. This is especially true for aquatic plants such as algae, which are considered to be the ancestors of land plants. The tendency to form polysaccharides is so pronounced that starch can even be synthesized *in vitro* in isolated chloroplasts. Thus, this principle, which evolved in a chemical context in primitive single cells, was further developed in higher biological species and used for new properties.

The different chemical stability of monomeric carbohydrates and the resulting polymers, which are directly or indirectly derived from D-glucose, are reflected in the spatial structure of wood (Fig. 4.67). Wood consists of a primary wall, which is surrounded by a middle lamella and the plasmalemma. The cellulose on the basis of the stable glucose is in the center. It is surrounded by hemicelluloses, which are already polymers of the first monomeric degradation products. At the edge of such structures, pectins are often found

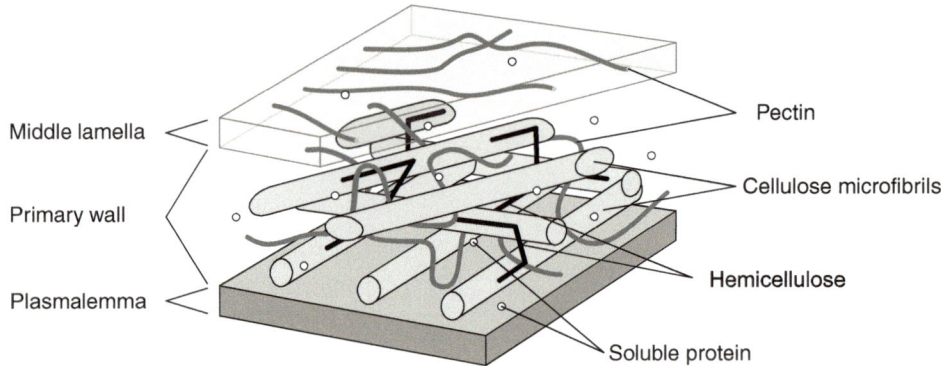

Middle lamella

Primary wall

Plasmalemma

Pectin

Cellulose microfibrils

Hemicellulose

Soluble protein

Fig. 4.67 The structure of wood

(Fig. 4.80), which represent oxidation products of monosaccharides with a terminal car-boxylic acid group, such as D-galactose.

Carbohydrates—a conclusion

D-Glucose has the highest chemical evolutionary potential of all carbohydrates due to unique structural properties, which materializes in a variety of biological structures. The formation of glycosides with sugar-free aglycones from monosaccharides via D-glucose is the beginning of a development towards increasingly complex structures. By linking several monosaccharides, disaccharides such as maltose or cellobiose are first formed. At the very end of this process are huge macromolecules such as hemicelluloses, dex-trans, cellulose, starch or glycogen, which are dominant in the living world. A prerequi-site for the formation of polysaccharides is the stability and uniformity of the monomeric sugar. In contrast, short-chain oligosaccharides often function as signal molecules on the surfaces of membranes, based on less stable and different monomers. They are usually bound to proteins (glycoproteins) or lipids (glycolipids) there. They have a "guardian function" about which molecules can pass through the membrane and which cannot. In this way, (in exceptional cases) even sugars can become information molecules.

With increasing molar mass of polysaccharides, stability also increases. It is rein-forced by inter- and intramolecular hydrogen bonds between the HO groups. At the same time, they prevent large amounts of water from nesting, which is essential for the cleav-age of the acetal bridges. Similar phenomena also affect modified polymers such as chi-tin. In contrast to anhydrous cellulose, the most important scaffold-forming substance in plants, storage carbohydrates starch and glycogen still contain considerable amounts of water, which explains their different functions and degradation accessibility in living organisms. In the case of wood, an additional effect is added by the community with the sugar-free polymer lignin. This makes the competing degradation pathway towards carbon dioxide energetically more expensive and, in terms of time, also longer. Thus, the entire sugar biochemistry is an illustrative example of the decelerated total oxidation of energy-rich carbon, which is a central guideline of this book. Each of the chemical structures and their concentration ratios to each other give the respective organisms unique biological properties. Ultimately, however, the biological applications in the form of scaffold building blocks (wood, chitin), energy storage (dextrans, starch or glycogen) and information molecules (glycoproteins, glycolipids) are "only" secondary effects based on the *a priori* of chemistry. Chemistry not only describes the material basis, but is also the origin of evolution.

Hemiaminals, imines and *N,O*-acetals

Hemiaminals are formed by the reaction of carbonyl compounds with amines (Fig. 4.68). The reaction is comparable to the reaction with water or alcohols as a nucleophile.

Hemiaminals are usually not stable. This corresponds to the properties of hydrates, for which the Erlenmeyer rule was explicitly formulated. In addition to the equilibrium with the reactant, the formation of a new functional group, the imino group, is generated by the release of water. The intermediate hemiaminal, despite its instability, plays a key role in the entire nitrogen chemistry of life, as an oxygen atom is exchanged for a nitrogen atom via this intermediate stage. The reason for this singular option is the redox variability of carbon and the ability to form stable double bonds between the (small) atoms of C and N.

Just as carbonyl compounds are hydrogenated to alcohols, imines are also reduced to amines.

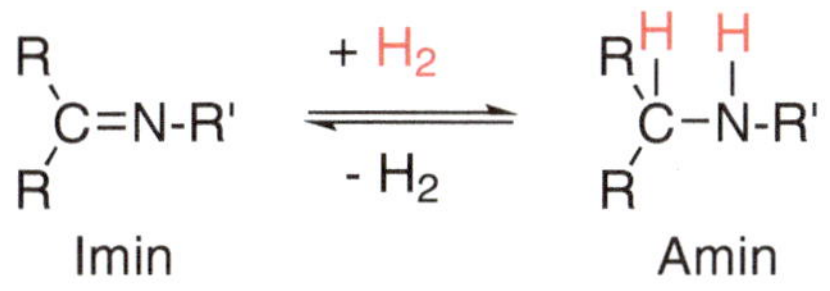

$$R_2C=N\text{-}R' \; \underset{-H_2}{\overset{+H_2}{\rightleftharpoons}} \; R_2\overset{H\,H}{\underset{}{C}}\text{-}\overset{}{N}\text{-}R'$$

Imin   Amin

If the redox reactions between alcohols/carbonyl compounds and amines/imines are coupled via the corresponding hemiaminals, a biochemically realizable reaction path results to convert alcohols into amines and back (Fig. 4.69).

As shown in Sect. 1.2.12, the direct exchange of NHR'/OH (nucleophilic substitution) is not possible under the conditions of biochemistry. The reasons for the hindrance lie in the low nucleophilicity of amines or water. At the same time, the tendency to replace the

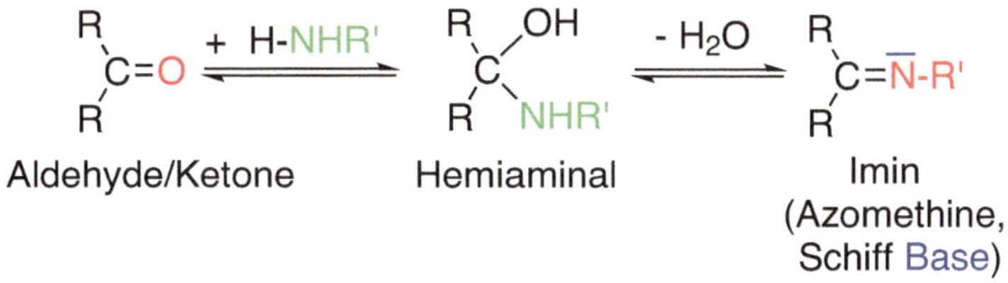

$$R_2C=O \; \overset{+\,H\text{-}NHR'}{\rightleftharpoons} \; R_2C(OH)(NHR') \; \overset{-\,H_2O}{\rightleftharpoons} \; R_2C=\overline{N}\text{-}R'$$

Aldehyde/Ketone   Hemiaminal   Imin
                                                          (Azomethine,
                                                          Schiff Base)

Fig. 4.68 The reaction of a carbonyl group with amines

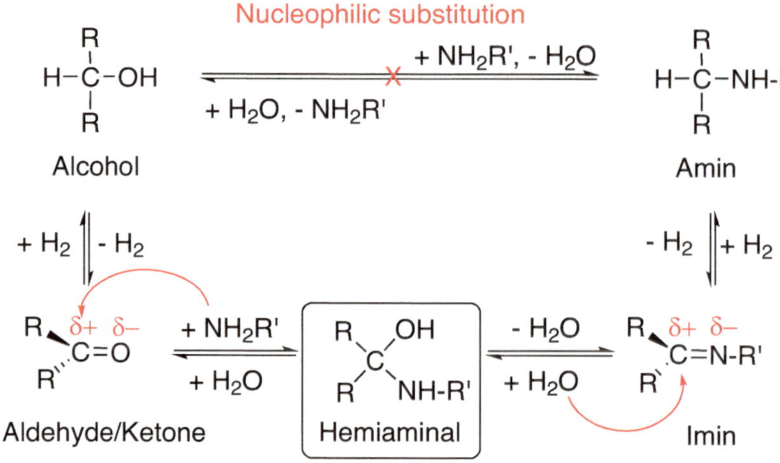

Nucleophilic substitution

$$\begin{array}{c} R \\ | \\ H-\overset{|}{\underset{|}{C}}-OH \\ R \end{array} \quad \xrightleftharpoons[+ H_2O,\ -\ NH_2R']{+\ NH_2R',\ -\ H_2O} \quad \begin{array}{c} R \\ | \\ H-\overset{|}{\underset{|}{C}}-NH-R' \\ R \end{array}$$

Alcohol   Amin

$+ H_2 \Updownarrow - H_2$ $- H_2 \Updownarrow + H_2$

$$\underset{\text{Aldehyde/Ketone}}{\overset{R}{\underset{R}{\diagdown}}{C}=O} \quad \xrightleftharpoons[+\ H_2O]{+\ NH_2R'} \quad \boxed{\underset{\text{Hemiaminal}}{\overset{R}{\underset{R}{\diagup}}{\overset{OH}{\underset{NH-R'}{C}}}} \quad \xrightleftharpoons[+\ H_2O]{-\ H_2O} \quad \underset{\text{Imin}}{\overset{R}{\underset{R}{\diagdown}}{C}=N-R'}$$

Fig. 4.69 A biochemically realizable reaction path for the conversion of alcohols into amines and back

corresponding leaving groups OH or NH_2 is only weakly pronounced. The "detour" via the carbonyl compound or via the imine requires a much lower activation energy. In both structures, the doubly bonded carbon atom is strongly polarized. At the same time, both structures are planar with only three substituents on the carbon, which makes the attack by a nucleophile easier from a steric point of view. On this "detour", numerous natural products with outstanding biological importance are formed.

It should be noted that the imine (also known as azomethine) changes the basicity in comparison to the carbonyl compound. Imines, like ammonia and many organic amines, are strong bases, as is also reflected in the name "Schiff's base". They are protonated in water, which increases the electron-withdrawing effect of the $C=NH^+$ group in comparison to the $C=O$ group (Fig. 4.70). This has significant consequences for adjacent bonds, which are now even more susceptible to cleavage.

For example, this transformation is important for the cleavage of D-fructose during glycolysis (Fig. 4.71).

Fig. 4.70 Imines have basic properties

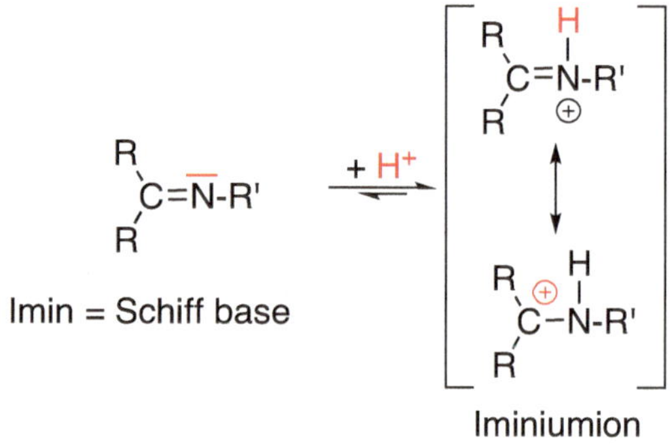

$$\underset{\text{Imin = Schiff base}}{\overset{R}{\underset{R}{\diagdown}}{C}=\overline{N}-R'} \quad \xrightleftharpoons{+\ H^+} \quad \left[\begin{array}{c} \overset{R}{\underset{R}{\diagdown}}\overset{H}{\underset{\oplus}{C}}=\overset{|}{N}-R' \\ \Updownarrow \\ \overset{R}{\underset{R}{\diagdown}}\overset{H}{\underset{}{\overset{\oplus}{C}}}-\overset{|}{N}-R' \end{array} \right]$$

Iminiumion

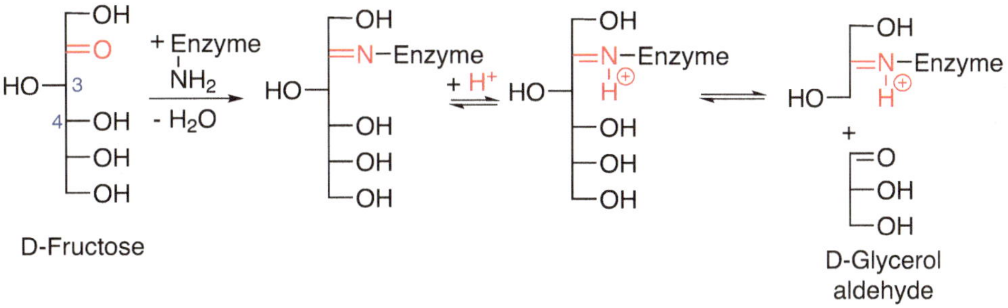

Fig. 4.71 Protonation of an imine supports C–C bond cleavage during glycolysis

The intrinsic electron-withdrawing effect of the carbonyl group is intermediarily increased by the condensation of the ketone with the terminal amino group of an amino acid (L-lysine) of an enzyme to the imine. The protonation of the imine leads to an iminium cation, which exerts a much stronger electron-withdrawing effect on the adjacent C^3–C^4 bond than the originally C=O group in fructose. In the end, the chain is cleaved.

At first, two different cleavage products arise. After deprotonation of the iminium group and hydrolysis of the imine, the enzyme is recycled and at the same time the original C=O group is regenerated; 1,3-dihydroxyacetone is formed (Fig. 4.72). A double tautomerism isomerizes the ketone to D-glyceraldehyde, whereby ultimately two identical cleavage products result and glycolysis can be continued to pyruvate.

The cleavage of the iminium cation is reversible, which has an important consequence: Not only can D-glyceraldehyde be bound again, but also other aldehydes can compete for the formation of a C–C bond. In this way, shorter, but also longer sugars with up to seven carbon atoms are formed in the pentose phosphate pathway (Fig. 4.73). A typical example is D-ribose, the basic sugar for all nucleosides. But here too it should be noted that the degradation pressure exerted by carbon dioxide on glucose is also a limiting factor for such chain extensions.

As mentioned, in biochemistry, hemiaminals are central intermediates between amines and alcohols. In principle, amines are poisons because they catalyze the cleavage of vital esters (soap effect!) due to their basic properties. At the same time, they react with all available carbonyl compounds in the cell. For this reason, amines are either neutralized (Sect. 3.2) or immediately converted into carbonyl compounds or alcohols.

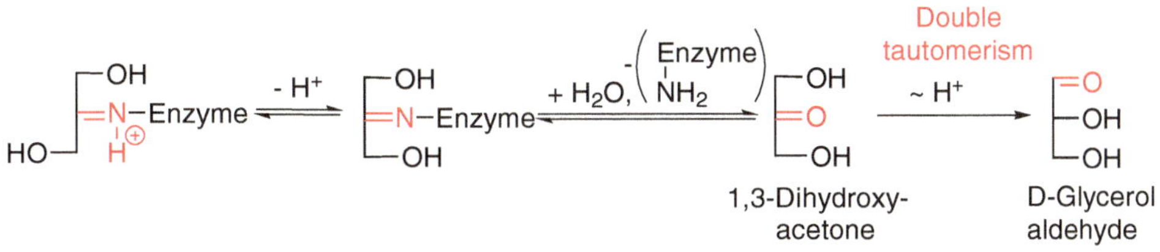

Fig. 4.72 Subsequent reactions of the iminium cation during glycolysis

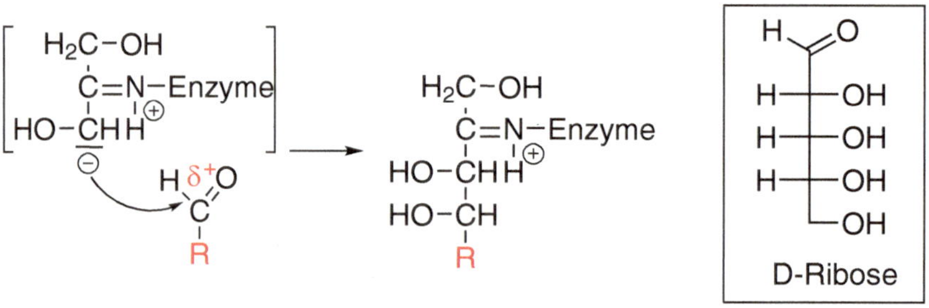

Fig. 4.73 Chain break *versus* chain extension during glycolysis

A variety of amines are derived from α-amino acids from which they arise by cleavage of the carboxylic acid group. Further reactions, such as oxidations of aromatic or aliphatic C–H bonds, may follow. Amines that are formed in this way from the proteinogenic amino acids are referred to as **biogenic amines**.

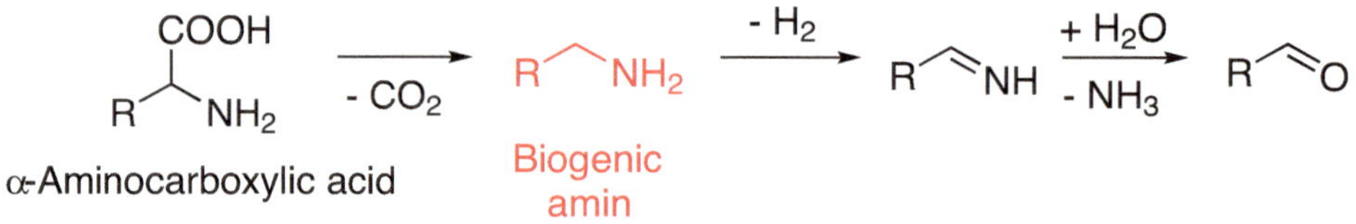

Biogenic amines have diverse effects, particularly as neurotransmitters, where they are involved in the transmission of stimuli at the postsynaptic gap. Examples are noradrenaline, dopamine, γ-aminobutyric acid and serotonin. After physiological effect, they are oxidized and the resulting imines are hydrolyzed. Since some biogenic amines have a stimulating effect, especially on animals, this results in a behavioral pressure to procure proteins. This is particularly evident in carnivores. This is an illustrative example of how chemical and biological evolution interact and feedback into each other up to behavior.

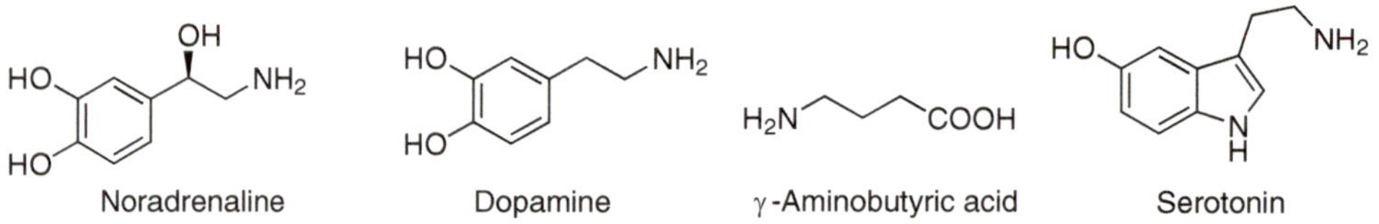

Alternatively, biogenic amines are found in highly complex structures, such as cysteamine, which acts as a substrate for CoA-SH as an acyl group transfer agent (Fig. 4.74). It is noteworthy that the amine in the form of a carboxamide amide has lost its basicity. This is also a variant of the inactivation of the life-threatening basic effect by a chemical modification and will play an outstanding role in proteins (Fig. 4.88).

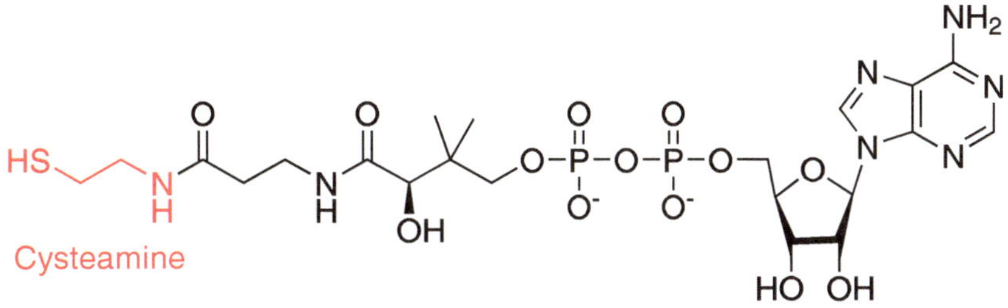

Fig. 4.74 The structure of CoA-SH, the major acyl group transmitter in biochemistry.

Other biogenic amines such as putrescine or cadaverine are formed at the end of decomposition processes. They are part of the ptomaines (corpse poisons) and thus have no meaning for the respective organisms. Whether the typical smell of decay, which attracts scavengers, can also be qualified as an ecologically useful property that goes beyond the biochemistry of the individual organism, should remain speculative at this point.

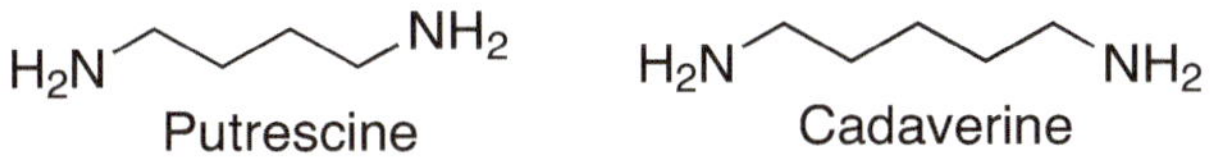

Similar to the formation of acetals from aldehydes or ketones via hemiacetals as intermediates, imines react with alcohols. This results in *N,O*-acetals. They are also more stable than the comparable hemiacetals.

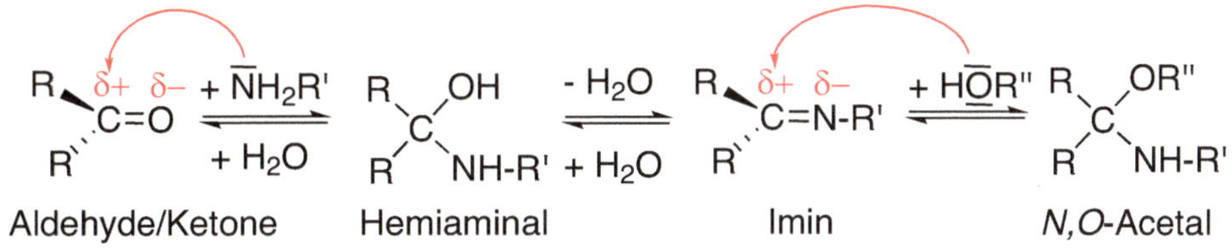

N,O-Acetals also have basic properties that disappear upon acylation with carboxylic acids (Fig. 4.75).

The best-known natural products with an *N,O*-acetal structure are the **nucleosides**, which are referred to as the corresponding 3'- or 5'-phosphoric acid esters as **nucleotides**. These include the compounds already described that play a role in energy storage, such as ATP or AMP (Fig. 1.7). In addition, a large number of nucleosides act as information carriers in RNA and DNA. Due to the *N,O*-acetal structure, there is a significant stability against water and oxygen, which is a prerequisite for lasting biochemical function. Selected reactions in the biochemical synthesis pathway for AMP are shown below (Fig. 4.76).

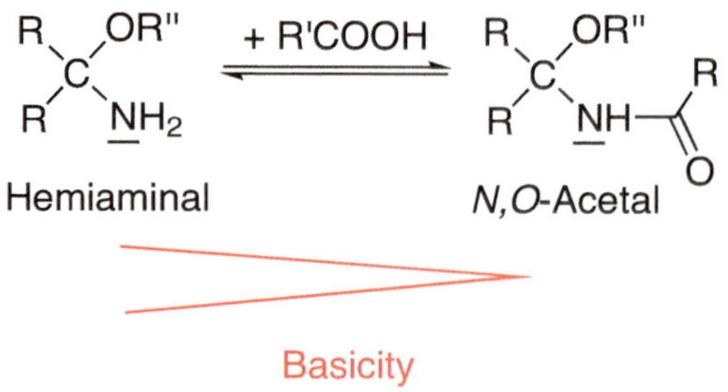

Fig. 4.75 Basicity of hemiaminals is decreased by acylation

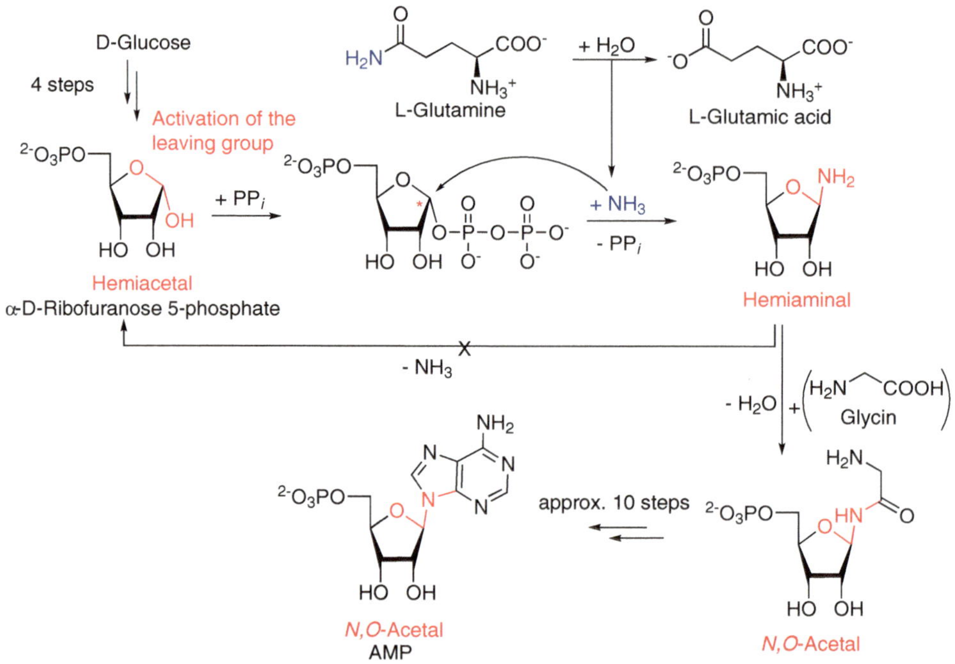

Fig. 4.76 Characteristic steps in the biosynthesis of AMP

In the first step, the hydroxy group at the anomeric center (hemiacetal!) of α-D-ribo-furanose 5-phosphate, which is formed from D-glucose, is esterified with pyrophosphate. This forms a well-substituted leaving group with the phosphate diester anhydride. In parallel, the nucleophile ammonia is generated from the hydrolysis of the amino acid L-glutamine to L-glutamic acid. The subsequent substitution leads to the formation of a hemiaminal. By carboxamide formation with the amino acid glycine to form an acy-lated N,O-acetal, it is stabilized, i.e. the reverse reaction of the unstable hemiaminal to the hemiacetal with release of ammonia is prevented. Numerous reactions follow, which complete the construction of the two heterocyclic rings of adenine.

In a similar way, other nucleosides are formed, which also play an excellent role in the construction of RNAs and DNAs.

4.1.6   Carboxylic Acids

Carboxylic acids are formed by oxidation of aldehydes (Fig. 4.77). The carbon atom in the carboxyl group has an oxidation state of +3. Thus, carboxylic acids are the last organic station before the formation of carbon dioxide, an inorganic compound with an oxidation state of +4. Ketones are not transformed in this way, since they lack the required C–H bond in which oxygen can insert.

The carboxyl group can be regarded as a composite of two functional groups, the hydroxy group and the carbonyl group. This represents a first step towards higher complexity. Compared to the two original functional groups, the properties change when the groups are linked. Alcohols are, with the exception of the electronic effect of an aromatic group on the HO group, generally not acidic. The electron-withdrawing effect of the adjacent carbonyl group polarizes the H–O bond more strongly and creates an acid, the namesake carboxylic acid.

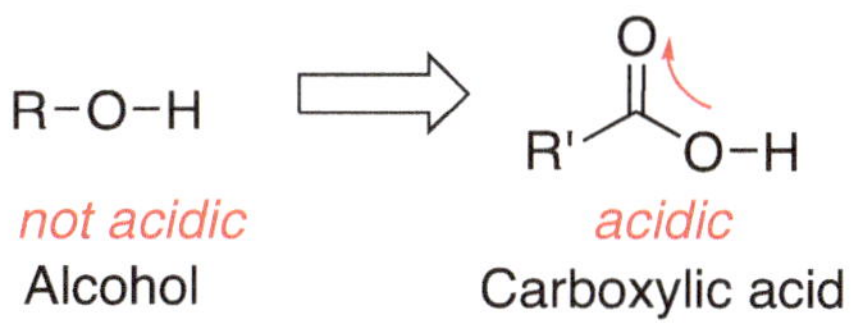

The proton in carboxylic acids is not only much more acidic than in alcohols, but carboxylic acids are also better hydrogen bond donors. At the same time, the C=O group has become a strong hydrogen bond acceptor. The proof of increased molecular interactions can be seen by comparing the physical data between ethanol and acetic acid: acetic acid has a higher melting point (mp.) or boiling point (bp.) than ethanol (Fig. 4.78). This difference not only affects the simple carboxylic acids, but also many of their nitrogen

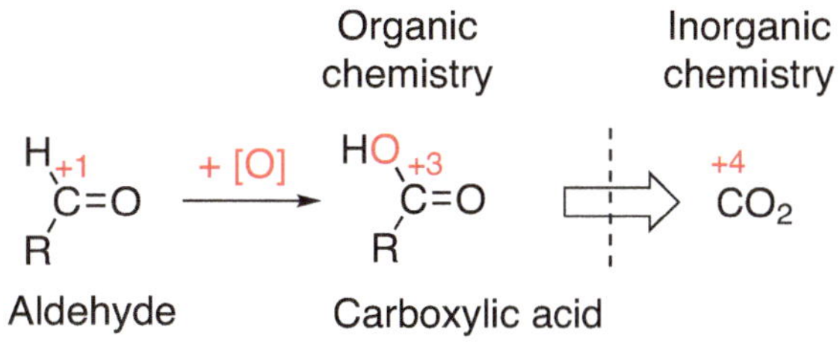

Fig. 4.77 Carboxylic acids as products of the oxidation of carbonyl compounds and precursors of carbon dioxide

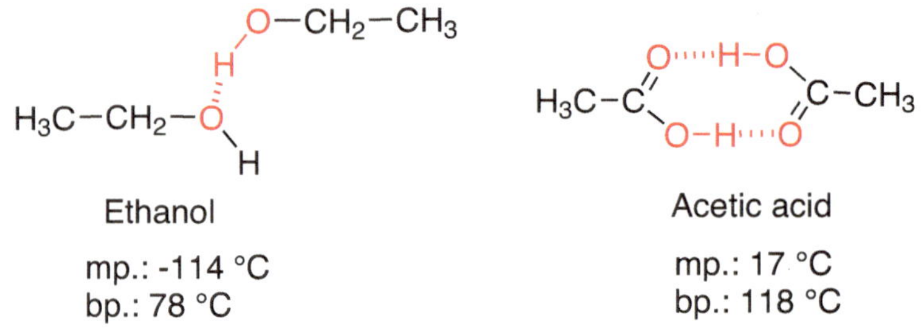

Fig. 4.78 Different physical properties of ethanol and acetic acid

derivatives, and ultimately leads to the hydrogen bonds in the very complex structures of proteins and polynucleic acids.

Acids are neutralized by bases. In the case of carbonic acids, a carboxylate anion is formed by the reaction with HO⁻. Two mesomeric boundary structures can be constructed for this. The driving force of the neutralization reaction is therefore not only the destabilization of the educt by the electron pull of the carbonyl group, but also the stabilization of the product.

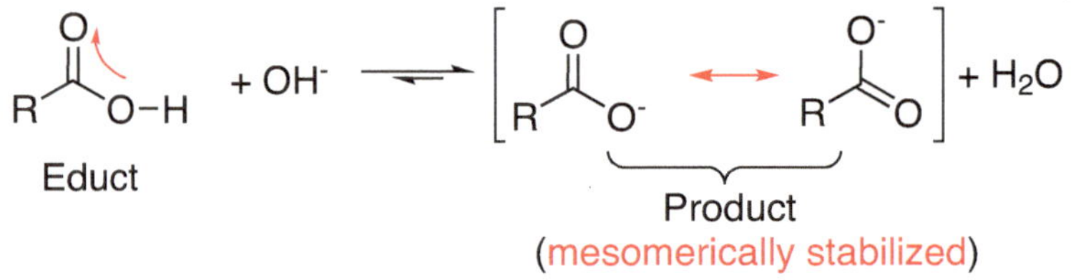

Since carboxylic acids—like all acids—catalyze the cleavage of acetals (polysaccharides!) with water and thus destroy their biological function, they occur in living systems almost exclusively in the form of their carboxylates. Typical examples can be found in glycolysis or in the citrate cycle with pyruvate, citrate, α-ketoglutarate, succinate, fumarate and L-malate (Fig. 4.79). Often, alkali or alkaline earth metal ions such as Na^+, K^+, Mg^{2+} or Ca^{2+} act as positively charged counterions.

Not only low-molecular natural substances with alcoholic groups or aldehyde functions are subjected to the permanent oxidation pressure of oxygen, but it is also relevant in polymers that are built from highly oxidized monomeric components. An example are pectins (Fig. 4.80). These are polysaccharides based on D-galacturonic acid. The original primary hydroxy group at C^6 of the D-galactose has been oxidized to the carboxylic acid in them. Pectins are not uniformly constructed. The composition varies depending on the producing plants and their age. Pectin chains are interrupted by other monosaccharides (rhamnose). They are additionally branched with oligomeric side chains consisting of up to 50 neutral sugars (arabinose, galactose or xylose). Pectins often occur in higher land

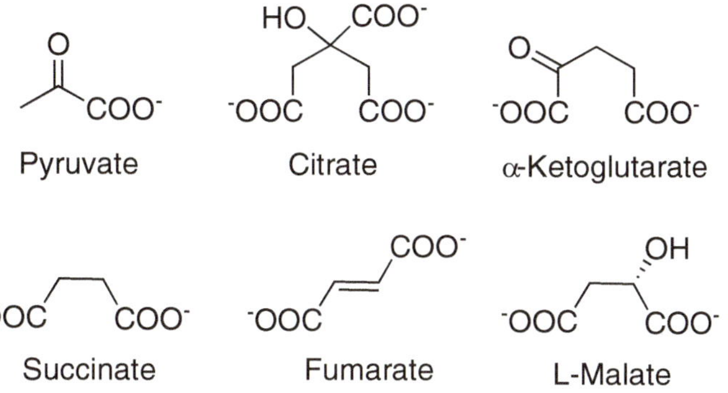

Fig. 4.79 Examples of salts of carboxylic acids

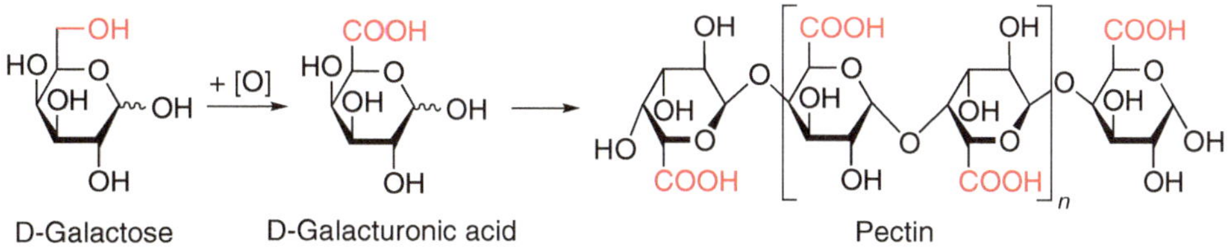

Fig. 4.80 Galacturonic acid and pectin as primary oxidation products of D-galactose

plants in the form of their methyl esters. They are contained in the middle lamellae and primary cell walls and take on a strengthening and water-regulating function from a biological perspective. As the peel of oranges, lemons and apples, they protect the fruits from mechanical environmental influences.

In addition to inorganic acids, carboxylic acids are often used to neutralize basic natural substances. Alkaloids such as nicotine, morphine, cocaine or sparteine based on amines are known, which have lost their basicity in the form of their carboxylates (Fig. 4.81). Important salts are derived from acetic acid (acetate), oxalic acid (oxalate) and L-malic acid (L-malate).

Neutralization can take place both intermolecularly, as in the case of alkaloids, and intramolecularly (Fig. 4.82). In the latter case, inner salts are formed. The most important of these, also known as **betaines**, are the amino carboxylic acids. In particular, α-amino carboxylic acids, usually referred to simply as amino acids, play a major role in the construction of proteins.

Regardless of the biological species, 20 amino acids are involved in the formation of proteins. With the exception of glycine, they are chiral. The chiral amino acids are homochiral, i.e. in accordance with the Fischer convention, they are all L-configured. The canonical amino acids only differ in terms of the organic moiety. Amino acids are roughly divided into four groups according to the hydrophobic, hydrophilic, acidic and basic side chains. L-Alanine, L-tyrosine, L-asparagine and L-lysine are typical representatives (Fig. 4.83).

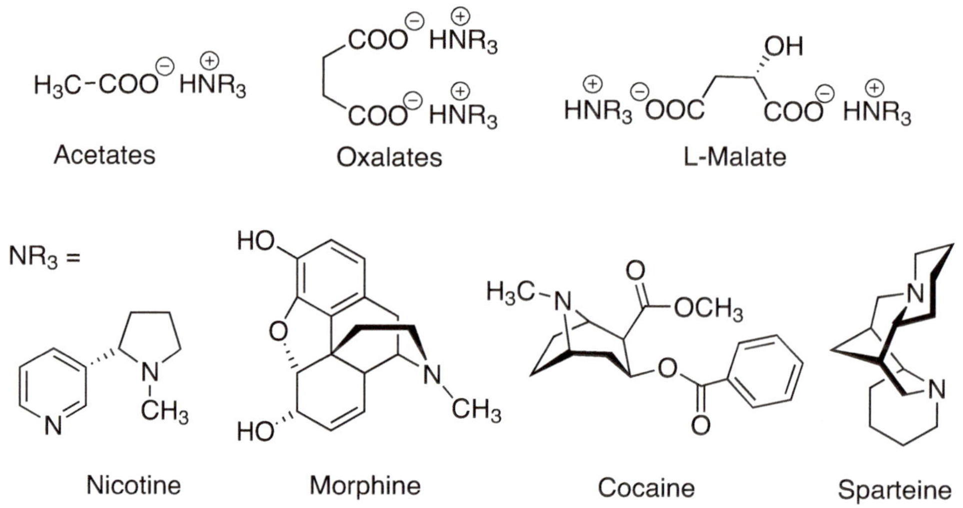

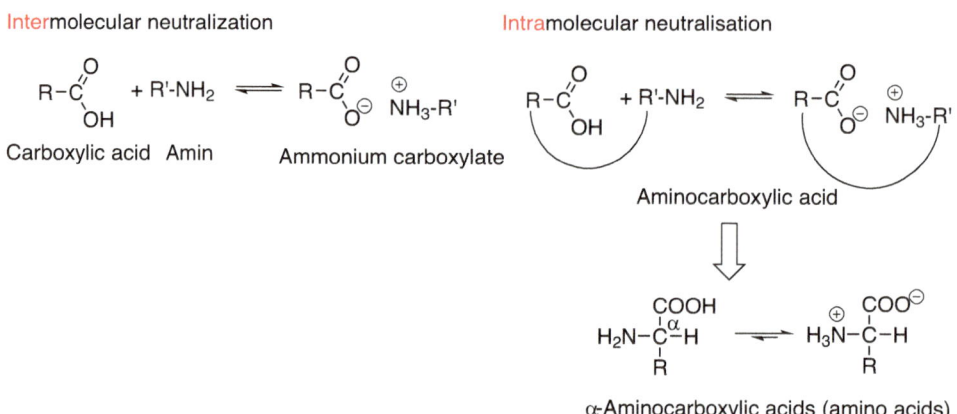

Fig. 4.81   Examples of ammonium salts of carboxylic acids

Fig. 4.82   Inter- and intramolecular neutralization

In principle, natural substances based on carboxylic acids are direct precursors of carbon dioxide with respect to the high oxidation state. However, under biochemical conditions, the direct transition is not possible. It is only realized by means of numerous organic compounds. From this situation, a delay effect on the otherwise rapid oxidation process results. The "slowing down" is materialized by the conversion of carboxylic acids into various derivatives, including carboxylic esters, thioesters, amides, imides and imidamides. They form a large number of partly very stable and highly biologically significant compounds. On the other hand, carboxylic acid phosphoric acid anhydrides are reactive structures and drivers of complex biochemical oxidation processes.

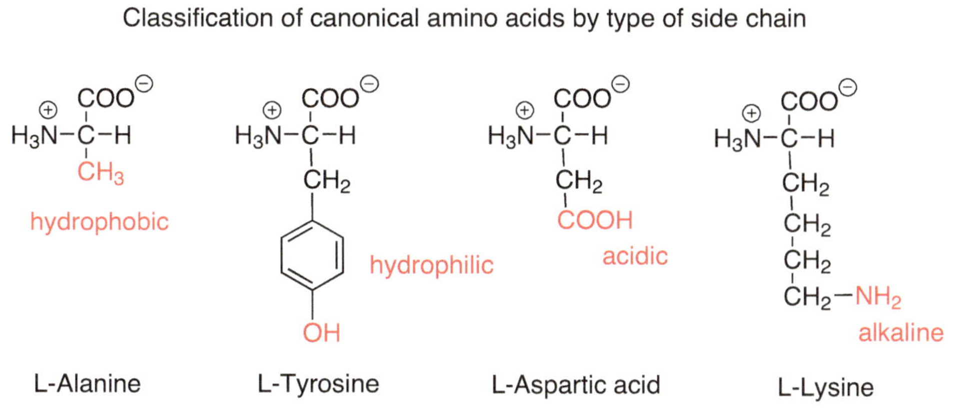

Classification of canonical amino acids by type of side chain

| L-Alanine | L-Tyrosine | L-Aspartic acid | L-Lysine |

Fig. 4.83 Basic types of proteinogenic amino acids

Carboxylic esters and thioesters

Similar to inorganic acids (Sect. 1.2.14), carboxylic acids react with alcohols to form esters (Fig. 4.84). The reaction is catalyzed by protons, which increase the carbonyl activity for the nucleophilic attack of an alcohol. In the first step, a carbocation is formed, which, after the release of the catalytically active proton, passes into an unstable compound with two HO and one R'O group. By the effect of the Erlenmeyer rule, water is split off and the stable carboxylic acid ester is formed.

Carboxylic esters are decomposed by water under the influence of bases. The driving force of the reaction is the attack of the hydroxide nucleophile on the ester bond and the formation of the mesomerically stabilized carboxylate anion. Since the action of alkaline soaps is due to this reaction, it is also referred to as as **saponification**.

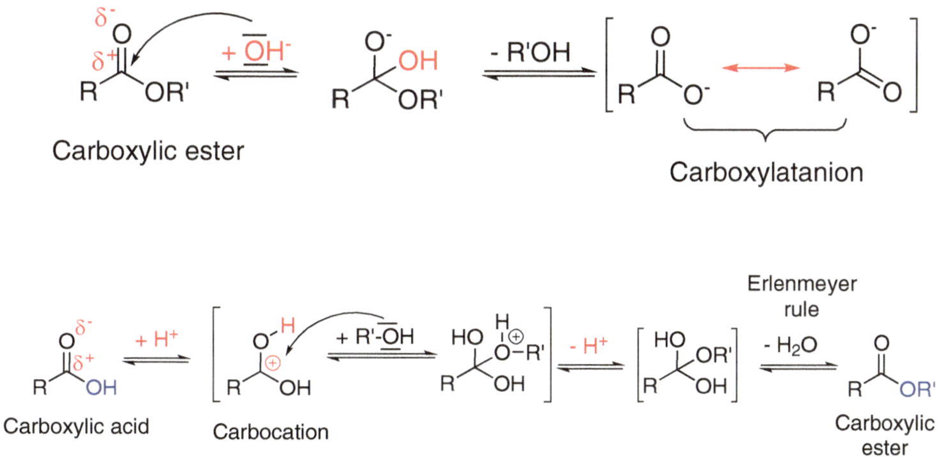

Fig. 4.84 The mechanism of esterification of a carboxylic acid

Just as with neutralization, the acidic effect is lost through esterification of acids. Acid esters are intermediates in biochemical metabolic pathways and at the same time play important biological functions in an aqueous environment. Above all, esters with long chains generate delimited spaces in water together with long-chain amphiphiles due to their **hydrophilic character**. This is the prerequisite for the formation of organelles and organisms. Triglycerides are prototypical, in which saturated or/and unsaturated fatty acids are esterified with the trivalent alcohol glycerol (Fig. 4.85).

In the presence of bases, e.g. strongly basic amines, these long-chain esters are also subject to saponification. This loses their boundary-forming effect, which is a prerequisite for any life. The result is the basic conclusion that not only acids, but also bases have a destructive influence and that life can develop only in an approximately neutral environment.

It should be remembered at this point to the comparison between *O*-esters and *S*-esters, the so-called thioesters. Thioesters are easier to cleave. A interplay of both types of functional groups is found in the transport of fatty acids through the mitochondrial membrane. Long-chain fatty acids are bound to coenzyme A (CoA) via the thioester group (Fig. 4.86). Due to the size of this molecular aggregate, they do not pass through the mitochondrial membrane. The fatty acid is taken over by the "shuttle" L-carnitine and an ester is formed. After passage through the membrane, a new CoA-SH molecule binds the fatty acid again via the thiol group. Subsequently, the fatty acid is subjected to β-oxidation.

In addition to intermolecular esterification, esters can also be formed by the reaction of a hydroxy and a carboxyl group in the same molecule. Intramolecular esters are referred to as lactones. They can be six-membered lactones, such as digitoxigenin from the series of cardenolides, which occurs in the foxglove *(Digitalis lanata)* (Fig. 4.87). Bufotalin is a five-membered lactone produced by a number of toad species. A five-membered lactone also represents vitamin C (L-ascorbic acid).

Carboxamides and lactams

Carboxamides (organic amides) are formed by the condensation of carboxylic acids with amines (Fig. 4.88). The neutralization reaction with the formation of ammonium carboxylate can be regarded as an intermediate on the way to the carboxamides. In a synthetic

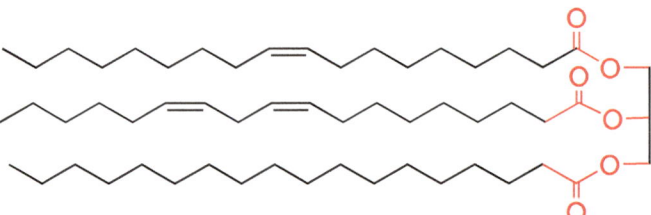

Fig. 4.85 Prototype of a triglyceride

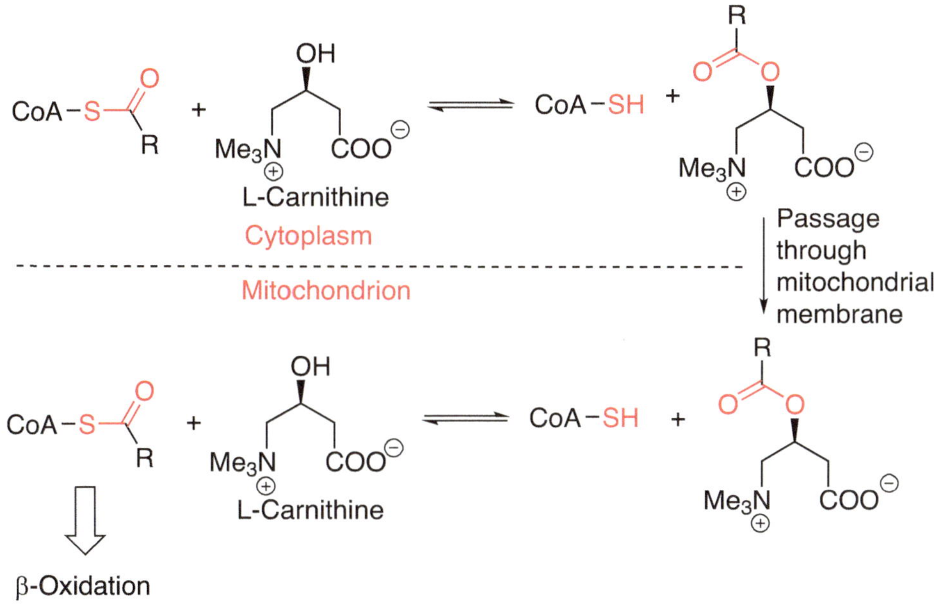

Fig. 4.86 The reversible change between an ester and a thioester group in the transport of fatty acids

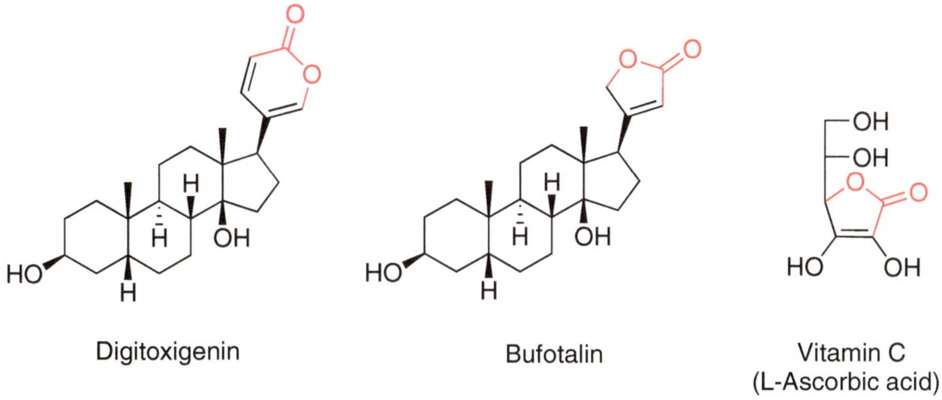

Digitoxigenin Bufotalin Vitamin C (L-Ascorbic acid)

Fig. 4.87 Examples of lactone groups in natural products

laboratory, the removal of water requires high temperatures. In contrast, the reaction in the presence of enzymes (synthetases) proceeds under mild conditions in biotic systems.

In addition to ammonium carboxylate, the formation of carboxamides also leads to neutral products (Fig. 4.89). *N*-Acetyl-D-glucosamine is a component of the cell wall of bacteria and a monomer in the construction of chitin (Fig. 3.16), which also occurs in many higher organisms. In melatonin, the primary amino group is also protected

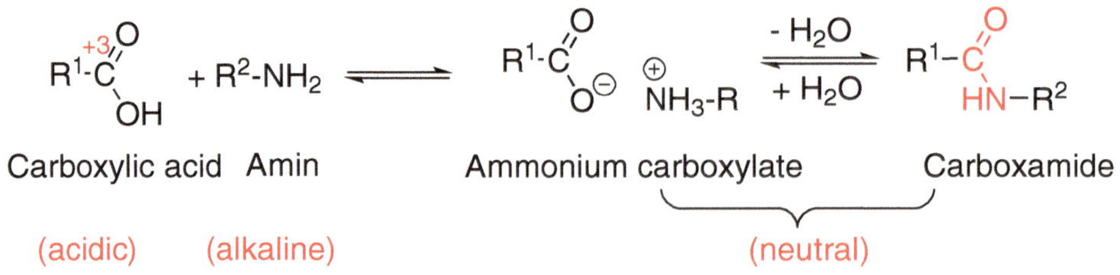

Fig. 4.88 Formation of carboxylic acids

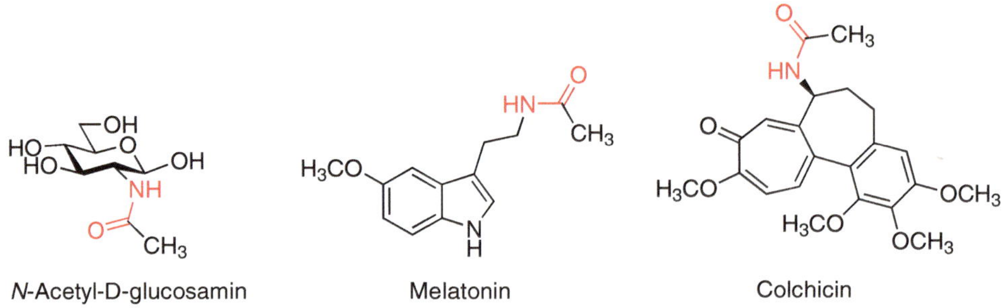

Fig. 4.89 Examples of carboxamides in natural products

from rapid degradation by oxidation by acetylation. Some alkaloids that are not based on tertiary amines, but on primary amines, such as colchicine, lose their basic properties by formation of carboxamides. Colchicine is the toxic alkaloid from autumn crocus (*Colchicum autumnale*) and occurs in its seeds, flowers, bulbs and leaves. Melatonin is formed from a proteinogenic amino acid (L-tryptophan) via a biogenic amine as an intermediate stage. This chemical degradation pathway is uniform in the animate world and even leads to a uniform biological effect: melatonin is responsible for the wake-sleep rhythm of numerous organisms from algae to humans. Colchicine, on the other hand, is a secondary natural product and thus a solitary in the plant world.

The formation of carboxamides can also take place intramolecularly, resulting in lactams. Lactams are *N*-analogs of lactones. For example, β-lactams from β-amino-carboxylic acids with a four-membered ring form the basic structure of penicillins (Fig. 4.90). They originally come from the secondary metabolism of various *Penicillium*, *Aspergillus*, *Trichophyton* and *Streptomyces* species and are now also produced synthetically. They are antibiotic. Cephalosporins are also used as antibiotics, some of which, like cephalosporin C, can be isolated from biogenic sources in the mold *Acremonium chrysogenum*.

The stability of lactams depends on the size of the ring; small rings and thus highly strained rings are not preferred. This is also the basis for the antibiotic effect of penicillins and cephalosporins. The β-lactam ring is opened and prevents the build-up of

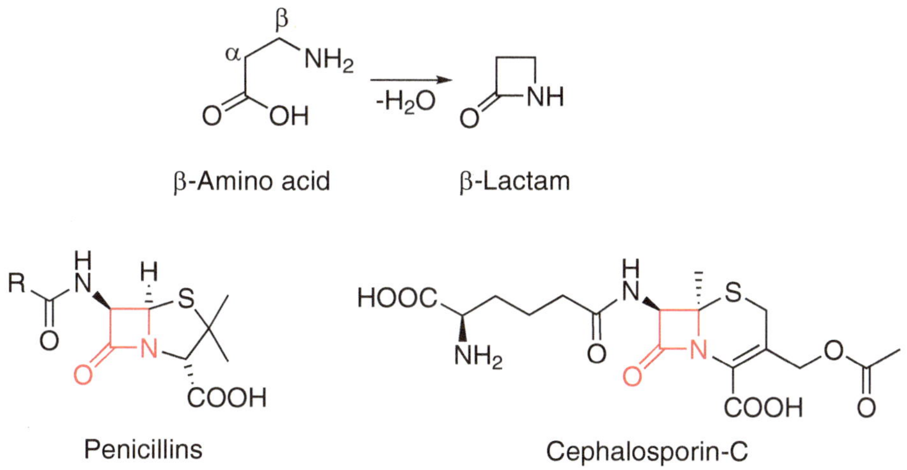

β-Amino acid   β-Lactam

Penicillins   Cephalosporin-C

Fig. 4.90 β-Lactams are formed from β-aminocarboxylic acids

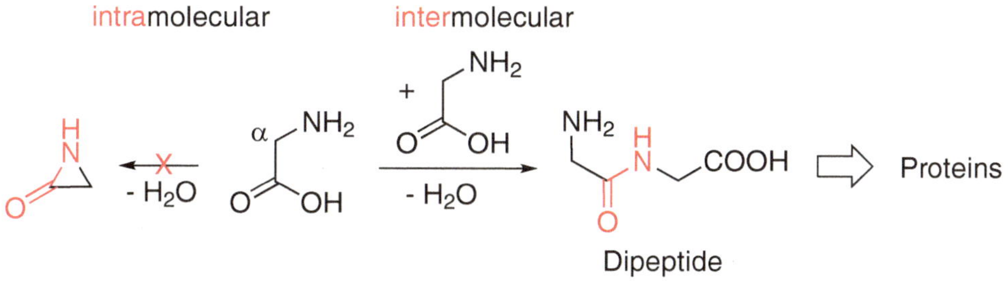

intramolecular   intermolecular

Dipeptide

Fig. 4.91 Intra- *versus* intermolecular carbonyl acid formation

membranes of multiplying bacteria. Penicillins not only act on bacteria, but also on other lower organisms such as algae or mosses. They have no effect on higher vascular plants.

Even more strained are three-membered ring lactams, which arise formally from the intramolecular condensation of α-amino acids. The hindrance of this reaction pathway is the reason why α-amino acids split off water intermolecularly, which is the prerequisite for the formation of dipeptides and longer chains, the proteins (Fig. 4.91).

Proteinogenic amino acids carry an organic residue (with the exception of glycine), which gives rise to two options for the linking of two different amino acids (Fig. 4.92). The two dipeptides A and B are chemically different and therefore have different biochemical effects. The condensation with further amino acids increases the number of isomers.

The formation of isomers is prevented in the biochemical context. The sequence and thus the decision, which amino or carboxyl group reacts, is ensured during protein synthesis in the ribosomes by successive translocation through the *t*RNAs (*transfer*RNA)

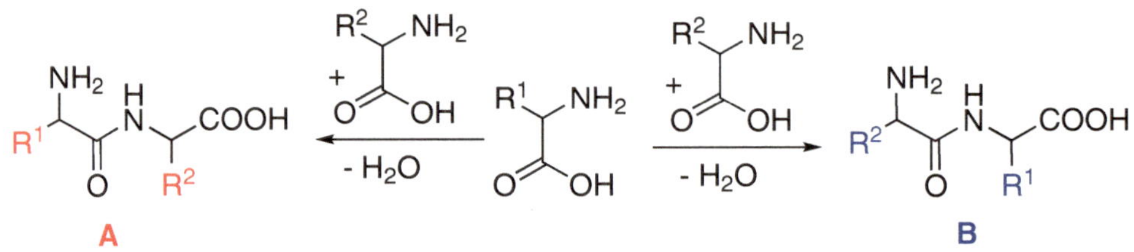

Fig. 4.92 Two possibilities in the formation of dipeptides from different amino acids

and the immediate linking with the growing protein chain. This uniqueness in biosynthesis already encodes information in the first linking step.

Not only the sequence of the individual amino acids in a chain, but also the properties of the amide bond lead to a new quality in the architecture of peptide chains and thus proteins (Fig. 4.93).

The protein chain, which actually consists entirely of single bonds, loses part of its conformational flexibility due to the restriction of the free rotational movement of the C(O)—N bond as a result of the partial double bond character (Sect. 3.3). Along the chain, freely rotatable bonds alternate with bonds that have a stable *trans*-**configuration** at normal temperature. Since all proteinogenic α-amino acids, with the exception of the achiral amino acid glycine, are homochiral and belong to the L-amino acids, the residues R^1, R^2, R^3 etc. alternate along the chain. Thus, the **primary structure** of proteins is already a highly ordered structure that receives specific information from the sequence of individual amino acids. At the same time, the primary structure determines the formation of higher associates. By the newly added property of the increased acidity of the N–H group due to the amide bond, protein chains either build up associates with themselves or with other proteins via hydrogen bonds. The **secondary structures** α-helix or sheet structure (Fig. 4.94) arise.

By further binding interactions, which take their origin in the additional functional groups in the side chains of the individual amino acids, e.g. further carboxylic acid amide and H-bonds, disulfide bridges or attractive interactions between ions or aromatics, the **tertiary structure** (Fig. 4.95) is formed.

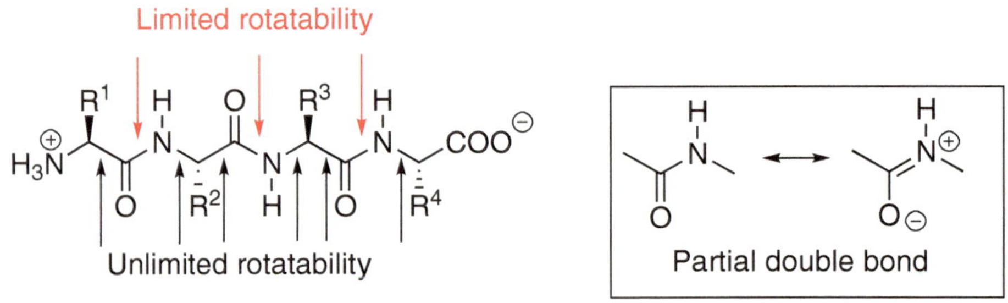

Fig. 4.93 Properties of a peptide chain

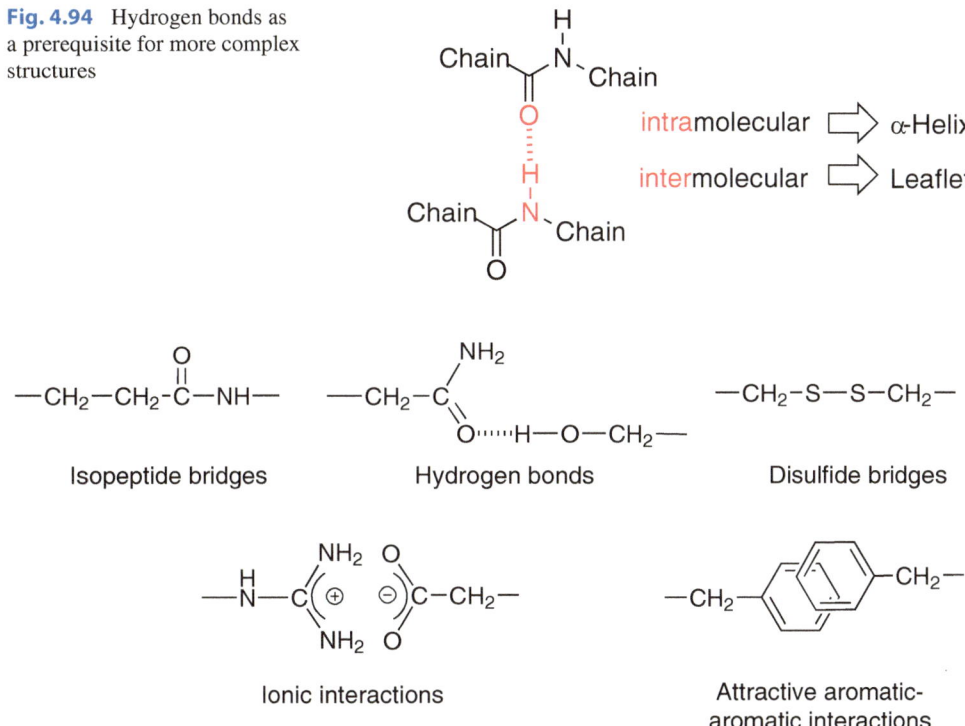

Fig. 4.94 Hydrogen bonds as a prerequisite for more complex structures

intramolecular ⟹ α-Helix

intermolecular ⟹ Leaflet

Isopeptide bridges

Hydrogen bonds

Disulfide bridges

Ionic interactions

Attractive aromatic-aromatic interactions

Fig. 4.95 Additional binding interactions stabilize complex protein structures

If several proteins join together to form a functional complex, this is called a **quaternary structure**. With the exception of the covalent isopeptide and disulfide bridges, the other binding interactions are responsible for the temporary binding between enzyme and substrate.

Proteins irreversibly denature to an increasing extent at temperatures above 40 °C, i.e. starting with the more complex architectures, the binding interactions within the chain are gradually destroyed. Therefore, any life based on proteins only takes place at moderate temperatures. In other words, only the moderate temperature regime on Earth is the prerequisite for life. Even small temperature fluctuations have an impact on natural product chemistry and thus biology. The biochemistry of fertilized eggs is particularly sensitive to changes in the environment. This form of reproduction was adopted by flightless ancestors and used for a different mode of locomotion. The temperature dependence of life processes is also shown in the development of the testes in male mammals, which are located outside the body to protect the sperm. Although this exposes the germ cells to the danger of external mechanical injuries, the chemical laws determine this exposed position.

Since most living organisms are only limited in their ability to synthesize amino acids by themselves, the intake in the form of proteins through food is vital for survival. Since, in particular, enzymes are synthesized for their function in their own organism according to the specifications of the DNA, the information contained therein is not useful for another organism and can even be counterproductive (e.g. snake venom based on proteins, allergies in humans). For this reason, proteins are first unfolded and then split into the individual amino acids after food intake. Denaturation begins with the cleavage of the hydrogen bonds in the acidic environment, for which an organ of its own has evolved in many animals with the stomach. Stomach cells are therefore particularly stressed, and the countermeasures already mentioned such as buffers in the mucous membrane and the continuous regeneration of the cells have developed. Both the laying of eggs by birds and the cleavage of proteins in the stomach are direct evidence of how biological phenomena have developed from their material basis, which can only be described by chemistry.

Proteins versus Polysaccharides

The similarities in the origin and principle structure of proteins with those of polysaccharides are obvious. But there are also significant differences. Both represent long chains and *a priori* achieve their stability through hydrogen bonds. In polysaccharides, these originate from the numerous alcoholic groups that are already present in the monomers from the beginning. In contrast, this property only evolves in proteins through the amide bond between two monomers. Polysaccharides always contain the same monomeric building blocks. In contrast, proteins can be made up of identical or different amino acids. Polysaccharides and proteins with a uniform structure are therefore referred to as scaffold or structural molecules. The function of polysaccharides, for example, as cellulose in wood, is already visible to the naked eye. This also applies to simple fibrous proteins such as keratin, the main component of mammalian hair, claws, hooves, horns and beaks.

In contrast, the transition from the secondary to the tertiary structure in proteins is associated with the emergence of new ordering forces (Fig. 4.96).

The inevitability of this results from the 20 proteinogenic amino acids. Hydrogen bridges alone would generate almost an infinite number of forms. Only through the

Fig. 4.96 The amount of information grows with the complexity of protein structures

assistance of the additional binding interactions arising from the differently functional-ized side chains of the amino acids do proteins acquire their unique tertiary structure. The tertiary and later also the quaternary structure is already determined in the sequence of the amino acids in the chain and thus in the primary structure. Thus, proteins with dif-ferent amino acids contain a much higher content of information than polysaccharides. An increase in information takes place at each level of organization. These information not only relate to the genesis and form of their own structure (e.g. exclusive selection of L-amino acids), but they are also passed on. The transfer takes place in the form of enzymes. In enzyme reactions, in the course of which substrates are converted into prod-ucts with certain rates and selectivities, these information are duplicated. The special structure of each enzyme is thus the material basis for its activity and selectivity. Since enzymes are always built from homochiral amino acids, they also contribute to the main-tenance and dissemination of homochirality in living nature in every catalytic reaction in which chiral compounds are formed.

With the exception of a few exceptions of short signal molecules mostly based on rare sugars, proteins extend the chemical evolution level enormously towards biologi-cal applications. This is comparable to the high information content of RNA and DNA, which in turn provide the matrix for the synthesis of proteins. In contrast to polysaccha-rides, the functions of enzyme proteins and polynucleic acids can only be understood in terms of their information properties. Therefore, the term information molecules is justi-fied. All biopolymers have in common that they represent stability islands and resist the rapid breakdown into lower molecular compounds, which is a prerequisite for life.

Carboximidamides

In the presence of an excess of the same or another amine, a carboximidamide is reversi-bly formed from a carboxamide under water splitting (Fig. 4.97).

It should be noted that in this transformation, the oxidation number of the carbonyl carbon atom does not change from +3 with respect to the underlying carboxylic acid.

Carboximidamide, which are derived from formic acid, are found in particular as sub-structures of numerous biologically important heterocycles. There is still a C–H bond that can be oxidized with oxygen (Fig. 4.98). If this C–H bond is missing (as in deriva-tives of higher carboxylic acids beginning from acetic acid), carboximidamides become

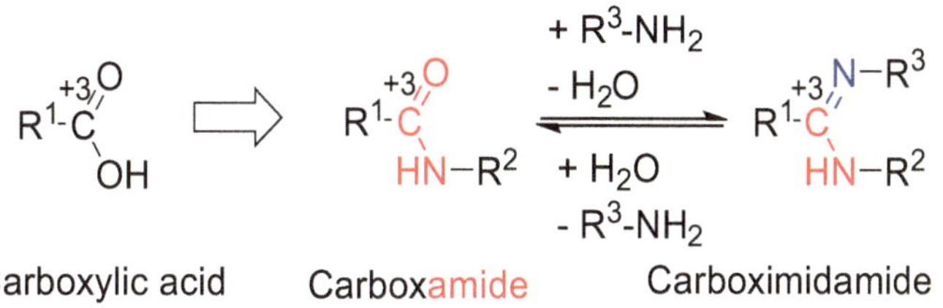

Fig. 4.97 The formation of carboximidamides

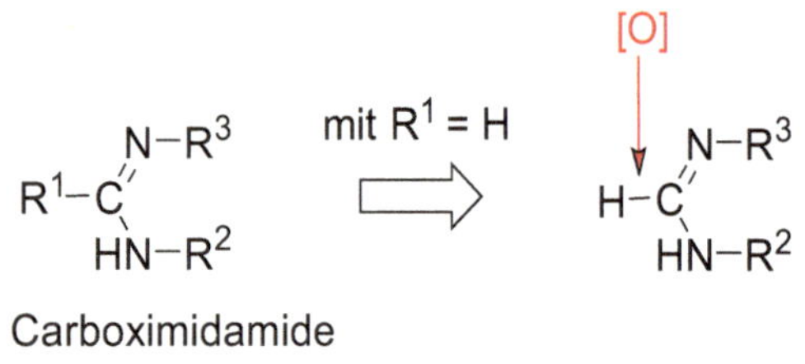

Fig. 4.98 Only carboximidamides based on formic acid are subject to oxidation

[O]

$$R^1-C\begin{array}{c}N-R^3\\\\HN-R^2\end{array} \quad \overset{mit\ R^1=H}{\Longrightarrow} \quad H-C\begin{array}{c}N-R^3\\\\HN-R^2\end{array}$$

Carboximidamide

extremely stable and can no longer be directly integrated into the dynamics of biochemical degradation processes.

Nucleosides, such as deoxyguanosine, have a carboximidamide structure (Fig. 4.99). They are subject to the usual oxidative degradation processes with oxygen radicals, which do not spare components of DNA, even if they are more strongly protected by embedding in chromosomes and the cell nucleus than other natural substances in the cell. The insertion of oxygen into the C–H bond of the five membered-ring heterocycle is followed by imidol-amide tautomerism to the more stable urea derivative. In stressful situations, where especially many free oxygen radicals are formed, this reaction occurs frequently and can lead to cancer in the corresponding organism. Therefore, the resulting 8-hydroxy-deoxyguanosine also serves as a medical stress marker in cancer therapy.

The same mechanisms are responsible for the oxidative degradation of the nucleobase adenine from AMP to uric acid (Sect. 3.2) (Fig. 4.100). The degradation starts during the hydrolytic separation of adenine from ribose 6-phosphate or D-ribose and proceeds via the nucleosides adenosine and inosine.

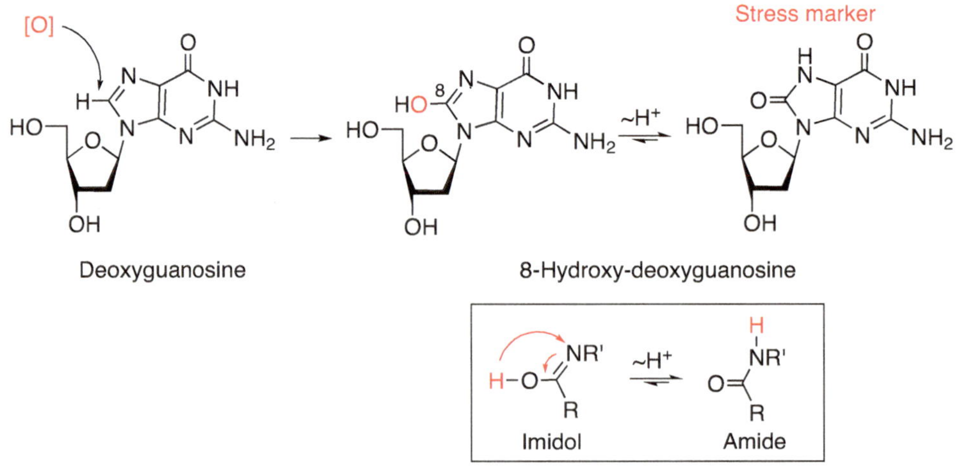

Deoxyguanosine   8-Hydroxy-deoxyguanosine

Imidol   Amide

Fig. 4.99 The oxidation of deoxyguanosine

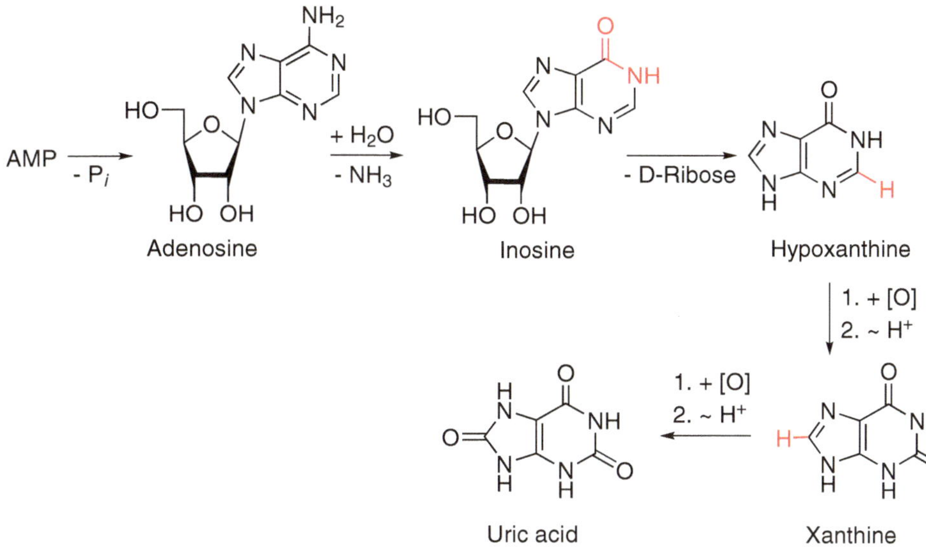

Fig. 4.100 The degradation from AMP to uric acid

Carboxylic anhydrides

In contrast to carboxylic esters, which arise from the reaction between a carboxylic acid and an alcohol, carboxylic anhydrides are formed by the condensation of a carboxylic acid with another acid (Fig. 4.101). In principle, both inorganic and organic acids can act as reactants. A similar situation was already discussed in the case of the anhydrides of inorganic oxoacids, such as phosphoric anhydride and its salts, the pyrophosphates (Fig. 1.5). Carboxylic anhydrides can only be synthesized in the laboratory and are only stable there in the absence of water. The hydrolysis reaction is highly exothermic and can, for example, lead to the boiling of water in the case of acetic anhydride. Dicarboxylic anhydrides do not exist in the living world.

Different acids can also react with each other, as the reaction between a carboxylic acid and phosphoric acid shows in the formation of a mixed anhydride (Fig. 4.102). The phosphoric acid residue is present in the physiological environment as an anion or dianion, which avoids the exothermicity during the solvation of the hydrolyzed acids and qualifies them for biochemical applications.

For example, a mixed anhydride provides the energy for the attachment of fatty acids to the acyl group carrier CoA-SH (Fig. 4.103). The negatively charged oxygen atom of the carboxylate first attacks the positively charged phosphorus atom in the phosphoric ester of ATP. The energy for the reaction comes from the parallel exergonic hydrolysis of pyrophosphate. At the same time, the carbonyl carbon atom of the reaction-inert carboxylate anion is activated by the electron-withdrawing effect of the phosphate. The nucleophilic sulfur atom in the anionic CoA-SH substitutes phosphate to form the "activated fatty acid", which is subsequently degraded by β-oxidation.

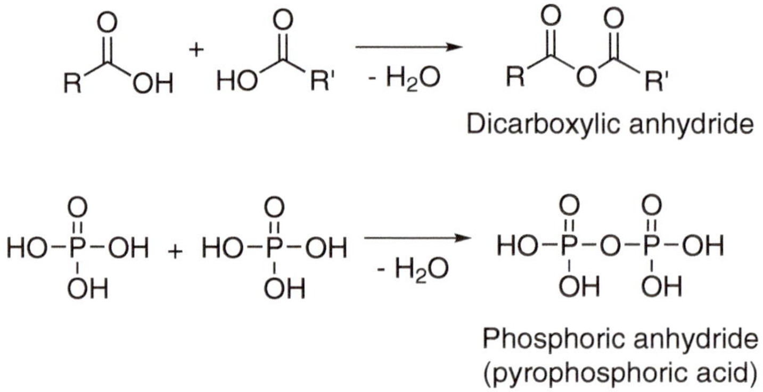

Fig. 4.101 Comparison of carboxylic and phosphoric anhydrides

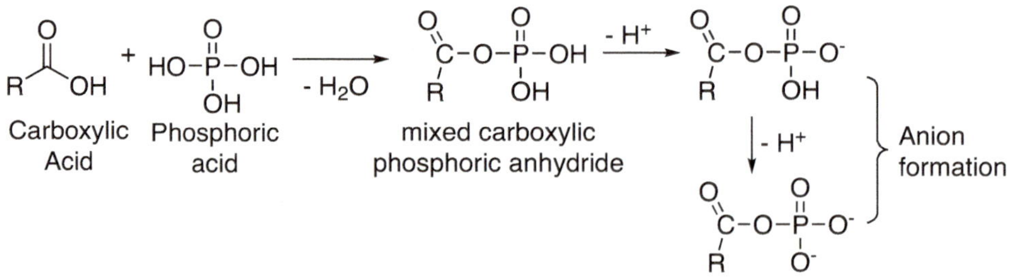

Fig. 4.102 Formation of mixed carboxylic acid phosphoric acid anhydrides

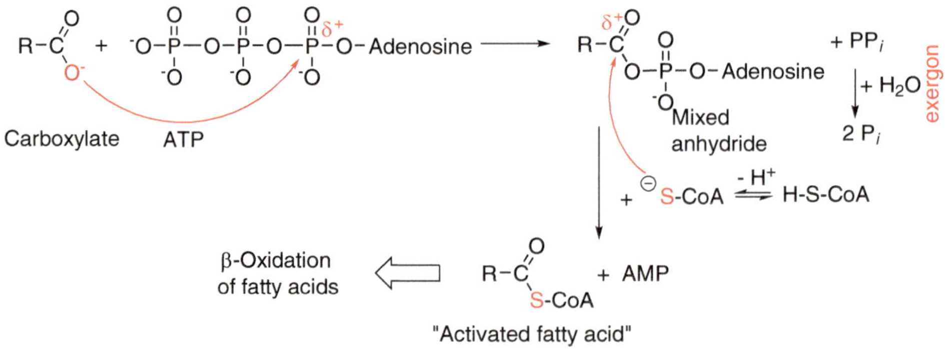

Fig. 4.103 ATP mediates the formation of an activated fatty acid from the inactive carboxylate

A similar hybrid structure consisting of a carboxylic acid and a hydrogen phosphate is generated during glycolysis. In the 6th step of this mechanism, 1,2-bisphosphoglycerate is formed from D-glyceraldehyde 3-phosphate by oxidation and phosphorylation, as the overall reaction of the biochemically relatively complex mechanism shows (Fig. 4.104).

The mixed anhydride is so energy-rich that its hydrolysis in the subsequent 7th step of glycolysis provides the energy for the synthesis of one equivalent ATP from ADP (Fig. 4.105).

The different stabilities of the two phosphoric acid functions are expressed in this reaction. While the (energy-rich) carboxylic phosphoric anhydride is cleaved and the released energy materializes in the form of ATP, the (energy-poor) phosphoric ester remains unchanged. A similar situation can be found in the hydrolysis of ATP to ADP or AMP already discussed, where only the phosphoric anhydride bonds are cleaved, while the phosphoric ester bond to D-ribose remains intact (Fig. 1.9).

Analogous to the reaction of phosphoric acid with ammonia (or amines in general) to phosphoramides, the condensation of carboxamides with phosphoric acid is also possible (Fig. 4.106). Often, the energy to build the associated biochemical structures comes from

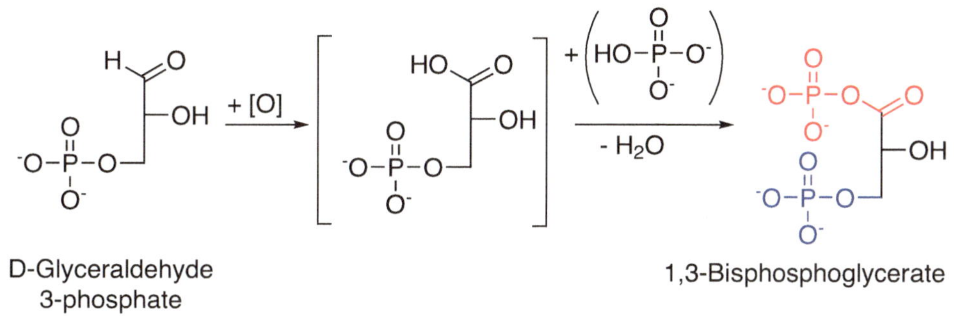

D-Glyceraldehyde
3-phosphate   1,3-Bisphosphoglycerate

Fig. 4.104   The formation of a phosphoric ester and a mixed anhydride during glycolysis

1.3-Bisphospho-
glycerat

3-Phosphoglycerate

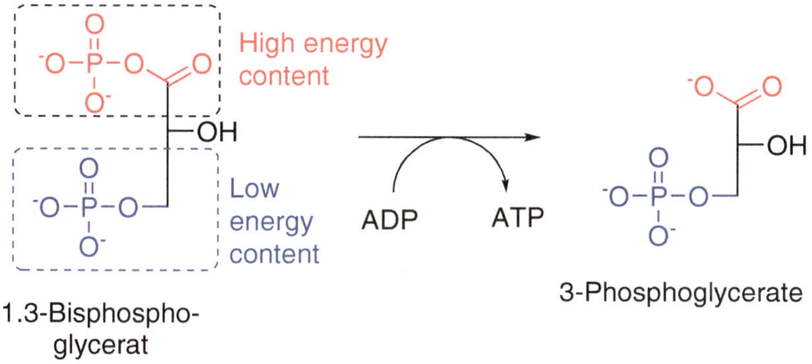

Fig. 4.105   Phosphoric ester *versus* mixed anhydride during glycolysis

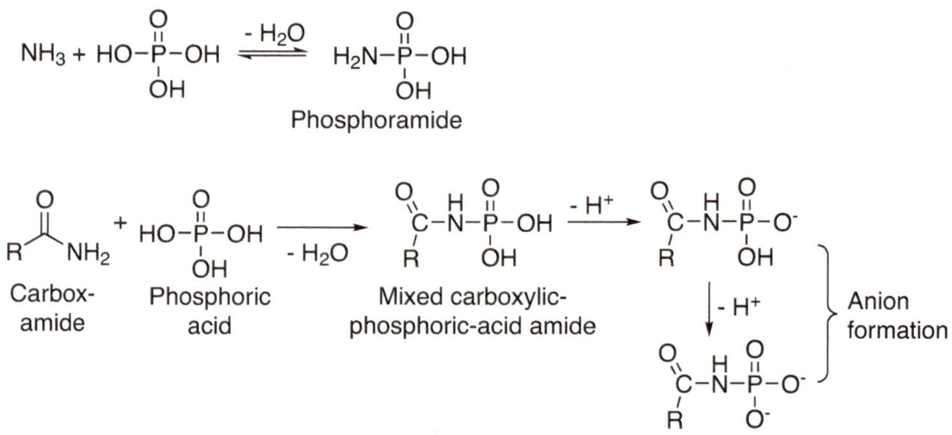

Fig. 4.106 The phosphorylation of carbamides

the reaction with ATP. The mixed carboxylic phosphoramides are much more reactive than carboxamides and allow substitution reactions at the carbonyl carbon atom (see urea cycle). Due to the physiological milieu, the products are present as anions.

The combination of the C=O group with other functional groups—a comparison of activity

By combining the carbonyl group with one or two other functional groups, new functional groups evolve that differ significantly in their properties from those of the parent functionalities. It is epistemologically advantageous to show the shifts in reactivity in the form of the terms "passive" and "active" in order to be able to assess acceleration or deceleration effects in the biochemical context (Fig. 4.107). HO-Groups in alcohols, which are in principle not or only slightly acidic, become acidic by combination with a carbonyl group and thus the formation of the carboxyl group in carboxylic acids, whereby an active principle is formed. Carboxylic acids are also passivated by esterification with alcohols. In contrast to the esters, carboxylic thioesters play a more active role in biochemistry.

By anhydride formation of a carboxylic acid with phosphoric acid, a hybrid of three functional groups is formed. With the carboxylic phosphoric anhydride or its salts, a very reactive structure is formed, which, for example, enables reactions at the carboxyl group, which are normally deactivated as a carboxylate or ester under biochemical conditions.

A reversal of the ratios in comparison to alcohols can be observed for amines (Fig. 4.108). Amines have fundamentally basic properties and thus influence biological structures in a partly destructive manner (e.g. saponification of esters in membranes). By acylation, carboxamides are formed which are no longer basic and thus withdraw the

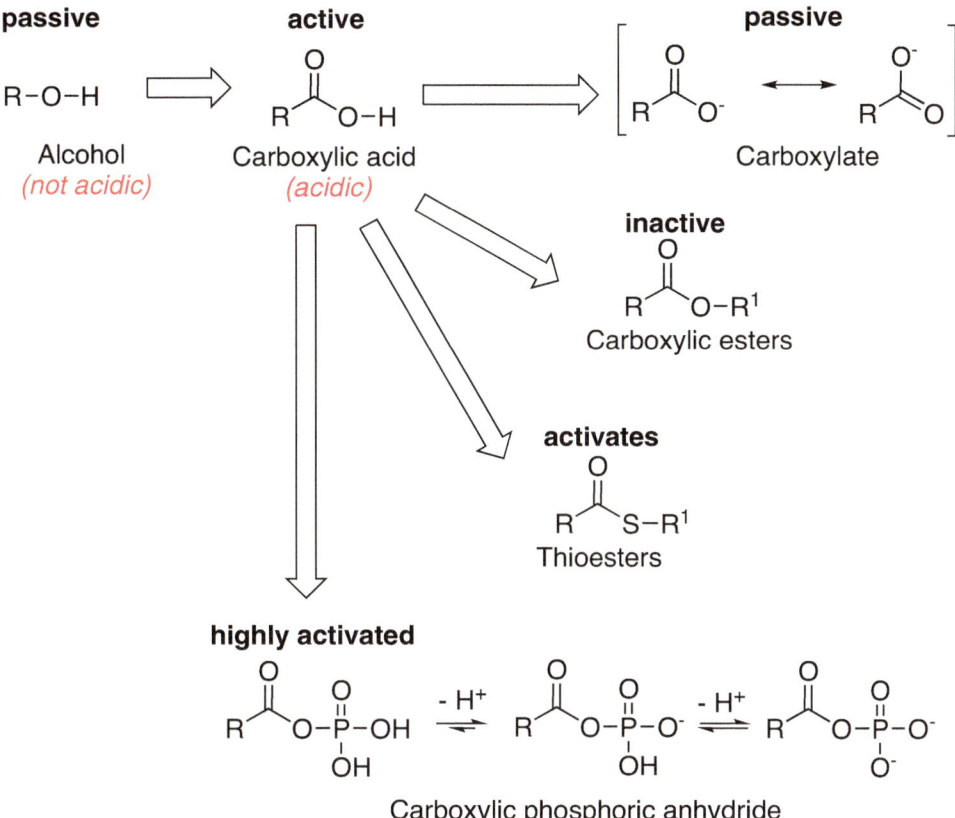

Fig. 4.107 Carboxylic acid derivatives—a comparison of activity

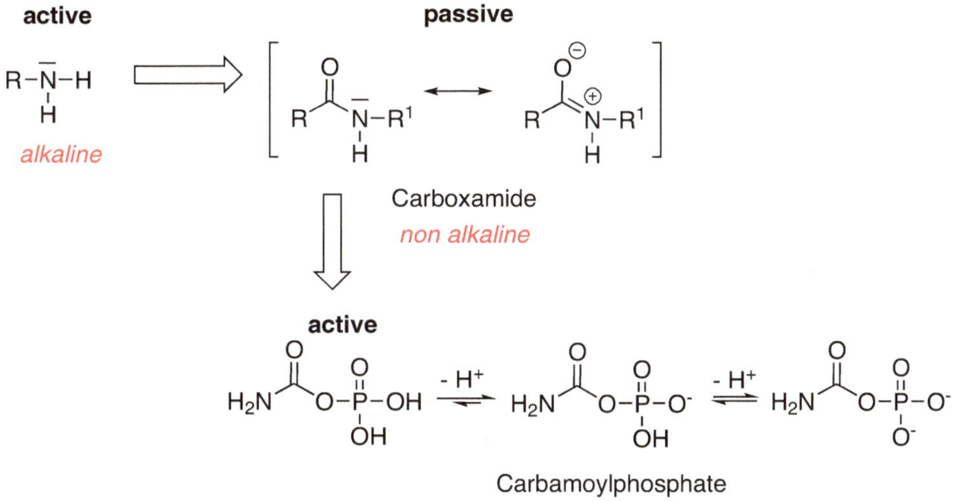

Fig. 4.108 Amine derivatives—a comparison of activity

relevant natural product from rapid transformation. In particular, phosphorylation of the carboxamide leads to another functional group which consists of three simple functional groups. In carbamoylphosphate or its salts, the originally passive carboxamide structure is activated for subsequent reactions.

Carbamoyl phosphate and carbamoyl phosphate thus take up the energy which was mainly built up in ATP, store it in chemical form and mediate coupling reactions.

Decarboxylation

The cleavage of CO_2 from a carboxylic acid always requires the support of an activating group in the vicinity. In comparison to an unfunctionalized alkane chain, high oxidation numbers of the participating groups are therefore the prerequisite for successful decarboxylation. Mesomeric stabilized carboxylates must first be converted into carboxylic acids by protonation. Most of the time, the activation of adjacent C=O groups takes place, which exert a significant cleavage pressure on nearby C–C bonds due to their electron-withdrawing effect. Already in the middle cleavage of D-fructose during glycolysis, the destabilizing effect of a carbonyl group was mentioned (Fig. 4.71). The decarboxylation of C_1-unit CO_2 is, for example, observed in α-ketocarboxylic acids. The product aldehyde is often subsequently oxidized to the carboxylic acid.

α-Ketocarboxylate   α-Ketocarboxylic acid   Aldehyde   Carboxylic acid

β-Ketocarboxylic acids are also unstable. The electron-withdrawing effect of the two C=O groups is also the prerequisite for C–C bond cleavage here. A hydrogen bond, which is energetically particularly favored due to the six-membered ring, supports the decarboxylation. After cleavage of CO_2, an enol is formed first. By tautomerism from the enol to the ketone, a product is formed which is no longer oxidizable.

β-Ketocarboxylic acid   Enol   Ketone

Since the shortening of carbon chains *is* the predominant reaction in degradation processes ultimately leading to carbon dioxide, decarboxylation is found in many catabolic processes.

An example for the biosynthesis of other monosaccharides is the transformation of D-glucose into D-ribose, as it is shown as a chemical gross reaction (Fig. 4.109). In the

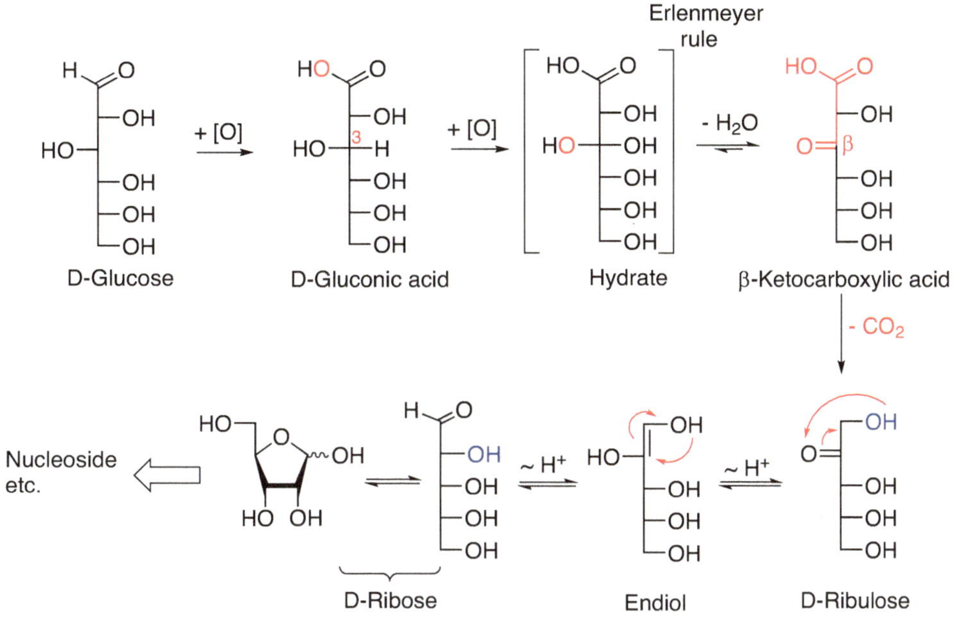

Fig. 4.109 The shortening of glucose by one C_1 unit

first step, the aldehyde is oxidized to the carboxylic acid. The D-gluconic acid undergoes further oxidation at C^3. The resulting unstable hydrate splits water off according to the Erlenmeyer rule, and the structure of a β-ketocarboxylic acid is formed. The subsequent decarboxylation shortens the original chain by one carbon unit. The formed D-ribulose undergoes a double tautomerism, which leads to D-ribose. The latter is a central component of nucleosides, which play, for example, a major role in ribonucleic acids (RNA, DNA).

Of course, decarboxylations are also in the center of the total degradation of D-glucose to CO_2. Already when entering the citrate cycle, pyruvate, which results from glycolysis, loses a C_1 unit (Fig. 4.110). As a product, the poisonous acetaldehyde should be formed formally. However, its formation is prevented in the biochemical context by the involvement of vitamin B_1 (thiamine) (see Figs. 4.9 and 4.10). Acetic acid is also not present in the free state, but is always bound to CoA via a thioester. Acetyl-CoA (Fig. 4.7) is also the end product of the degradation of fatty acids in β-oxidation. The juxtaposition of chemical gross reaction and biochemical mechanism shows the increase in complexity from purely chemical to biochemical processes in a vivid way. The chemical perspective with a focus on functional groups allows the reduction of this complexity.

Neither the acetyl group in acetyl-CoA nor free acetic acid tend to go over into two CO_2 equivalents due to lack of activation. Activation and total oxidation take place in the biochemical context of the citrate cycle (Fig. 4.111). The logic and inevitability of

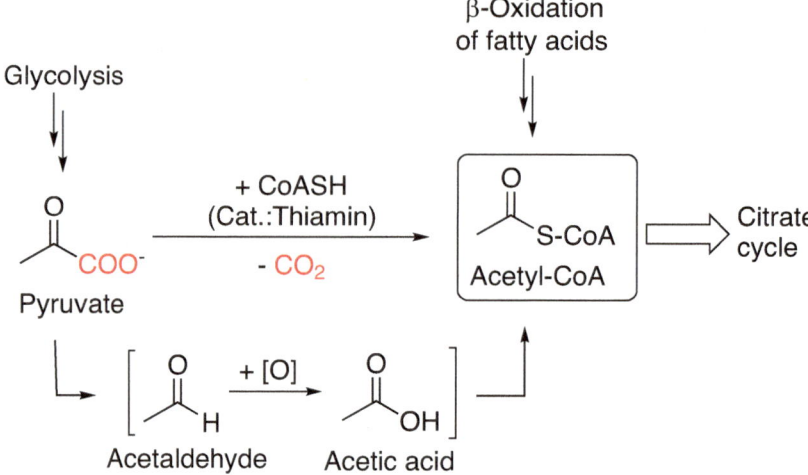

Fig. 4.110 Comparison of a biochemical and a purely chemical pathway using pyruvate degradation as an example.

this complex cycle result from the structure of the individual intermediate stages and the sequence of their conversions.

The citrate cycle is named after the salt of lemon acid (citric acid), which provides the activating "platform" for the generation of CO_2. Individual steps have already been discussed in previous chapters and should be included in the overall consideration below. Citrate is formed by C–C bond between oxaloacetate and the acetyl group bound to CoA (Fig. 4.7). Citrate is a tertiary alcohol and therefore not oxidizable. Only by isomerization to a secondary alcohol an oxidizable compound is formed, the isocitrate (Fig. 4.33). The isomerization takes place over the *cis*-aconitate. The oxidation of isocitrate to oxalosuccinate provides the activating keto group for the two following decarboxylations.

Oxalosuccinate contains three COO^- groups that are exposed to different decarboxylation pressures, as partially illustrated in Fig. 4.112. The carboxylate group marked with C^1 has no tendency to split off as CO_2. Of the other two, the one marked with C^3 is particularly activated in direct proximity to the keto group. However, after its loss, a β-ketocarboxylate would be formed, which has a lower tendency to decarboxylate than an α-ketocarboxylate. This could lead to a stall in the further breakdown of the carbon chain. However, this reaction path is not taken in the citrate cycle, since the particularly sensitive α-carboxylate group is part of a stable magnesium(II) complex (compound A). The anionic carboxylate oxygen atom is fixed on the metal via a salt formation (red) and the keto oxygen atom via a coordinative bond (green). The existence of the mesomeric boundary structure A' with partial salt character shows the increased stability.

For this reason, the less reactive carboxylate group with C^2 is first split off as CO_2 (Fig. 4.111). In parallel, a coordinative bond is converted into a salt, with magnesium salt

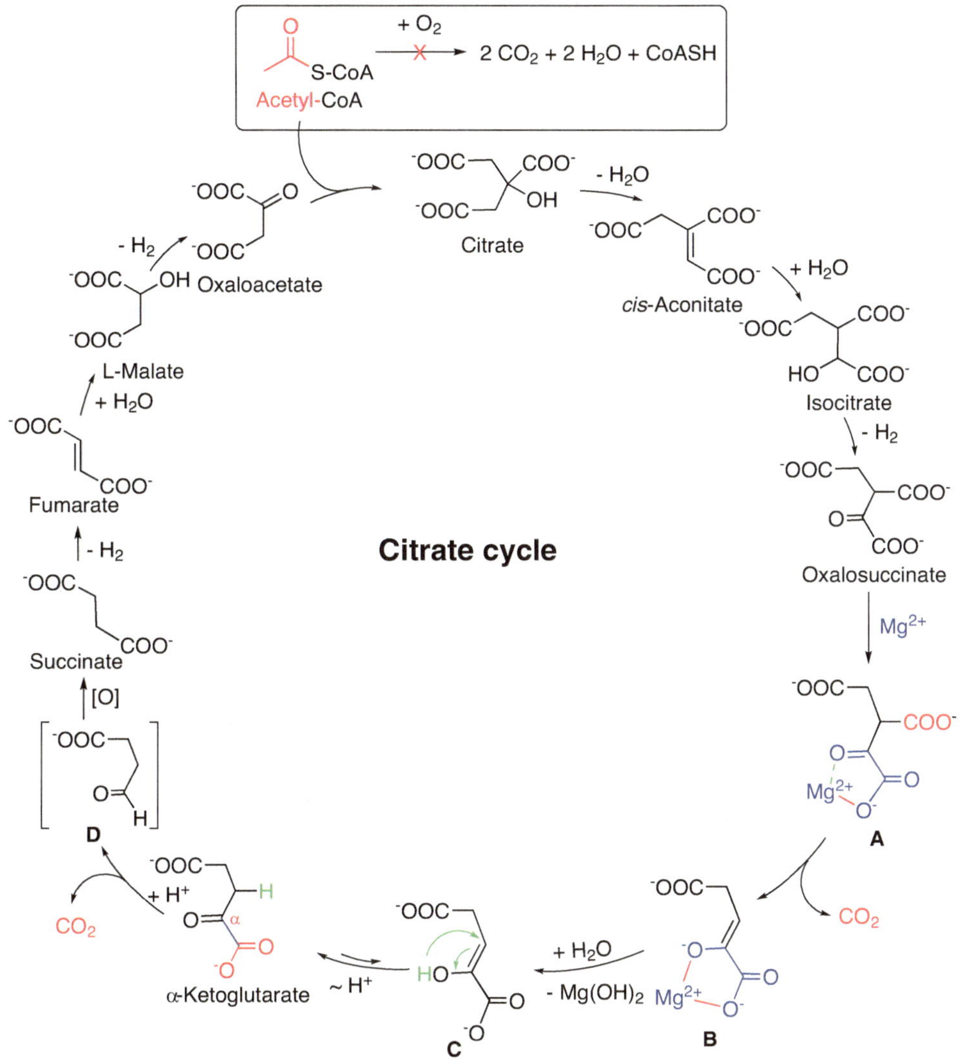

Fig. 4.111 Comparison of a biochemical and a purely chemical pathway using the example of the citrate cycle

B being formed from the original coordinative bond. The compound is hydrolyzed and the unstable enol C tautomerized to α-ketoglutarate. This clears the way for the release of the second equivalent of CO_2, now involving C^3. The aldehyde group in the only formally existing intermediate D is oxidized, and succinate is formed, which reacts with fumarate and L-malate to oxaloacetate. The citrate cycle begins anew. In parallel, the resulting hydrogen is transferred to molecular oxygen via NAD^+ or FAD and the electron

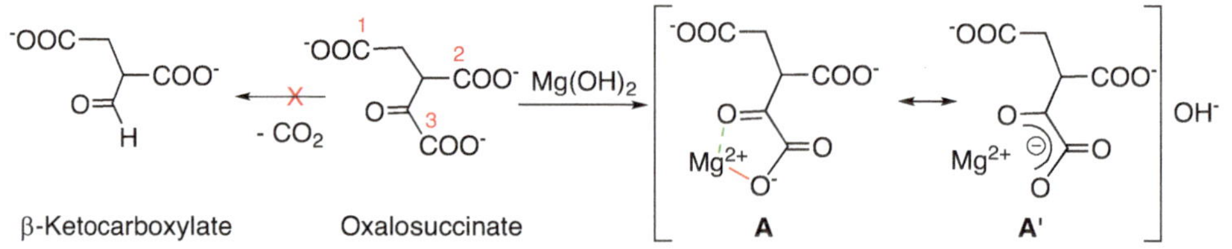

Fig. 4.112 Two decarboxylation options, only one is realized

transport chain, ultimately resulting in water. The energy released by this "slowed down" explosion reaction is used to generate ATP from AMP via chemiosmotic coupling.

When considering the entire cycle, it becomes clear that the formation of gaseous carbon dioxide is the driver. The citrate cycle is not only a biochemical way to oxidize chemically unreactive acetic acid to carbon dioxide, but many intermediates can come from degradation mechanisms of other natural substances (kataplerotic reactions). At the same time, the intermediates of the cycle form educts for the synthesis of amino acids, fatty acids, nucleotides or porphyrins in anabolic metabolism (anaplerotic reactions). This leads to the conclusion that the citrate cycle is not one of many other options for converting acetic acid to carbon dioxide, but it is also the only one in the biochemical context.

4.1.7   Carbonic Acid and Derivatives

The oxidation of aldehydes leads to carboxylic acids, which cannot be further oxidized due to a missing C–H bond. Formic acid is the only carboxylic acid that is oxidized directly to carbonic acid without C–C bond cleavage (Fig. 4.113). Carbonic acid decomposes in accordance with the Erlenmeyer rule to carbon dioxide and water. Carbon dioxide is, as repeatedly emphasized, an extremely stable gas, which further explains the shift of the chemical equilibrium to the right.

As with all other carboxylic acids, there are corresponding derivatives of carbonic acid. In particular, in their case, they prevent its rapid decomposition. Hydrogencarbonate, carbonate, carbamic acid, carbamate, urea and guanidine are important in the biochemical context (Fig. 4.114). The reactive salts of the mixed anhydride

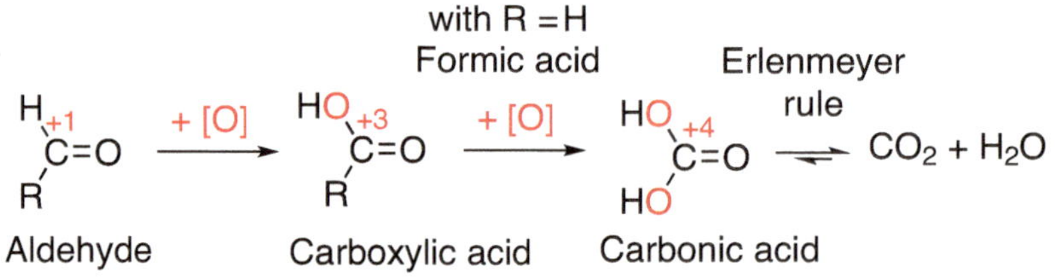

Fig. 4.113 The formation of carbon dioxide as the terminus of all biochemical processes

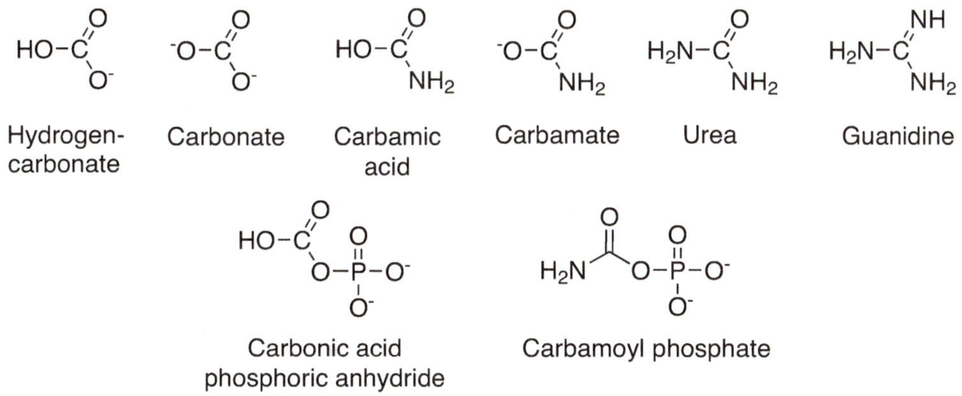

Fig. 4.114 Inorganic derivatives of carbonic acid

of carbonic acid and carbamic acid are formed by phosphorylation. All compounds are assigned to inorganic chemistry, an indication that the total oxidation of the energy-rich carbon with its countless organic compounds and thus the concept underlying this book finds its conclusion.

The neutralization of the unstable carbonic acid leads to hydrogencarbonate or carbonate.

$$H_2CO_3 + OH^- \rightleftharpoons HCO_3^- + H_2O$$

$$HCO_3^- + OH^- \rightleftharpoons CO_3^{2-} + H_2O$$

Buffer: HCO_3^-/CO_3^{2-}

The HCO_3^-/CO_3^{2-} buffer plays an important role in maintaining an approximately neutral pH value, for example in blood, a proof that even final inorganic degradation products play a role in the biochemical context. Buffers are characterized by the fact that the pH value only changes slightly when a limited amount of acid or base is added. In water-dissolved hydrogencarbonate serves in the chloroplasts of green plants as an inorganic substrate for the construction of D-glucose in the context of photosynthesis.

Carbamic acids and the urea cycle

The decomposition of carbonic acid into carbon dioxide and water is prevented by amidation. A typical example is urea, whose basicity and toxicity are reduced compared to ammonia. Urea is able to form hydrogen bonds, which is the reason for its good water solubility. Wherever ammonia is not directly released into water as the final breakdown product of N-containing natural substances such as amino acids or nucleobases and thus diluted, it is converted into urea. It is collected in the form of its aqueous solution in the kidneys of vertebrates and excreted in the urine. Urea is formed from ammonia as part of the urea cycle.

The aim of the urea cycle is the conversion of the cytotoxic and neurotoxic ammonia into the non-toxic and namesake urea (Fig. 4.115). The direct way would be a purely inorganic transformation of ammonia with carbon dioxide dissolved in water. But this reaction does not lead to urea under physiological conditions, but remains on the level of ammonium hydrogen carbonate or ammonium carbonate. The blockade of this reaction pathway has led to the evolution of the urea cycle based on organic compounds. The cycle represents a sequence of nucleophilic substitution reactions on organic derivatives of carbonic acid, namely on those of carbamic acid, urea and guanidine.

The participating partners and the individual reaction steps, like all other biochemical cycle processes, follow a strict logic that can only be understood by looking at the chemical formulas and the dependencies on other biochemical mechanisms.

At the beginning of the urea cycle is the reaction of ammonia with an activated carbonic acid derivative (Fig. 4.116). Mixed anhydrides of carbonic acid, like other carboxylic anhydrides, are highly reactive compounds that react with nucleophiles. The first step in the formation of carbamoyl phosphate is the formation of a carbonic phosphoric acid anhydride from hydrogen carbonate and ATP. By nucleophilic substitution of the phosphate by ammonia, which results from the equilibrium with NH_4^+, carbamic acid is formed. The latter is also loaded with a phosphate group by the action of ATP, resulting in carbamoyl phosphate.

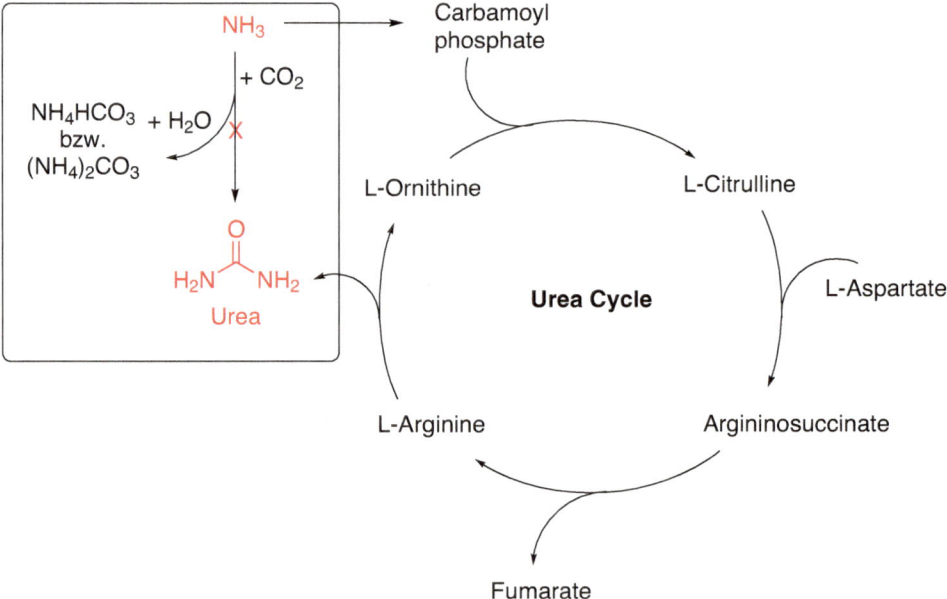

Fig. 4.115 Contrasting a biochemical and a purely chemical pathway using the urea cycle as an example

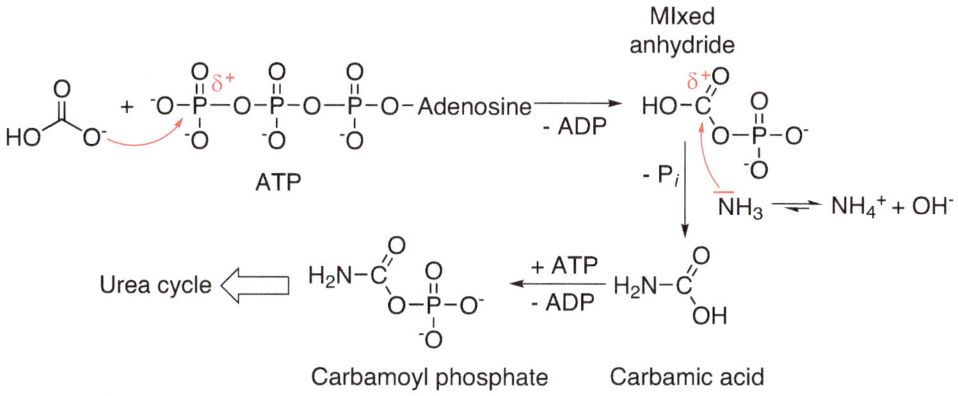

Fig. 4.116   Activation of ammonia

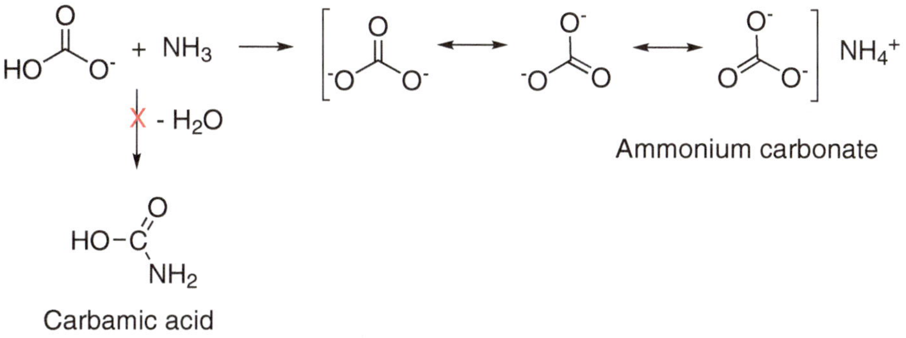

Fig. 4.117   The formation of carbamic acid *versus* ammonium carbonate

The alternative-less determinism of this two-step sequence results from the consideration of the theoretical alternatives (Fig. 4.117): The reaction of hydrogen carbonate with ammonia in the absence of ATP does not lead to carbamic acid under physiological conditions, but to the mesomeric-stabilized and therefore unreactive carbonate anion.

However, the intermediate binding of the activating phosphate makes the formation of carbamic acid possible. However, carbamic acid is just as unstable as carbonic acid and would either decompose directly into carbon dioxide and ammonia or be neutralized into the slightly more stable ammonium carbamate (Fig. 4.118), depending on the ammonia concentration. Carbamic acid is therefore not suitable for urea synthesis under physiological conditions.

Only another phosphorylation step converts carbamic acid into a reactive compound; the carbonyl carbon atom is activated for a nucleophilic attack. Carbamoyl phosphate opens the urea cycle.

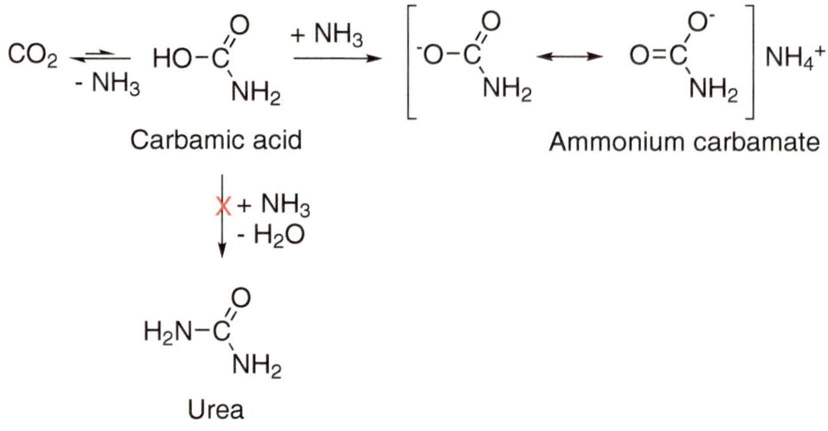

Fig. 4.118 The formation of urea *versus* ammonium carbamate

All reactions of the urea cycle take place at the terminal end of L-ornithine, so to speak a higher molecular weight carbon platform (Fig. 4.119). Ornithine is at the beginning and at the end of the cycle. The terminal amine group attacks carbamoyl phosphate nucleophilically and L-citrulline is formed.

L-Citrulline is already an N-substituted urea, but is not suitable for the cleavage of urea, since this requires a CH_2–N bond cleavage, which is not realizable under biochemical conditions (see also degradation of amines, Fig. 3.10). Only the introduction of another amino function creates the prerequisite. For this purpose, L-aspartate condenses with citrulline to argininosuccinate (Fig. 4.120). The energy for this endergonic reaction comes from the parallel hydrolysis of ATP or pyrophosphate.

Argininosuccinate is a guanidinate derivative in which—in contrast to urea—it is a very strong base (Fig. 4.121). The imino group is therefore always protonated in water, which causes a strong electron pull on the environment and provokes the follow-up reaction that drives the cycle forward. L-Aspartate is an amino acid with four carbon atoms, which is converted into fumarate by elimination of the substituted amine. The structural

Fig. 4.119 The opening reaction in the urea cycle

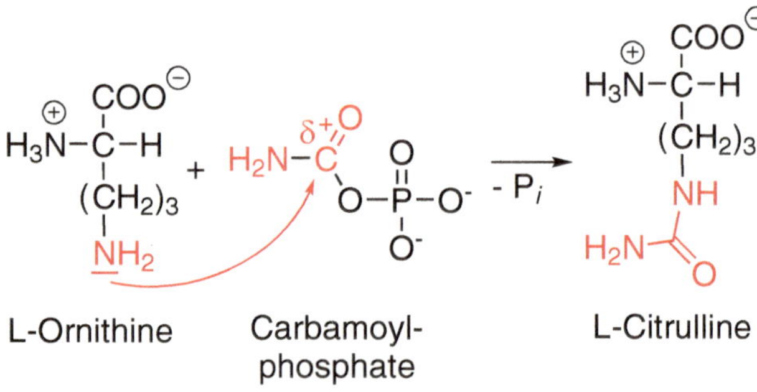

L-Ornithine   Carbamoyl-   L-Citrulline
                     phosphate

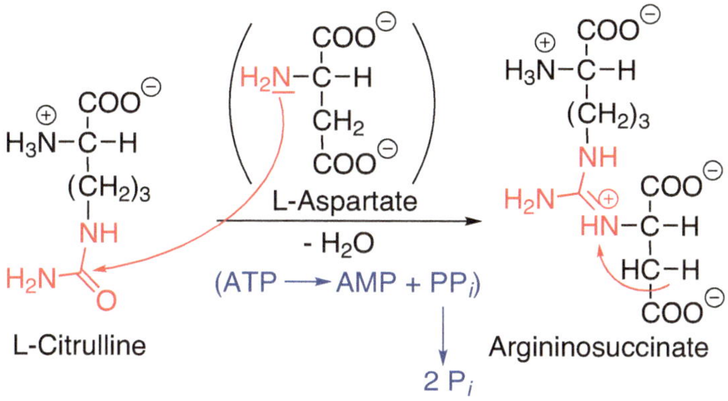

Fig. 4.120   The generation of an alkaline guanidinium structure

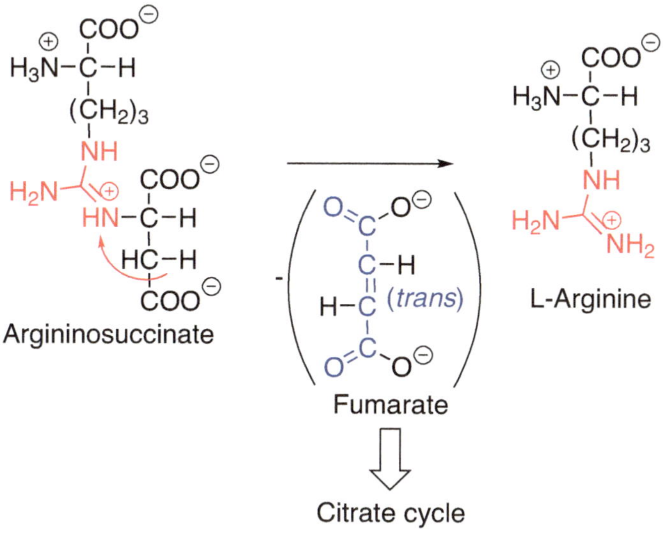

Fig. 4.121   The cleavage of the "helper reagent"

features of fumarate are the energetically favorable *trans* configuration of the two flanking carboxylate groups at the C=C double bond as well as the three conjugated double bonds extending over the entire molecule. This favors its elimination from argininosuccinate and the formation of L-arginine. Fumarate enters the citrate cycle and thus forms a bridge between the two biochemical mechanisms. This also makes the citrate cycle influence the urea cycle.

In L-arginine, the protonated guanidine group also continues to exert its strong electron pull on the relevant carbon atom (Fig. 4.122). This is the prerequisite for water to nucleophilically attack one of the two tautomeric forms and hydrolyze the C=N⁺ bond

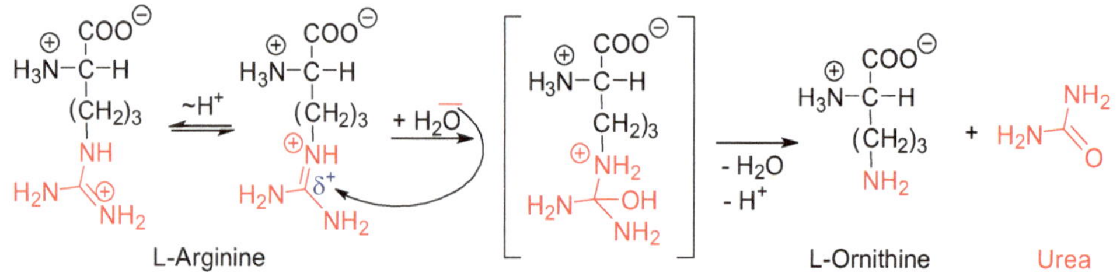

Fig. 4.122 The generation of urea from L-arginine

(formally via a hemiaminal equivalent). L-Ornithine and urea are formed. The former opens the next catalytic cycle and urea—and thus ultimately two equivalents of ammonia—leaves the organism dissolved in water.

It is striking that with L-ornithine and L-citrulline, in addition to the two proteinogenic amino acids L-aspartate and L-arginine, two non-proteinogenic amino acids are involved in the urea cycle. They are homologues of L-lysine and L-glutamine, respectively. This fact suggests that a mechanism with participants evolved here, which are partly outside the close correlation between DNA and the proteinogenic amino acids and thus the "self-cannibalization" is prevented. The consumption of ATP for activation reactions makes it clear that the "detoxification" of ammonia is an energy-consuming process. This is also reflected in the marginal energy gain, which is to be expected when amino acids are degraded in comparison to the oxidation of the main energy suppliers carbohydrates and fats.

Overall, the urea cycle provides an impressive evidence of how "dead-ends" of inorganic chemistry are avoided by means of structures and mechanisms of organic chemistry. Only this increase in complexity makes it possible to circumvent energetic sinks that would interrupt the dynamics of biochemical reactions.

β-Oxidation of fatty acids

As a functionalized C_1 building block, carbonic acid derivatives are also involved in the elongation of carbon chains during β-oxidation of fatty acids. In this biochemical mechanism, long-chain fatty acids, such as oleic acid or linoleic acid, each with 18 carbon atoms, are broken down into acetyl residues, i.e. C_2 fragments, which are further processed in the citrate cycle. When fatty acids with an odd number of carbon atoms are degraded, a C_3 fragment remains at the end. There is no connection mechanism for this. The blockade is "solved" by combining this C_3 unit with a C_1 unit to form a compound with four carbon atoms.

First, the already mentioned mixed phosphoric anhydride is bound to the basic side chain of vitamin H (biotin), which is part of an enzyme (Fig. 4.123). This results in *N*-carboxybiotin, an anionic carboxamide.

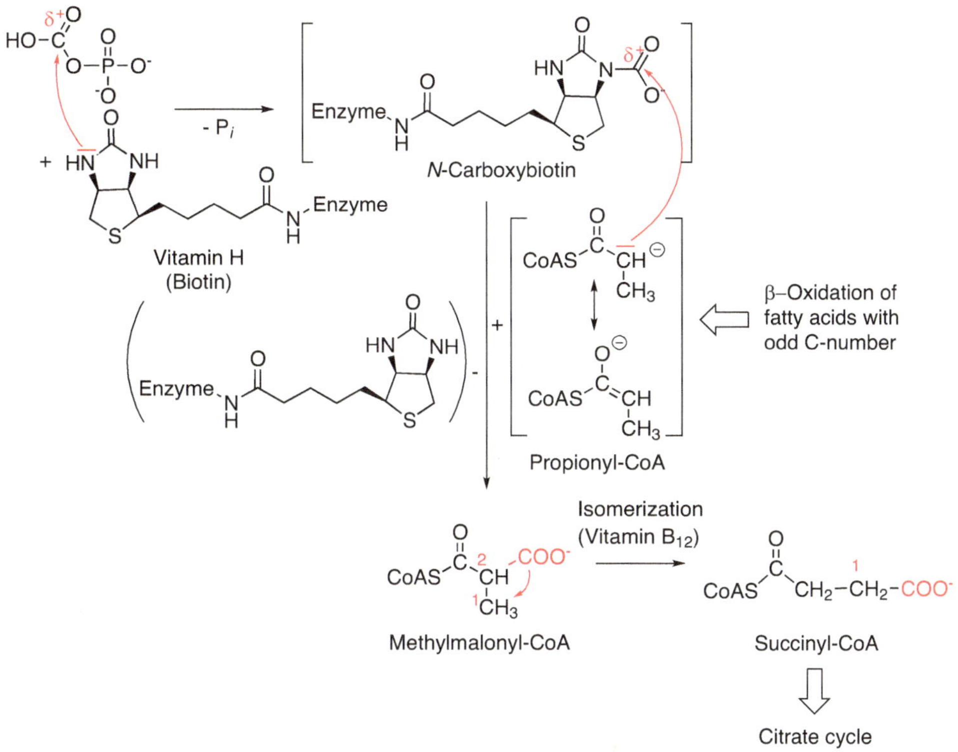

Fig. 4.123 The extension of a C_3 unit by a C_1 unit

While the phosphate group is still being cleaved off, the carboxamide is attacked by the mesomeric stable anion of propionyl-CoA, which results from the β-oxidation of the odd-numbered carbon fatty acid. The enzyme is released after the C–C bond has been formed. The product methylmalonyl-CoA is a compound with four carbon atoms, but with a branched chain, for which no degradation mechanism exists. By isomerization of the carboxylate group at the end of the chain, which is catalyzed radically by vitamin B_{12} (Sect. 2.2.3), succinyl-CoA is formed. The latter is a component of the citrate cycle. In this way, fragments of different lengths of fatty acids are introduced into a single degradation process, which prevents the "uneconomic" passage through several mechanisms with the corresponding separate enzyme equipment.

Even in the transformation shown in Fig. 4.123, the purely organic-based way proves its superiority over theoretical alternatives. The reaction of non-activated hydrogen carbonate with propionyl-CoA would reverse the deprotonation of the extremely strong base propionyl-CoA and the C–C coupling would not take place.

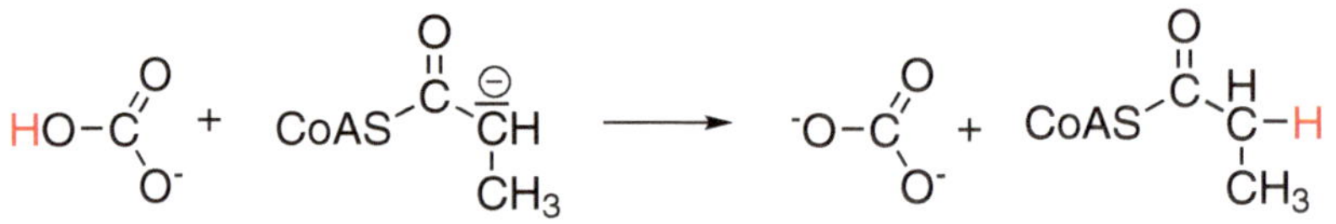

Incorporation of highly oxidized carbon derivatives into complex natural products: nucleobases

In addition to the occurrence of reactive carboxylic acid derivatives in degradation and rearrangement reactions, there are numerous stable and complex natural products that are based on carbonic acid, formic acid, and other carboxylic acids. The identification of the relevant structures can be difficult because the rules for the derivation of oxidation numbers are tailored to oxygen, which is often not present in such structures. However, the determination of the oxidation number is important to illustrate the thesis of this book that the majority of all life can be traced back to the slowed-down total oxidation of energy-rich carbon in all its biochemical and biological consequences. However, by applying fewer theoretical transformations to the corresponding oxygen compounds, the oxidation number can be assigned to each carbon atom.

The procedure should be explained below using the nucleobase adenine (Fig. 4.124). In adenine there are two C–H bonds. By formal exchange of N- for O-functionalities, the rules for the determination of the oxidation levels become valid again. In this formal procedure, no change in oxidation numbers must occur. As a result, it becomes clear that the carbon atoms considered can be traced back to formimidamide and finally to formic acid. In formic acid, the oxidation number, namely +2, can be clearly assigned.

The chemical affinity between the two nucleobases uracil and cytosine can only be recognized by this method (Fig. 4.125). After exchange of the amino group in the imidamide substructure of cytosine and subsequent imide-amide-tautomerism, uracil is formed. The oxidation numbers do not change in this transformation, both bases are compounds that can be traced back to the same carboxylic acid. If one takes into account that the escaping ammonia is a gas, the shift of the equilibrium in the reaction of cytosine to uracil with water becomes plausible. The RNA base uracil is thus a hydrolysis product of the DNA base cytosine from a formal chemical point of view. This relationship only arises when considering the chemical structures, since both nucleobases have individual synthesis pathways under biochemically evolved conditions. The relationship

Fig. 4.124 Oxidation numbers make chemical relationships visible

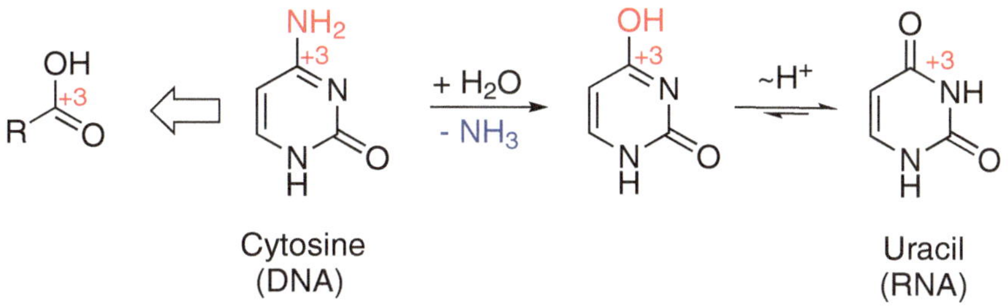

Cytosine
(DNA)

Uracil
(RNA)

Fig. 4.125   The chemical affinity between cytosine and uracil

between the two bases depends on the stability of the corresponding polynucleic acid. While the imidamide structure of cytosine is stabilized as a nucleobase of a DNA chain by hydrogen bonds with adenine of a second chain (Fig. 4.128), this stabilization does not occur in the single RNA chain, and the hydrolysis of cytosine to uracil is inevitable.

If the same formalism is applied to the nucleobase adenine, it becomes clear that it is also a precursor of another nucleobase, namely guanine (Fig. 4.126). By hydrolysis and tautomerism, the carboximidamide is converted into a carboxamide. In the vicinity is a C–H bond, into which oxygen is inserted. After exchange of the formed HO group with ammonia, guanine is formed. It follows that adenine is a derivative of formic acid, while guanine can be traced back to carbonic acid. Only the determination of the oxidation number on the respective carbon reveals these relationships. Guanine is thus already an oxidative degradation product of adenine, although both bases are central components of DNA and are formed in different ways. The paradoxical situation arises that precisely the biomolecules with the highest information content contain mainly structures in which carbon atoms with the highest oxidation state are gathered.

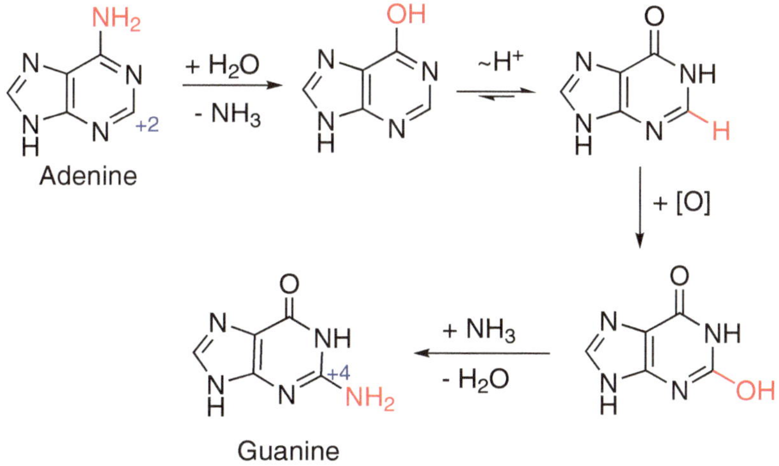

Adenine

Guanine

Fig. 4.126   The chemical link between adenine and guanine

Adenine not only has a C–H bond, but also two (Fig. 4.127). In addition to the oxidative route to guanine, the further insertion of oxygen into the C–H bond of the five-membered ring is presented. This results in hypoxanthine and then xanthine. Via the oxidation of the C–H bond in the six-membered ring, uric acid is formed via a tautomeric structure. Uric acid is decomposed into several steps to urea or ammonia and carbon dioxide, with glyoxylic acid remaining as the organic residue.

The oxidative degradation, associated with the reaction with water or ammonia, thus not only generates the variability in the nucleobases of RNA or DNA, but also leads to the total degradation and thus to the loss of the genetic function. It should be remembered that the methylation products of xanthine, that is caffeine, theophylline and theobromine, are defensive substances for the producing plants (Sect. 3.2). This results in an unexpected chemical connection between two biologically completely different functions, which can only be accessed via the formula language.

The examples show that there are numerous ways in which carboxylic acids can be converted into carbon dioxide, with the biochemical way always being the more complex and the only possible one under physiological conditions. The direct way is the oxidation of formic acid and its derivatives. In contrast, the decarboxylation of higher carboxylic acids requires numerous preparatory steps that aim at the activation of the environment of the carboxyl group. Many carboxylic acids, in particular carbonic acid derivatives, are components of high molecular weight and very central natural substances, which from a biochemical perspective slows down the rapid conversion into CO_2.

Hydrogen bonds in polynucleic acids
Carboxylic acids and many derivatives thereof are capable of forming hydrogen bonds. This property was already highlighted in Fig. 4.78 in the comparison between ethanol and acetic acid. Hydrogen bonds are particularly effective if they originate from

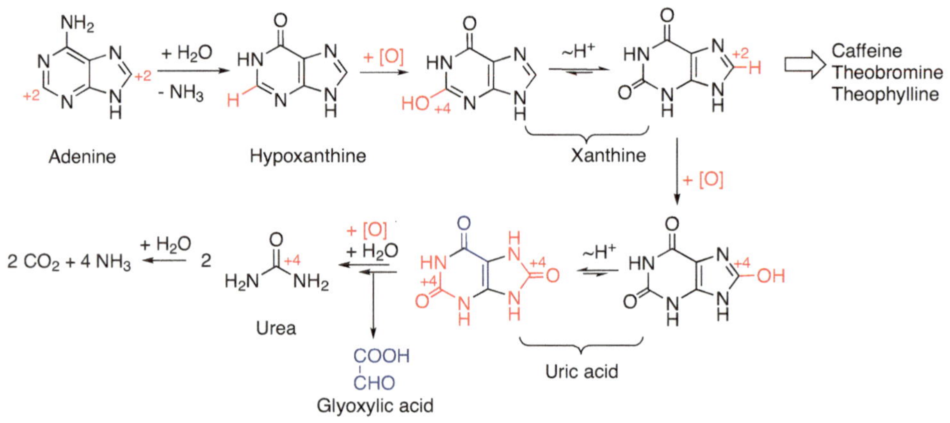

Fig. 4.127 The total degradation of adenine

carboxamides. Let us remember the H-bonds in the secondary structures of proteins or in the binding interactions between chitin chains. Multiple hydrogen bonds are also a key characteristic of polynucleic acids, in particular DNA. The complementary pairing between the "common" nucleobases thymine (T) and adenine (A) or between cytosine (C) and guanine (G) is based on such binding interactions (Fig. 4.128). In the thymine-adenine pair, two hydrogen bonds are formed between a carboxamide and an imidamide structure. In the cytosine-guanine pairing, three H-bonds are set up, in addition to the two aforementioned ones, between a carboxylic acid amide and a guanidine fragment. This strict complementarity determines the 1:1 concentration ratio between thymine and adenine or cytosine and guanine in any DNA. The formation of such attractive interactions is linked to the singular structure of the participating bases, i.e. the amide or guanidine substructure must be located in the "right" places in the molecule. In biology, the term *fitness* is used for such relationships. The compulsion to complementarity simultaneously makes it possible to almost losslessly transfer the information from an original DNA chain to RNAs or second DNA strands in the context of protein synthesis or identical replication, resulting in another adaptation pressure, this time on external partners.

In guanine (G), the six-membered ring is not aromatic. However, aromatization occurs by an amide-imidol tautomerism to the isomeric structure G* (Fig. 4.129).

This has the consequence that cytosine is no longer accepted as a partner for the tautomerized guanine*, but thymine, now over three hydrogen bridges (Fig. 4.130). Such changes in the genetic code are, inter alia, the starting points for hereditary diseases, but also determine the genetic variability of a species.

Nucleosides with structurally even more strongly modified bases do not lead to pairing. This is one reason why most RNAs only occur as single-stranded molecules and only form double-stranded domains where the "common" bases face each other. Only by removing the so-called "rare" bases during chemical evolution, such as dihydrouracil (D), pseudouridine (Ψ), ribothymidine (T) or inosine (I), is a higher level of complexity in the form of DNA achieved (Fig. 4.131).

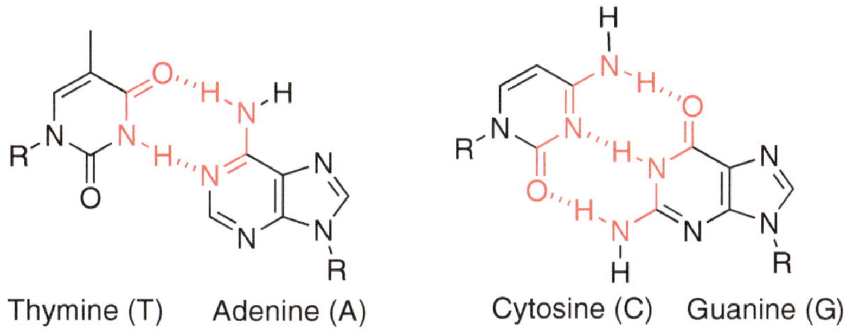

Thymine (T)   Adenine (A)   Cytosine (C)   Guanine (G)

Fig. 4.128   Hydrogen bonds determine base pairing in DNA

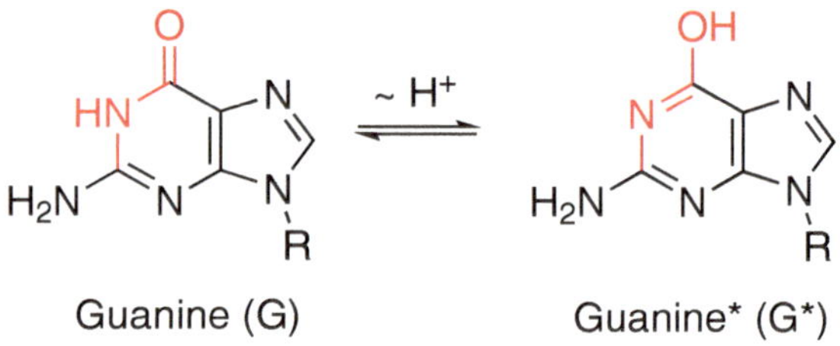

Fig. 4.129 Tautomerism
between guanine and guanine*

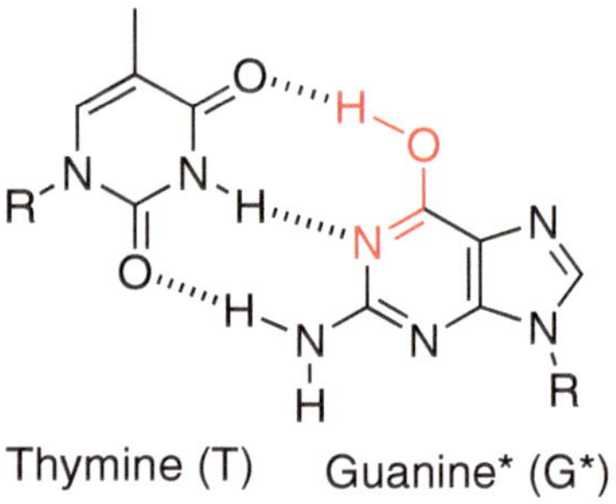

Guanine (G)   Guanine* (G*)

Fig. 4.130 Guanine* forces
thymine as the complementary
base

Thymine (T)   Guanine* (G*)

The extremely balanced selection of nucleobases with regard to their suitability in the double helix of DNA is also reflected in the exchange of uracil for thymine during the transition from RNA to DNA. In thymine there is a CH_3 group where a hydrogen atom is bound in uracil (Fig. 4.132). The exchange of the (small) hydrogen atom for the bulky methyl group hinders the [2+2] cycloaddition, in which two alkenes react to form a conformationally rigid cyclobutane ring. This reaction can occur when irradiating between two adjacent uracil residues. Since high-energy light (hv) has always accompanied the development of organisms on Earth, this small chemical modification contributes to the enduring functionality of DNA. At the same time, the tetrahedral structure of the connecting phosphoric acid leads to the fact that the base pairs are not completely flat, but slightly inclined. This not only leads to the formation of the double helix, but also the [2+2] cycloaddition is additionally restricted.

In addition to the causes that were elaborated in Sect. 1.2.14 for the higher evolutionary potential of phosphoric acid compared to other inorganic acids or for the superiority

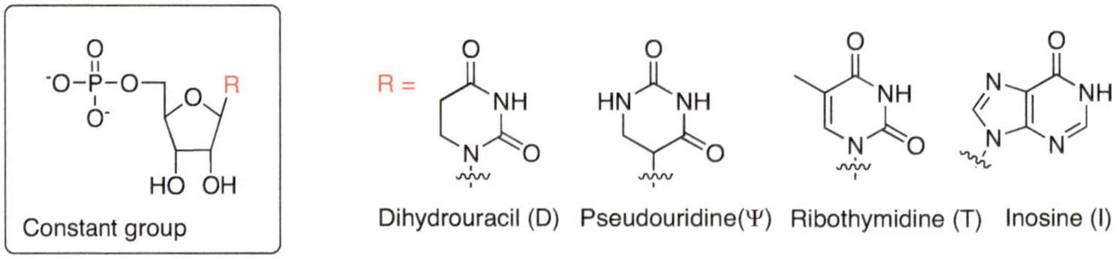

Constant group

R =

Dihydrouracil (D) Pseudouridine(Ψ) Ribothymidine (T) Inosine (I)

Fig. 4.131 "Rare" nucleosides as a structural property of RNA

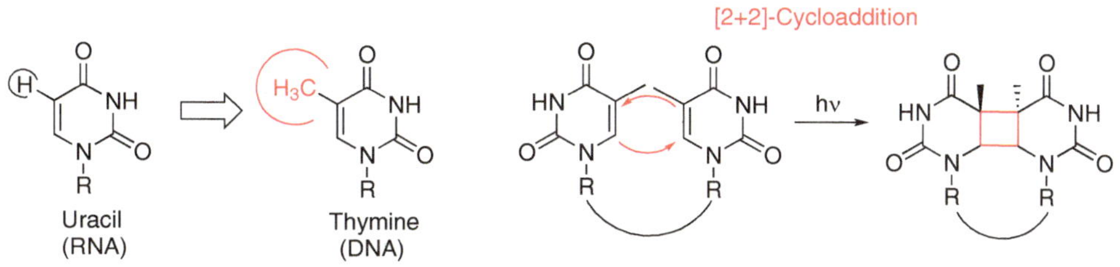

Fig. 4.132 Thymin *versus* uracil

of deoxyribose over ribose, the following overall picture results for the chemical evolution of polynucleic acids via RNA to DNA, with modifications in both inorganic and organic substructures contributing (Fig. 4.133):

On the lower two levels of evolution, which are about the potential and stability of inorganic oxoacids to form diester bridges, sulfuric acid, arsoric acid and phosphoric acid prevail. Carbonic acid is too unstable to form diesters under biotic conditions, and nitric acid does not meet this selection criterion as a monobasic acid. When comparing sulfuric acid and phosphoric acid on the next higher level of ability to bridge binding, the high acidity of sulfuric acid prevents the formation of diesters in the nearly neutral biotic environment. The tribasic arsoric acid does not form lasting esters under physiological conditions. At the end of this selection process, the evolutionary *fitness* of phosphoric acid becomes apparent. The anionic oxygen in the phosphoric diesters causes the formation of attractive ionic interactions with the basic side chains of amino acids (histones). This is the prerequisite for proteins in chromosomes to control the reading process on DNA as part of epigenetic processes. By reducing the HO group at $C^{2'}$ of ribose and the formation of 2'-deoxyribose, esterification processes of bridging phosphoric diesters are excluded, which lead to the cleavage of the polynucleic acid chain. In this way, the chain stability is increased in the form of DNA. The sorting out of "rare" bases in an RNA single strand that do not have the potential for complementary base pairing *necessarily* leads to the low-energy storage of a second DNA chain. The bonds are not realized by

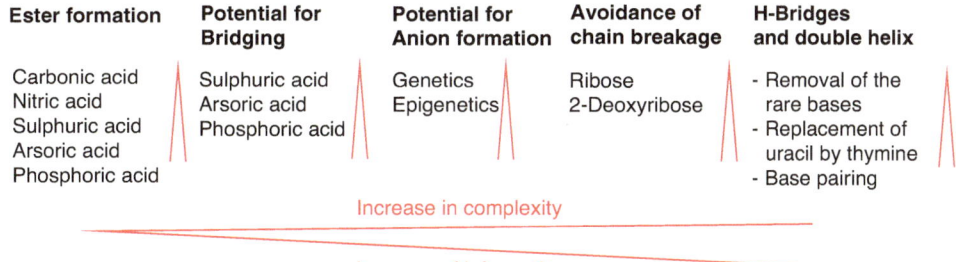

Ester formation	Potential for Bridging	Potential for Anion formation	Avoidance of chain breakage	H-Bridges and double helix
Carbonic acid Nitric acid Sulphuric acid Arsoric acid Phosphoric acid	Sulphuric acid Arsoric acid Phosphoric acid	Genetics Epigenetics	Ribose 2-Deoxyribose	- Removal of the rare bases - Replacement of uracil by thymine - Base pairing

Increase in complexity

Increase of information

Fig. 4.133 DNA as a complex and logical evolutionary product of its components

strong covalent bonds, but by flexible hydrogen bonds, which are temporarily dissolved during the reading process. Thus, the mechanism of reading determines the binding form in DNA. The exchange of uracil for thymine also restricts destabilizing side reactions in a chain under biotic conditions, namely light. These interlocking structures and processes are further stabilized by biochemical repair mechanisms.

Overall, with each step of chemical selection, the overall stability of polynucleic acids is improved. At the same time, the amount of information increases with each modification. The less stable RNAs are always shorter than DNA, regardless of their biochemical function. The overall ensemble of a DNA is the sum of the individual levels of evolution, all of which are feedback to each other. However, this also means that fewer and fewer options are possible. The biochemical mechanisms and associated biological organisms are less flexible in changing environmental conditions, but at the same time guarantee the generation-spanning stability of biological organisms.

The fact that, with the exception of RNA viruses, both RNAs and DNA occur in higher organisms, proves that during biochemical evolution, conservative structures do not have to be sorted out in every case. On the contrary, natural products at different levels of evolution contribute to the overall structure of biology.

From Chemical Structures and Individual Reactions to Complex Biochemical Networks

5

The aim of the methodology presented here is to trace biochemical processes and biological evolutionary phenomena back to characteristic chemical substructures and overall structures of natural products and their interactions with each other. In this context, functional groups in particular are in the focus, which represent starting and attack points for chemical transformations. By focusing on the biologically relevant elements of the periodic system, chemical compounds and their reactions with each other, a logical and objective approach to the evolution of life processes results.

The most important biological conditions under which life takes place on Earth are the aqueous, mostly neutral environment and the presence of reactive oxygen. The physical framework is moderate temperatures and atmospheric pressure. They establish an overarching context that is reflected in the structures of natural products, their reactions and reaction relationships. Finding such relationships can be difficult. Individual biochemical reactions always take place in cells parallel or successively to other reactions. This results in dependencies that cannot be represented in their entirety so far due to their complexity. It is questionable whether this will succeed at all if one wants to analyze the dynamics of all biochemical transformations in a single organism at a certain point in time and then fix them quantitatively.

Explanations for the occurrence of certain natural substances in a variety of biochemical sub-structures or applications are not trivial. A good example is the entire phosphate metabolism, which is basically dependent on whether soluble or insoluble phosphate compounds are considered (Fig. 5.1). Another differentiation results from the focus on specific structures, such as adenosine-5'-monophosphate (AMP) and its derivatives. AMPs occur in biochemistry in a variety of chemical forms and there in different functions. On the one hand, this is an indication of their stability and uniqueness. On the other hand, it is also proof of their great functional variability. Of particular note is the AMP in RNA. After reduction of the HO group at $C^{2'}$ of the sugar, it becomes part of

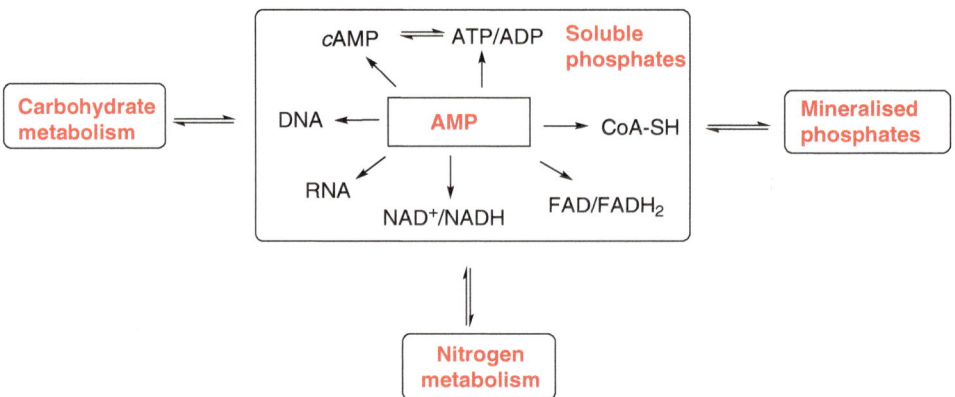

Fig. 5.1   Biochemical key compounds using AMP as an example

the DNA. Adenosine is also a component of ATP, the most important energy storage in the cell. As cyclic AMP (cAMP), which is in equilibrium with AMP, ADP and ATP, it serves as a signal transducer at the cellular level. In CoA-SH, AMP is located at the other end of the reactive thioether group. Also at a great distance from the reactive center is an AMP unit in the redox systems NAD+/NADH and in the H_2-acceptor/donor system FAD/FADH$_2$. Most of these structures are found in both carbohydrate and nitrogen metabolism.

One reason for the properties and the ubiquitous occurrence of AMP-based structures as nodes in biochemical networks in a variety of organism types and functional contexts lies in the frequency of occurrence of phosphorus, specifically the weak trivalent phosphoric acid and its salts, on Earth. The structure of adenine with two annular aromatic (pyrimidine and pyrrole) also contributes to the high stability of this part of AMP. However, the partial structure of D-ribose on the basis of a five-membered ring is surprising from this point of view, one would rather expect the more stable glucose at this point. However, a plausible explanation arises when the higher-level structure of the polynucleic acids RNA and especially DNA and their genesis are analyzed for the answer. In the context of information storage and forwarding, an overall system has evolved that only allows limited degrees of freedom with respect to the variability of individual molecule structures. This includes, in this context, the D-glucose. If one bases this thesis on the answer to the question of the widespread distribution of the AMP structure, it becomes clear that not only the structure of the individual monomers of RNA and DNA is determined by the superstructure, but that also many other "applications" outside of genetics benefit from it. From this point of view, RNA, like the insoluble phosphate of bones and teeth, can ultimately be considered as an AMP source or storage, from which different mechanisms of information processing and energy gain "draw".

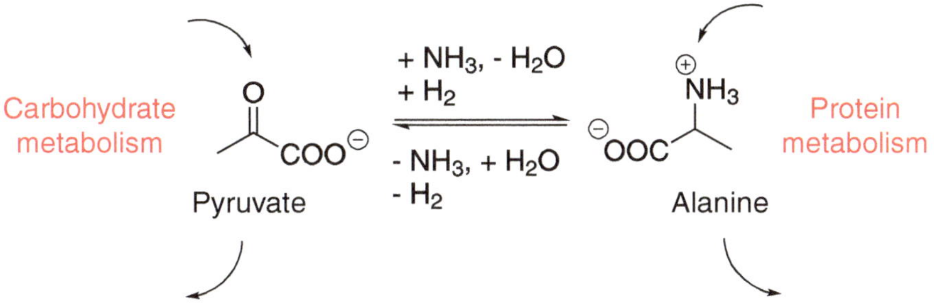

Fig. 5.2 Chemical transitions between complex mechanisms using the example of pyruvate/alanine

In biochemistry, it has become customary to represent chemical processes in cycles. In this way, the greatest possible visual clarity is sought. However, in living cells, the simultaneous presence of numerous chemical compounds and their reactions with each other and thus the actually prevailing complexity is lost due to the cycle representations. In order to capture parts of this complexity, the chemical formula language is of invaluable cognitive advantage. Only through the knowledge of the structures with their functional groups and their reactions with each other are cross-relationships between the individual cycles visible.

An example concerns the synthesis of amino acids from intermediates of glycolysis and the citrate cycle. It was shown in Sect. 4.1.7 that the release of carbon dioxide from organic compounds is usually initiated by the structural activation of adjacent carbonyl groups. Typical are α-ketocarboxylic acids such as pyruvate or α-ketoglutarate. By reaction with ammonia and subsequent hydration, α-aminocarboxylic acids are formed, as the example of alanine from pyruvate shows (Fig. 5.2).

In this way, mechanisms for the breakdown of sugars are linked to amino acid metabolism. Since amino acids in the form of histones also control the reading of DNA and thus the processes during protein synthesis in ribosomes as part of epigenetic processes, another reaction context becomes clear. At the same time, proteins are involved as enzymes in almost all biochemical transformations.

Another example concerns acetyl-CoA, which is both a product of the breakdown of glucose via glycolysis and a final product of β-oxidation of fatty acids (Fig. 5.3).

Fig. 5.3 Acetyl-CoA as a bridge between central biochemical mechanisms

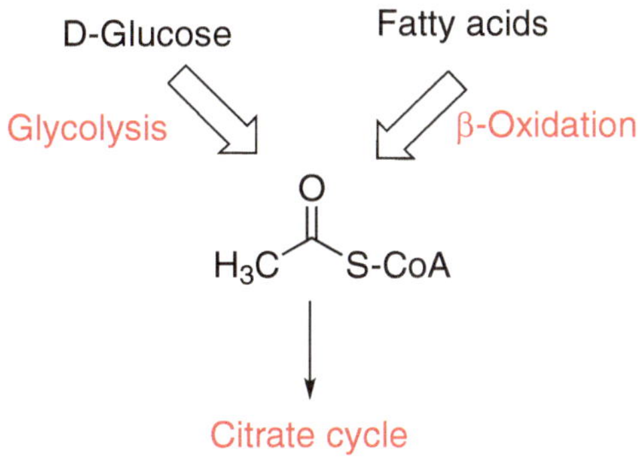

Fig. 5.4 Chemical transitions
between complex mechanisms
using the example of the
fumarate.

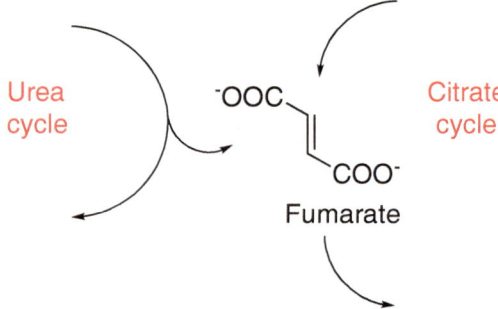

Another reaction relationship can be found in the breakdown of nitrogen-containing compounds with the breakdown pathway of carbohydrates and fats. Fumarate is
formed as a by-product in the urea cycle and is also an intermediate of the citrate cycle
(Fig. 5.4).

Determinism, Flexibility, and Contingency in Biochemistry

6

There are numerous natural substances and biochemical mechanisms that are found in organisms of different developmental stages, such as viruses, bacteria, archaea, plants, and animals alike. This is evidence that certain chemical structures and processes were and are extremely favored over alternatives in evolution. From this, a tendency towards convergence and ultimately unification can be derived.

At the same time, the existence of different biological species and individual organisms proves that deviations from uniform processes are possible. This biological fact is caused by the occurrence of partially very different natural substances. The diversity is thus a strong evidence that a considerable scope exists even under the limited conditions on Earth. However, this scope is not predictable, but is contingent (random) through evolutionary processes. The chaotic action of free oxygen radicals is particularly noteworthy from a chemical point of view.

Between the two poles of determinism and flexibility, the evolution corridors of biology develop. Chemical compounds and reactions form the basis. The converging of different degradation mechanisms of natural substances through the use of common intermediates, enzymes and reaction pathways is an aspect of evolution that constantly forms a counterpole to diversification.

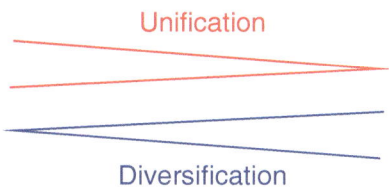

Unification

Diversification

The most important prerequisite for structured chemical reactions is the formation of distinct spaces in which reactants meet and react with each other without immediately

A. Börner and J. Zeidler, *The Chemistry of Biology*, https://doi.org/10.1007/978-3-662-66521-3_6

being diluted to infinity. It is assumed that at the beginning of biological evolution on Earth, cells made of inorganic materials, e.g. calcium carbonate, formed distinct spacess on the sea floor. This is where the predecessors of biochemical reactions took place. Low-molecular organic compounds were formed, which later evolved into higher-molecular natural products as part of increasingly complex mechanisms through a selection process. The necessary energy was supplied by the migration of protons through inorganic membranes against a concentration gradient. This concentration gradient built up between the milieu of the carbonic acid seawater and the basic interior of the inorganic cell. The energy was stored in chemical form by phosphoric anhydrides. The principle with ATP at the center has been preserved to this day.

However, higher life is linked to mobility, which static inorganic reaction spaces cannot offer. The generation of dynamic boundaries and thus mobile organisms in an aqueous environment is thus linked to organic structures, namely hydrophobic alkane chains (e.g. fatty acid esters, sulfatides, phospholipids). Unpolar carbon chains form ordered associates by means of van der Waals forces. They lead to compartmentalization, i.e. "microreactors" form in the water environment and set a resistance to the constant tendency to dilute. Membranes are organized by the incorporation of proteins or rare carbohydrates into these boundary layers, which control the inflow or outflow of certain molecules by means of chemical recognition mechanisms. They represent the first information carriers. Dynamic organic systems emerged from static inorganic structures through continuous exchange processes with the environment. In biology, they are referred to as organelles and cells.

In addition to creating distinct spaces, successful reactions between connections with each other are another fundamental condition for life. The reactants are dependent on the environmental conditions. On Earth, molecular oxygen and its radicals react with those elements that are oxidizable in an aqueous, approximately neutral environment. This results in water-soluble inorganic salts via oxides and hydroxides. Frequently occurring metal salts are able to decompose oxygen radicals. The rapid decomposition is slowed down when organic ligands coordinate with the metals. The resulting complex compounds evolve into "oxygen carriers" and subsequently catalyze selective oxidation reactions as enzymes.

Not only non-metals and some semi-metals, but all organic natural products are subject to a continuous oxidation pressure, which is both directed and undirected. Those oxidation products and their derivatives, which are produced by enzymatic and thus directed processes, form the conservative basis for a variety of biological species. In addition, compounds are formed that can never be assigned a direct biological function in the present or future. They are intermediates in biochemical mechanisms.

Oxidation products that are formed by the unselective attack of free oxygen radicals act against the determinism of biochemical processes and are responsible for part of the contingency (randomness) in the development of new biochemical mechanisms and biological forms. Changes in the chemical environmental conditions contribute to evolution as extrinsic factors.

Due to the limitation of the material basis to relatively few elements of the PSE and the limitation of the conditions to water as a reaction medium and oxygen as the primary reaction partner, life also runs in narrow channels. This is reflected in the limited number of natural substances, typical single reactions and uniform mechanisms. For example, the growth and cleavage of polysaccharides, polynucleic acids and proteins are based on the same principles. Polynucleic acids, regardless of whether DNA or RNA, represent conservative information carriers that not only pass on their information within the biological individual, but also across generations. Proteins that are formed according to the specifications of the polynucleic acids are the basis for almost all enzymes, which materialize information storage and multiplication.

The numerical limitation of biochemical reactions and natural substances is reinforced by the numerous interactions within the framework of self-organizing processes. Thus, self-organization is also a principle of selection. An illustrative example is the existence of "only" 20 proteinogenic amino acids. Their number is, inter alia, limited by the link with the information storage in the polynucleic acids (RNA, DNA). At the same time, the synthesis of amino acids and proteins is catalyzed by other proteins in the form of enzymes, from which further feedback loops arise.

Limiting feedback effects on chemistry and biochemistry also come from biological levels. For example, the nucleic acids in the cell nucleus are separated from the mitochondria. The latter are sites of the formation of free oxygen radicals. This creates an additional, biologically based protection against the rapid oxidative degradation of these central information molecules. Something similar applies to the separately existing chloroplasts in green plants.

Another example of biological feedback involves uric acid. It is excreted directly by birds. It is not soluble in water, which is reflected in their anatomy: birds do not have a urinary bladder. Uric acid in mammals, on the other hand, is usually subject to further transformation into urea. Urea is soluble in water, which makes the existence of the urinary bladder necessary. Fish and tadpoles excrete the ultimate breakdown product ammonia. It is immediately diluted by the aqueous environment and no longer poses a danger to these animals. There is therefore a direct correlation between the living environments of organisms, anatomical features and the chemical properties of breakdown compounds. The chemical properties of breakdown products decisively influence the biology of the corresponding organisms and vice versa.

Life is not only limited, but also diverse. This diversity is just as much as the limitedness conditioned by the material basis and the framework conditions. Small differences in chemistry always have consequences for the associated biochemistry and ultimately for the organism. This makes it clear that a "blindly acting" chemistry is the prerequisite for all biological existence. The compatibility of biochemical and biological functions with chemical structures is the criterion for further evolutionary selection.

Differences in the chemical basis form the basis for biological systematics with their division into domains, kingdoms, phyla, classes, orders, families, genera and species. Smaller chemical deviations within a species characterize individuals and can even depend on sex and age. Individual and short-term changes in chemistry are usually "caught" by established biochemical processes. Longer exposures to foreign elements or chemical compounds can cause larger changes in the long term and are ultimately drivers of biological evolution.

"Synthetic Chemistry" *versus* Biochemistry

As stated in the preface to this book, a "felt" contradiction is constructed between chemistry and biology in modern knowledge societies. This contradiction does not exist on a chemical level. The synthetic chemistry that is carried out in academic laboratories and the chemical industry's technical facilities is a chemistry that is *not* subject to the strong limitations of biochemistry. Or, to put it another way, biochemistry is in principle also a synthetic chemistry, but under very limited conditions. The "man-made" synthetic chemistry has a much larger parameter space available than biochemistry, which opens up an incomparable potential for the number and variety of chemical compounds. The parameters include, for example, the temperature and pressure under which chemical reactions take place. Restrictions in the technical field only arise from the limitations of the technical apparatus and facilities. If biochemistry is only limited to the solvent water, then in the technical field there are hundreds of solvents with different properties. If one adds those reactions in which only the reactant or the product is used as a solvent, the choice of solvent is almost unlimited. As in biochemical systems, catalysts also play a central role in reducing the activation energy in technical synthetic chemistry. Due to the short optimization time, the catalysts are structurally much simpler, but some already match enzymes in activity and productivity that have evolved over millions of years.

The most important distinction between technical chemistry and biochemistry concerns complexity. While in synthesis chemistry the "purification" of the system from all interfering compounds is sought, biochemistry is characterized by a side-by-side and one after the other of reactions. In such reaction cascades, substrates or products influence each other. There are positive or negative feedback effects, which are also referred to as **competitive inhibition** and are often characterized by the competition of substrates or products at a catalyst. Such feedback processes are the prerequisite for the optimization and expansion of biochemical mechanisms. On this chemical basis, biological evolution takes place over more or less long periods of time. Only by separating reactions

© The Author(s), under exclusive license to Springer-Verlag GmbH, DE, part of Springer Nature 2023
A. Börner and J. Zeidler, *The Chemistry of Biology*,
https://doi.org/10.1007/978-3-662-66521-3_7

in different organelles or cells, or by connecting reaction partners to membranes, does a more or less ordered structure arise, which, however, cannot be compared with the strictly separated reactors and processing modules of chemical engineering.

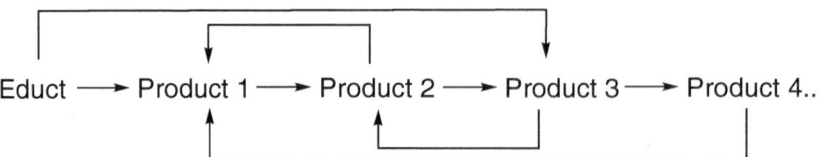

The derivation of simple if-then relationships is not possible in biochemistry due to the variety of parallel or consecutive reactions and the countless reaction participants. This is one of the most important differences to synthetic chemistry in the laboratory. There, such relations are sought and often also realized. Only on this basis is an optimal reaction possible.

In addition, there are no higher-level regulating units in biochemical systems, such as people or computers in a technical plant or in a research laboratory, which plan and monitor the reaction course and intervene in case of deviations. The apparent chaos of biochemical mechanisms, in which countless educts, products and enzymes are involved, turns out to be a self-organizing (**autopoietic**) system in a living organism, in which almost everything is connected with everything else. Only through this structure is every organism transformed into a living, homeodynamic entity.

In addition to these aspects, another phenomenon can be observed when comparing "classical" synthetic chemistry and biochemistry. The former mainly takes its starting point in unfunctionalized petroleum or natural gas products or coal, from which high-molecular and highly functionalized compounds are built up in a *bottom-up* process to complex structures (e.g. pharmaceuticals). Reactive starting compounds with very low molecular weights are generated via cracking processes or C-C bond formation reactions (e.g. Fischer-Tropsch synthesis) and then transformed. Biochemical transformations usually take place in the reverse direction, i.e. already highly functionalized compounds (especially glucose) are converted into others in a *top-down* process on the basis of the decelerated oxidation of the energy-rich carbon and subsequent reactions. A trend that is becoming increasingly strong in modern synthetic chemistry is to convert natural products (*renewable resources*) into chemicals. Due to the highly functionalized structures of the starting products, such selective transformations are still associated with considerable challenges. Nevertheless, the underlying chemistry does not change. The distinction between "synthetic chemistry" and biochemistry must therefore be made on the basis of other, in this case social, specifically political, criteria.

These differences between "man-made" synthetic chemistry and biochemistry may be responsible for the fact that chemical education in schools and basic chemistry at universities are still focused on the former, which has negative consequences for the acceptance of the natural science chemistry in society.

In principle, the generalization of synthetic compounds as "unnatural" or even poisonous is always wrong. Synthetic compounds that are produced in laboratories and chemical plants are subject to the laws of nature in their origin and structure. Considering that the most poisonous compounds known to date are natural products, this criterion also becomes obsolete. Through the strict application of the chemical formula language, such misunderstandings can be excluded and a deeper understanding of so-called "natural" processes is possible.

Further Reading

General and Inorganic Chemistry

Großmann G, Fabian J, Kammer H-W (1973) Struktur und Bindung – Atome und Moleküle, 2. Aufl. VEB Deutscher Verlag für Grundstoffindustrie, Leipzig
Hollemann AF, Wiberg N (2007) Anorganische Chemie, 102. Aufl. de Gruyter, Berlin
Pscheidl H (1975) Allgemeine Chemie. Grundkurs, Teile 1 und 2, Berlin

Organic Chemistry

Hart H, Crane LE, Hart DJ (2002) Organische Chemie, 2. Aufl. Wiley-VCH, Weinheim
Uhlig E, Domschke G, Engels S, Heyn B, Walther D (1973) Reaktionsverhalten und Syntheseprinzipien, 1. Aufl. VEB Deutscher Verlag für Grundstoffindustrie, Leipzig
Walter W, Francke W (1998) Lehrbuch der Organischen Chemie, 23. Aufl. Hirzel, Stuttgart

Biological Chemistry

Biesalski HK (2015) Mikronährstoffe als Motor der Evolution. Springer, Berlin
Behr A, Seidensticker T (2018) Chemistry of Renewables. An Introduction. Springer, Berlin
Follmann H, Grahn W (1999) Chemie für Biologen. Teubner, Stuttgart
Fox MA, Whitesell JK (1995) Organische Chemie. Grundlagen, Mechanismen, bioorganische Anwendungen. Spektrum Akademischer Verlag, Heidelberg
Kaim W, Schwederski B (2004) Bioanorganische Chemie. Zur Funktion chemischer Elemente in Lebensprozessen, 3. Aufl. Teubner, Wiesbaden
Latscha HP, Kazmaier U (2008) Chemie für Biologen, 3. Aufl. Springer, Berlin
Lehmann J (1996) Kohlenhydrate, 2. Aufl. Thieme, Stuttgart

Natural Product Chemistry and Biochemistry

Berg JM, Tymoczko JL, Gatto GJ, Stryer L (2018) Biochemie, 8. Aufl. Springer Spektrum, Berlin

Epple M (2003) Biomaterialisation und Biomineralisation. Eine Einführung für Naturwissenschaftler, Mediziner und Ingenieure. Teubner, Stuttgart

Harborne JB (2013) Ökologische Biochemie. Eine Einführung. Springer, Heidelberg

Karlson P, Doenecke D, Koolman J (1994) Kurzes Lehrbuch der Biochemie für Mediziner und Naturwissenschaftler, 14. Aufl. Thieme, Stuttgart

Lehninger AL (1985) Grundkurs Biochemie, 2. Aufl. de Gruyter, Berlin

Müller-Esterl W (2004) Biochemie. Eine Einführung für Mediziner und Naturwissenschaftler. Elsevier, München

Nuhn P (1997) Naturstoffchemie. Mikrobielle, pflanzliche und tierische Naturstoffe, 3. Aufl. Hirzel, Leipzig

Voet D, Voet JG (2019) Biochemie, 3. Aufl. Wiley-VCH, Weinheim

Special Aspects

Chai QY, Yang Z, Lin HW, Han BN (2016) Alkynyl-Containing Peptides of Marine Origin: A Review. Mar Drugs 14:216. https://doi.org/10.3390/md14110216

Hofmann H-J, Cimiraglia R (1988) Conformation of 1,4-dihydropyridine – planar or boat-like? FEBS Lett 241:38–40. https://doi.org/10.1016/0014-5793(88)81026-3

Hunter T (2012) Why nature chose phosphate to modify proteins. Phil Trans R Soc B 367:2513–2516. https://doi.org/10.1098/rstb.2012.0013

Jung M, Kim H, Lee K, Park M (2003) Naturally occurring peroxides with biological activities. Mini Rev Med Chem 3:159–165. https://doi.org/10.1039/c0np00024h

Kertesz MA (2000) Riding the sulfur cycle - metabolism of sulfonates and sulfate esters in Gram-negative bacteria. FEMS Microbiol Rev 24:135–175. https://doi.org/10.1016/S0168-6445(99)00033-9

Lane N (2016) Oxygen: The molecule that made the world, Oxford Landmark Science

Lane N (2017) Der Funke des Lebens: Energie und Evolution. Theiss WBG, Darmstadt

Metcalf WW, van der Donk WA (2009) Biosynthesis of Phosphonic and Phosphinic Acid Natural Products. Ann Rev Biochem 78:65–94. https://doi.org/10.1146/annurev.biochem.78.091707.100215

Parry R, Nishino S, Spain J (2011) Naturally-occurring nitro compounds. Natural Prod Rep 28:152–167

Spenser ID, White RL (1997) Die Biosynthese von Vitamin B1 (Thiamin) ein Beispiel für biochemische Vielfalt. Angew Chem 109:1096–1111. https://doi.org/10.1002/ange.19971091005

Thauer RK, Kaster A-K, Goenrich M, Schick M, Hiromoto T, Shima S (2010) Hydrogenases from Methanogenic Archaea, Mickel, a Novel Cofactor, and H2 Storage Ann Rev Biochem 79: 5507–536. https://doi.org/10.1146/annurev.biochem.030508.152103

Wächtershäuser G (1988) Before enzymes and templates: theory of surface metabolism. Microbiol Molecul Biol Rev 52:452–484